T0336040

Multi-Fractal Traffic and Anomaly Detection in Computer Communications

This book provides a comprehensive theory of mono- and multi-fractal traffic, including the basics of long-range dependent time series and $1/f$ noise, ergodicity and predictability of traffic, traffic modeling and simulation, stationarity tests of traffic, traffic measurement and the anomaly detection of traffic in communications networks.

Proving that mono-fractal LRD time series is ergodic, the book exhibits that LRD traffic is stationary. The author shows that the stationarity of multi-fractal traffic relies on observation time scales, and proposes multi-fractional generalized Cauchy processes and modified multi-fractional Gaussian noise. The book also establishes a set of guidelines for determining the record length of traffic in measurement. Moreover, it presents an approach of traffic simulation, as well as the anomaly detection of traffic under distributed-denial-of service attacks.

Scholars and graduates studying network traffic in computer science will find the book beneficial.

Multi-Fractal Traffic and Anomaly Detection in Computer Communications

Ming Li

CRC Press
Taylor & Francis Group
Boca Raton London New York

CRC Press is an imprint of the
Taylor & Francis Group, an **informa** business

First edition published 2023
by CRC Press
6000 Broken Sound Parkway NW, Suite 300, Boca Raton, FL 33487-2742

and by CRC Press
2 Park Square, Milton Park, Abingdon, Oxon, OX14 4RN

CRC Press is an imprint of Taylor & Francis Group, LLC

ISBN: 978-1-032-40846-0 (hbk)
ISBN: 978-1-032-40851-4 (pbk)
ISBN: 978-1-003-35498-7 (ebk)

DOI: 10.1201/9781003354987

Typeset in Minion
by KnowledgeWorks Global Ltd.

To Dr. Yonglan Zhang and Joanna Jiayue Li – for making it both possible and worthwhile.

Contents

Preface

This monograph is a sequel to my previous book titled *Fractal Teletraffic Modeling and Delay Bounds in Computer Communications* published by CRC Press with the link www.routledge.com/9781032212869 or https://doi.org/10.1201/9781003268802. In this monograph, we address the theory of multi-fractal traffic and its applications to anomaly detection of traffic under distributed denial-of-service attacking in computer communications. The monograph consists of 4 Parts with 14 chapters plus an Appendix.

Part I: Fundamentals is discussed in Chapters 1–5 and the Appendix. The basics of fractal time series are addressed in Chapter 1. We narrate $1/f$ noise in Chapter 2 and power laws with respect to fractal traffic in Chapter 3. We introduce the ergodicity and the predictability of long-range-dependent processes in Chapters 4 and 5, respectively. The convergence of sample autocorrelation function of long-range-dependent traffic is elaborated in the Appendix.

The second part is about traffic modeling and traffic data processing which are discussed in Chapters 6–8. Traffic modeling is described in Chapter 6, stationarity test of traffic is discussed in Chapter 7, and the record length requirement of traffic in measurement is addressed in Chapter 8.

Part III is on multi-fractal models of traffic, including the multi-fractional generalized Cauchy process discussed in Chapter 9, the modified multi-fractional Gaussian noise described in Chapter 10, and multi-scale traffic simulation described in Chapter 11.

The last part is about anomaly detection of traffic under distributed denial-of-service attacking which is discussed in Chapters 12 and 13. Three open problems are given in Chapter 14.

The main highlights of the book are as follows:

1. Concise descriptions of the fundamental theory of fractal traffic (Chapters 1–3).

2. Proving the ergodicity of long-range-dependent processes (Chapter 4).

3. Giving the proof of the predictability of long-range-dependent series (Chapter 5).

4. Showing the convergence of sample autocorrelation function of traffic with long-range dependence (Appendix).

5. Exhibiting that fractal properties of traffic from 1980s to the current time remain the same (Chapter 6).

6. Presenting correlation matching method for the weak stationarity test of traffic (Chapter 7).

7. Establishing a reference guideline for measurement of traffic (Chapter 8).

8. Proposing a novel multi-fractal traffic model called multi-fractional generalized Cauchy (mGC) process (Chapter 9).

9. Introducing a new multi-fractal traffic model called modified multi-fractional Gaussian noise (Chapter 10).

10. Putting forward a simulation method of traffic (Chapter 11).

11. Bringing forward a reliable anomaly detection approach of distributed-denial-of-service attacking (Chapter 12).

12. Revealing the variation trend of the Hurst parameter of traffic under distributed denial-of-service attacking (Chapter 13).

This monograph will be a reference for computer scientists, engineers, statisticians, researchers, and graduate students.

Prof. Ming Li
Ocean College, Zhejiang University, China

Acknowledgments

The author appreciates the ACM SIGCOMM for the real traffic data. Sincere thanks go to Will Leland (wel@bellcore.com) and Dan Wilson (dvw@bellcore.com) for the traffic traces measured by Bellcore, Jeff Mogul (mogul@pa.dec.com) of Digital's Western Research Lab (WRL), Digital Equipment Corporation (DEC) for the measured traffic traces, the WIDE Internet (MAWI) Working Group traffic archive of the WIDE Project (Japan) for the real traces, and MIT Lincoln Laboratory for the traffic data in network security research. I thank Ms. Ying Dong for her assistance in collecting the traffic data from the Measurement and Analysis on the real traces in the MAWI Traffic Archive and also Mr. Yong Chen for improving the resolution of drawings.

I am grateful to Prof. Vern Paxon (University of California, Berkeley) for his instructions to the traffic data in the ACM SIGCOMM traffic archive and the stationarity issue of traffic. Helpful discussions with Prof. Patrick Flandrin and Prof. Pierre Borgnat (école normale supérieure de lyon) in signal processing, fractal properties of traffic in particular, are highly appreciated. Particular thanks go to Prof. Swee Cheng Lim (Multimedia University) for his help in fractal time series. Prof. Yuming Jiang (the Norwegian University of Science and Technology) is appreciated for his helpful work in network calculus and traffic engineering. Prof. YangQuan Chen (University of California, Merced) is sincerely appreciated for his support and discussions about fractional calculus, fractional noise and systems. I am grateful to Prof. Jian-Xing Leng (Zhejiang University) for providing me with facilities on the Zhoushan Islands, where this monograph was peacefully written during the period of time when disastrous outbreaks of COVID-19 caused havoc in my hometown, Shanghai.

This work was partly supported by the National Natural Science Foundation of China under the project grant number 61672238. The views and conclusions contained in this book are those of the author and should not be interpreted as representing the official policies, either expressed or implied, of the Chinese government.

I

Fundamentals

CHAPTER 1

Fractal Time Series

Fractal time series substantially differs from conventional one in its statistic properties. For instance, it may have a heavy-tailed probability density function (PDF), a slowly decayed autocorrelation function (ACF), and a power spectrum density (PSD) function of $1/f$ type. It may have the statistical dependence, either long-range dependence (LRD) or short-range dependence (SRD), and global or local self-similarity. This chapter will give the basics of fractal time series for the theory of fractal traffic. Note that a conventional time series can be regarded as the solution to a differential equation of integer order with the excitation of white noise in mathematics. In engineering, such as mechanical engineering or electronics engineering, people may usually consider it as the output or response of a system or filter of integer order under the excitation of white noise. In this chapter, a fractal time series is taken as the solution to a differential equation of fractional order or a response of a fractional system or a fractional filter driven with a white noise.

1.1 BACKGROUND

Denote the n-dimensional Euclidean space for $n \in \mathbf{Z}_+$ by $\mathbf{R}^n$, where $\mathbf{Z}_+$ is the set of positive integers. Then, things belonging to $\mathbf{R}^n$ for $n = 1, 2, 3$ are visible, such as a curve for $n = 1$, a picture for $n = 2$, and a three-dimensional object for $n = 3$.

Denote an element belonging to $\mathbf{R}^n$ by $f(x_1, \ldots, x_n)$ and $x_n \in \mathbf{R}^1 = \mathbf{R}$. Denote a regularly orthogonal coordinate system in $\mathbf{R}^n$ by $\{e_1, e_2, \ldots, e_n\}$. Then, the inner product (e_l, e_m) is given by:

$$(e_l, e_m) = \begin{cases} 1, l = m, \\ 0, l \neq m. \end{cases}$$

Then,

$$f = \sum_{l=1}^{n} (f, e_l) e_l. \tag{1.1}$$

In the domain of the Hilbert space, $n \to \infty$ is allowed (Griffel [1], Liu [2]). Unfortunately, due to the limitation of human eyes, a high dimensional image of f, for example, $n > 4$, is

DOI: 10.1201/9781003354987-2

invisible unless some of its elements are fixed. One can only see an image f for $n > 4$ partly. For example, if we fix the values of x_n for $n \geq 3$, $f(x_1, x_2, x_3, ..., x_n)$ is visible. Luckily, humans have a nimbus such that people are able to think about high-dimensional objects in $\mathbf{R}^n$ even in the case of $n \rightarrow \infty$.

As we know, the nature is very rich and colorful (Mandelbrot [3], Korvin [4], Peters [5], Bassingthwaighte et al. [6]). Spaces of integer dimension are not enough. As a matter of fact, there exist spaces with fractional dimension, such as $\mathbf{R}^{n+d}$, where $0 < d < 1$ is a fraction. Therefore, even in the low-dimensional case of $n = 1, 2, 3$, things in $\mathbf{R}^{n+d}$ are not completely visible.

We now turn to time series. Intuitively, we say that $x(t)$ is a conventional series if $x(t) \in \mathbf{R}$. On the other side, $x(t)$ is said to be a fractal time series if it belongs to $\mathbf{R}^{1+d}$ for $0 < d < 1$. A curve of $x \in \mathbf{R}^{1+d}$ we usually see by eye, such as a teletraffic (traffic for short) series, is only its integer part belonging to $\mathbf{R}$. However, it is the fractional part of $x(t)$ that makes it substantially different from a conventional series in the aspects of PDF, ACF, and PSD, unless d is infinitesimal.

The theory of conventional series is relatively mature, see e.g., Fuller [7], Box et al. [8], Mitra and Kaiser [9], Bendat and Piersol [10], but the research regarding fractal time series is quite academic. However, its applications to various fields of sciences and technologies, ranging from physics to computer communications, are increasing, for instance, coast-lines, turbulence, geophysical record, economics and finance, computer memories (see Mandelbrot [11]), network traffic, precision measurements (Beran [12]), electronics engineering, chemical engineering, image compression (see Levy-Vehel et al. [13]), physiology (see Bassingthwaighte et al. [6]), just to name a few. The goal of this chapter is to provide the basics with respect to fractal time series.

The chapter is organized as follows: In Section 1.2, the concept of fractal time series from the point of view of systems of fractional order will be addressed. The basic properties of fractal time series are explained in Section 1.3. Some models of fractal time series are discussed in Section 1.4 and Section 1.5 gives the summary.

1.2 FRACTAL TIME SERIES: A VIEW FROM FRACTIONAL SYSTEMS

A time series can be taken as a solution to a differential equation. In terms of engineering, it is often called signal while a differential equation is usually termed as system or filter. Therefore, without confusion equation, system, or filter are taken as synonyms in the following text.

1.2.1 Realization Resulted from a Filter of Integer Order

A stationary time series can be regarded as the output $y(t)$ of a filter under the excitation of white noise $w(t)$. Denote the impulse function of a linear filter by $g(t)$. Then,

$$y(t) = \int_0^t g(t-\tau)w(\tau)d\tau. \tag{1.2}$$

On the other side, a nonstationary random function can be taken as the output of a filter under the excitation of nonstationary white noise. In general, filters with different $g(t)$s may yield different series under the excitation of $w(t)$. Hence, conventionally, one considers $w(t)$ as the headspring or root of random series (see, e.g., Press et al. [14]). In this chapter, we only consider stationary series.

A stochastic filter can be written as

$$\sum_{i=0}^{p} a_i \frac{d^{p-i} y(t)}{dt^{p-i}} = \sum_{i=0}^{q} b_i \frac{d^{q-i} w(t)}{dt^{q-i}}. \tag{1.3}$$

Denote the Fourier transforms of $y(t)$, $g(t)$, and $w(t)$ by $Y(\omega)$, $G(\omega)$, and $W(\omega)$, respectively, where ω is angular frequency. Then, according to the theorem of convolution, one has

$$Y(\omega) = G(\omega)W(\omega). \tag{1.4}$$

Denote the PSDs of $y(t)$ and $w(t)$ by $S_{yy}(\omega)$ and $S_{ww}(\omega)$, respectively. Then, when one notices that $S_{ww}(\omega) = 1$ if $w(t)$ is the normalized white noise [9, 10], one has

$$S_{yy}(\omega) = |G(\omega)|^2. \tag{1.5}$$

Denote the Laplace transform of $g(t)$ by $G(s)$, where s is a complex variable. Then (Lam [15]),

$$G(s) = \frac{1 + \sum_{i=1}^{q} b_i s^i}{1 + \sum_{i=1}^{p} a_i s^i}. \tag{1.6}$$

If the system is stable, all poles of $G(s)$ are located on the left of s plan. For a stable filter, therefore, one has (Papoulis [16])

$$G(\omega) = F[g(t)] = G(s)|_{s=j\omega}, \tag{1.7}$$

where F stands for the operator of the Fourier transform. A basic property of a linear stable system of integer order is stated as follows.

Note 1.1. Taking into account $b_0 = 1$ and (1.7), one sees that $|G(\omega)|^2$ of a stable system of integer order is convergent for $\omega = 0$ and so is $S_{yy}(\omega)$. □

In the discrete case, the system function is expressed by the z transform of $g(n)$. That is,

$$G(z) = Z[g(n)] = \sum_{n=0}^{\infty} g(n) z^{-n} = \frac{1 + \sum_{i=1}^{q} b_i z^{-i}}{1 + \sum_{i=1}^{p} a_i z^{-i}}, \tag{1.8}$$

where Z represents the operator of z transform. There are two categories of digital filters (Harger [17], Vegte [18]). One is in the category of infinite impulse response (IIR) filters, which correspond to the case of $a_i \neq 0$. The other is in the category of finite impulse response filters (FIR), which imply $a_i = 0$ [9, 15] (Harger [17], Vegte [18]). In the FIR case, one has

$$G(z) = \sum_{n=0}^{q} g(n)z^{-n} = 1 + \sum_{i=1}^{q} b_i z^{-i}. \tag{1.9}$$

Thus, an FIR filter is always stable with a linear phase.

Note 1.2. A realization $y(t)$ resulted from an FIR filter of integer order under the excitation of $w(t)$ is linear. It belongs to **R**. □

1.2.2 Realization Resulted from a Filter of Fractional Order

Let $v > 0$ and $f(t)$ be a piecewise continuous on $(0, \infty)$ and integrable on any finite subinterval of $[0, \infty)$. For $t > 0$, denote the Riemann-Liouville integral operator of order v by ${}_0D_t^{-v}$ [19, p. 45]. It is given by

$$_0D_t^{-v} f(t) = \frac{1}{\Gamma(v)} \int_0^t (t-u)^{v-1} f(u)du, \tag{1.10}$$

where Γ is the Gamma function. For simplicity, we write ${}_0D_t^{-v}$ by D^{-v} below.

Let $v_p, v_{p-1}, \ldots, v_0$ and $u_q, u_{q-1}, \ldots, u_0$ be two strictly decreasing sequences of nonnegative numbers. Then, for the constants a_i and b_i, we have

$$\sum_{i=0}^{p} a_{p-i} D^{v_i} y(t) = \sum_{i=0}^{q} b_{q-i} D^{u_i} w(t), \tag{1.11}$$

which is a stochastically fractional differential equation with constant coefficients of order v_p. It corresponds to a stochastically fractional filter of order v_p. The transfer function of this filter, expressed by using the Laplace transform, is given by (Ortigueira [20])

$$G(s) = \frac{1 + \displaystyle\sum_{i=1}^{q} b_{q-i} s^{-u_i}}{1 + \displaystyle\sum_{i=1}^{p} a_{p-i} s^{-v_i}}. \tag{1.12}$$

In the discrete case, it is expressed in z domain by (Ortigueira [21, 22], Chen and Moore [23], Vinagre et al. [24])

$$G(z) = \frac{1 + \displaystyle\sum_{i=1}^{q} b_{q-i} z^{-u_i}}{1 + \displaystyle\sum_{i=1}^{p} a_{p-i} z^{-v_i}}. \tag{1.13}$$

Denote the inverse Laplace transform and the inverse z transform by L^{-1} and Z^{-1}, respectively. Then, the impulse response of the filter expressed by (1.11) in the continuous and discrete cases are given by

$$g(t)=L^{-1}[G(s)], \tag{1.14}$$

and

$$g(n)=Z^{-1}[G(z)], \tag{1.15}$$

respectively.

Without loss of the generality to explain the concept of fractal time series, we reduce (1.11) to the following expression:

$$\sum_{i=0}^{p} a_{p-i}D^{v_i} y(t)=w(t). \tag{1.16}$$

Consequently, (1.12) and (1.13) are reduced to

$$G(s)=1+\sum_{i=1}^{q} b_{q-i}s^{-u_i} \tag{1.17}$$

and

$$G(z)=1+\sum_{i=1}^{q} b_{q-i}z^{-u_i}. \tag{1.18}$$

Recall that the realization resulted from such a class of filters can be expressed in the continuous case by

$$y(t)=w(t)*g(t), \tag{1.19}$$

where $*$ implies the operation of convolution. The realization in the discrete case can be expressed by

$$y(n)=w(n)*g(n). \tag{1.20}$$

Hence, we have the following notes.

Note 1.3. A realization $y(t)$ resulted from a stochastically fractional differential equation may be unbelonging to **R**. ☐

Note 1.4. For a stochastically fractional differential equation, Note 1.1 may be untrue. ☐

We shall further explain Note 1.4 in the next section. As an example to interpret the point in Note 1.3, we consider a widely used fractal time series called the fractional Brownian motion (fBm) introduced by Mandelbrot and van Ness [25].

Replacing v with $H + 0.5$ in (1.10) for $0 < H < 1$, where H is the Hurst parameter, the fBm defined by using the Riemann-Liouville integral operator, which is denoted by $B_{H, \mathrm{RL}}(t)$, is given by

$$B_{H,\mathrm{RL}}(t) = {}_0D_t^{-(H+1/2)}B'(t) = \frac{1}{\Gamma(H+1/2)}\int_0^t (t-u)^{H-1/2}\,dB(u), \tag{1.21}$$

where $B(t)$, $t \in (0, \infty)$, is the Wiener Brownian motion (see, e.g., Hida [26] for Brownian motions). The differential of $B(t)$ is in the domain of generalized functions over the Schwartz space of test functions (see, e.g., Gelfand and Vilenkin [27] for generalized functions). Taking into account the definition of the convolution used by Mikusinski [28], we have the impulse response of a fractional filter given by

$$\frac{(-t)^{H-1/2}}{\Gamma(H+1/2)}. \tag{1.22}$$

Consequently, the fBm $B_{H,\mathrm{RL}}(t)$ can be taken as an output of the filter (1.22) under the excitation $\frac{dB(t)}{dt}$ in the form (Li and Chi [29])

$$B_{H,\mathrm{RL}}(t) = \frac{dB(t)}{dt} * \frac{(-t)^{H-1/2}}{\Gamma(H+1/2)}. \tag{1.23}$$

Therefore, Note 1.5 comes.

Note 1.5. The fBm $B_{H,\mathrm{RL}}(t)$ is a special case as a realization of a fractional filter (1.22) driven by $\frac{dB(t)}{dt}$. $\square$

Other articles discussing the fBm from the point of view of systems or filters of fractional order can be seen in Daw [30], Buchmann and Klüppelberg [31], Zeinali and Pourdarvish [32], and Podlubny [33]. In the end of this section, I use another equation to interpret the concept of fractal time series. The fractional oscillator or fractional Ornstein–Uhlenbeck process is the solution of the fractional Langevin equation given by

$$({}_aD_t + A)^\alpha\, y(t) = w(t), \quad \alpha > 0, \tag{1.24}$$

where A is a positive constant and $w(t)$ is the white noise (Lim et al. [34, 35]). Obviously, the fractal time series $y(t)$ in (1.24) is a realization resulted from a fractional filter under the excitation $w(t)$. More about this will be discussed in Section 1.4.

1.3 BASIC PROPERTIES OF FRACTAL TIME SERIES

Fractal time series has its particular properties in comparison with the conventional one. Its power laws in general are closely related to the concept of memory. A particular point, which has to be paid attention to, is that mean and or variance may usually not exist in such a series. This may be the main reason why measures of fractal dimension and the Hurst parameter play a role in the field of fractal time series.

1.3.1 Power Laws in Fractal Time Series

Denote the ACF of $x(t)$ by $r_{xx}(\tau) = E[x(t)x(t + \tau)]$. Then, $x(t)$ is called SRD if r_{xx} is integrable (Beran [12]), i.e.,

$$\int_0^\infty r_{xx}(\tau)d\tau < \infty. \tag{1.25}$$

On the other side, $x(t)$ is LRD if r_{xx} is non-integrable, i.e.,

$$\int_0^\infty r_{xx}(\tau)d\tau = \infty. \tag{1.26}$$

A typical form of such an ACF for r_{xx} being non-integrable has the following asymptotic expression:

$$r_{xx}(\tau) \sim c|\tau|^{-\beta} \ (\tau \to \infty), \tag{1.27}$$

where $c > 0$ is a constant and $0 < \beta < 1$. The above expression implies a power law in the ACF of LRD fractal series.

Denote the PSD of $x(t)$ by $S_{xx}(\omega)$. Then,

$$S_{xx}(\omega) = \int_{-\infty}^\infty r_{xx}(t)e^{-j\omega t}\,dt. \tag{1.28}$$

In the LRD case, the above $S_{xx}(\omega)$ does not exist as an ordinary function but it can be regarded as a function in the domain of generalized functions. Since

$$F\left(|\tau|^{-\beta}\right) = 2\sin(\pi\beta/2)\Gamma(1-\beta)|\omega|^{\beta-1}, \tag{1.29}$$

(see, e.g., Gelfand and Vilenkin [27] and Li and Lim [36, 37]) the PSD of LRD series has the property of power law. It is usually called $1/f$ noise or $\frac{1}{f^\alpha}$ ($\alpha > 0$) noise (Mandelbrot [38]). Thus, comes Note 1.6.

Note 1.6. The PSD of an LRD fractal series is divergent for $\omega = 0$. This is a basic property of LRD fractal time series which substantially differs from that as described in Note 1.1. □

Denote the PDF of $x(t)$ by $p(x)$. Then, the ACF of $x(t)$ can be expressed by

$$r_{xx}(\tau) = \int_{-\infty}^\infty x(t)x(t+\tau)p(x)dx. \tag{1.30}$$

Considering that r_{xx} is non-integrable in the LRD case, one sees that a heavy-tailed PDF is an obvious consequence of LRD series (see, e.g., Li [39, 40] and Abry et al. [41]).

Denote the mean of $x(t)$ by μ_x. Then,

$$\mu_x = \int_{-\infty}^{\infty} xp(x)dx. \tag{1.31}$$

The variance of $x(t)$ is given by

$$\mathrm{Var}(x) = \int_{-\infty}^{\infty} (x-\mu_x)^2 p(x)dx. \tag{1.32}$$

One thing remarkable in LRD fractal time series is that the tail of $p(x)$ may be so heavy that the above integral either (1.31) or (1.32) may not exist. To explain this, we recall a series obeying the Pareto distribution that is a commonly used heavy-tailed distribution. Denote the PDF of the Pareto distribution by $p_{\mathrm{Pareto}}(x)$. Then,

$$p_{\mathrm{Pareto}}(x) = \frac{ab}{x^{a+1}}, \tag{1.33}$$

where $x \geq a$. The mean and variance of $x(t)$ that follows $p_{\mathrm{Pareto}}(x)$ are respectively given by

$$\mu_{\mathrm{Pareto}} = \frac{ab}{a-1}, \tag{1.34}$$

and

$$\mathrm{Var}(x)_{\mathrm{Pareto}} = \frac{ab^2}{(a-1)^2(a-2)}. \tag{1.35}$$

It can be easily seen that μ_{Pareto} and $\mathrm{Var}(x)_{\mathrm{Pareto}}$ do not exist for $a = 1$. That fractal time series with LRD may not have its mean and or variance is one of its particular points [6].

Note that μ_x implies a global property of $x(t)$ while $\mathrm{Var}(x)$ represents a local property of $x(t)$. For an LRD $x(t)$, unfortunately, in general, the concepts of mean and variance are inappropriate to describe the global property and the local one of $x(t)$. We need other measures to characterize the global property and the local one of LRD $x(t)$. Fractal dimension and the Hurst parameter are utilized for this purpose.

1.3.2 Fractal Dimension and the Hurst Parameter

In fractal time series, one respectively uses the fractal dimension and the Hurst parameter of $x(t)$ to describe its local property and the global one ([3], Li and Lim [37, 42]). In fact, if r_{xx} is sufficiently smooth on $(0, \infty)$ and if

$$r_{xx}(0) - r_{xx}(\tau) \sim c_1 |\tau|^\alpha \ \text{ for } |\tau| \to 0, \tag{1.36}$$

where c_1 is a constant and α is the fractal index of $x(t)$, the fractal dimension of $x(t)$ is expressed by

$$D = 2 - \frac{\alpha}{2}, \tag{1.37}$$

see, for example, Kent and Wood [43], Hall and Roy [44], and Adler [45].

On the other side, expressing β in (1.27) by the Hurst parameter $0.5 < H < 1$ yields

$$\beta = 2 - 2H. \tag{1.38}$$

Therefore,

$$r_{xx}(\tau) \sim c|\tau|^{2H-2} \ (\tau \to \infty). \tag{1.39}$$

Different from those in conventional series, we use D and H to characterize the local property and the global one of LRD $x(t)$, respectively, rather than mean and variance (Gneiting and Schlather [46], and Lim and Li [47]).

In passing, we mention that the estimation of H and/or D becomes a branch of fractal time series as can be seen from refs [11, 12]. Various methods regarding the estimation of fractal parameters are reported (see, e.g., Taqqu et al. [48]) such as methods based on ACF regression (Li [49–53]), periodogram regression method (Raymond et al. [54]), generalized linear regression (Beran [55, 56]), scaled and rescaled windowed variance methods [57–59], dispersional method (Raymond and Bassingthwaighte [60]), maximum likelihood estimation methods (Caccia et al. [61], Guerrero and Smith [62]), methods based on wavelet [63–67], and fractional Fourier transform (Chen et al. [68]).

In the end of this section, we note that self-similarity of a stationary process is a concept closely relating to fractal time series. Fractional Gaussian noise (fGn) is an only stationary increment process with self-similarity (Samorodnitsky and Taqqu [69]). In general, however, a fractal time series may not be globally self-similar. Nevertheless, a series that is not self-similar may be locally self-similar [45].

1.4 SOME MODELS OF FRACTAL TIME SERIES

Fractal time series can be classified into two classes from the view of statistical dependence: One is LRD and the other SRD. It can be also classified into Gaussian series or non-Gaussian ones. I shall discuss the models of fractal time series of Gaussian type in Section 1.4.1–1.4.4 and Section 1.4.6. Series of non-Gaussian type will be described in Section 1.4.5.

1.4.1 Fractional Brownian Motions

FBms are commonly used in modeling nonstationary fractal time series. They are Gaussian (Sinai [70, 71]). The definition of the fBm described in (1.21) is called the Riemann-Liouville type since it uses the Riemann-Liouville integral (see, e.g., [25], Sithi and Lim [72], Muniandy and Lim [73], and Feyel and Pradelle [74]). Its PSD is given by

$$S_{B_H, \text{RL}}(t, \omega) = \frac{\pi \omega t}{\omega^{2H+1}} \left[J_H(2\omega t) \mathbb{H}_{H-1}(2\omega t) - J_{H-1}(2\omega t) \mathbb{H}_H(2\omega t) \right], \tag{1.40}$$

where J_H is the Bessel function of order H (Korn and Korn [75]), $\mathbb{H}_H$ is the Struve function of order H, and the subscript on the left side implies the type of the Riemann-Liouville integral (see [72] for details). The ACF of the fBm in the form of $B_{H,\,RL}(t)$ is given by

$$r_{B_H,\,RL}(t,s) = \frac{t^{H+1/2}s^{H-1/2}}{(H+1/2)\Gamma(H+1/2)^2} \, _2F_1(1/2-H,1,H+1/2,t/s), \qquad (1.41)$$

where $_2F_1$ is the hypergeometric function.

Note that the increment process of $B_{H,\,RL}(t)$ is nonstationary (Lim and Muniandy [76]). Therefore, another definition of the fBm based on the Weyl integral [25] is usually used when considering the stationary increment process of that fBm.

The Weyl integral of order v is given for $v > 0$ by

$$W^{-v}f(t) = \frac{1}{\Gamma(v)} \int\limits_t^\infty (u-t)^{v-1} f(u)du. \qquad (1.42)$$

Thus, the fBm of the Weyl type, denoted by $B_{H,\,Weyl}(t)$, is defined by

$$B_{H,\,Weyl}(t) - B_{H,\,Weyl}(0)$$

$$= \frac{1}{\Gamma(H+1/2)}\left\{ \int\limits_{-\infty}^0 \left[(t-u)^{H-0.5} - (-u)^{H-0.5}\right]dB(u) + \int\limits_0^t (t-u)^{H-0.5}\,dB(u) \right\}. \qquad (1.43)$$

It has stationary increment. Its PSD is given by (Flandrin [77])

$$S_{B_H,\,Weyl}(t,\omega) = \frac{V_H}{|\omega|^{2H+1}}\left(1 - 2^{1-2H}\cos 2\omega t\right). \qquad (1.44)$$

Its ACF is expressed by

$$r_{B_H,\,Weyl}(t,s) = \frac{V_H}{(H+1/2)\Gamma(H+1/2)}\left[|t|^{2H} + |s|^{2H} - |t-s|^{2H}\right], \qquad (1.45)$$

where V_H is the strength of $B_{H,\,Weyl}(t)$ and it is given by

$$V_H = \mathrm{Var}[B_{H,\,Weyl}(1)] = \Gamma(1-2H)\frac{\cos \pi H}{\pi H}. \qquad (1.46)$$

The basic properties of fBms are listed below.

Note 1.7. Either $B_{H,\,RL}(t)$ or $B_{H,\,Weyl}(t)$ is nonstationary as can be seen from (1.40) or (1.44). □

Note 1.8. Both $B_{H,\,RL}(t)$ and $B_{H,\,Weyl}(t)$ are self-similar because they have the property expressed by

$$B_{H,RL}(at) \equiv a^H B_{H,RL}(t), \quad a > 0,$$
$$B_{H,Weyl}(at) \equiv a^H B_{H,Weyl}(t), \quad a > 0, \qquad (1.47)$$

where $\equiv$ denotes equality in the sense of probability distribution. □

Note 1.9. The PSDs of two types of fBms are divergent at $\omega = 0$, exhibiting a case of $1/f$ noise. □

Note 1.10. Two types of fBms reduce to the standard Brownian motion when $H = 1/2$, as can be seen from (1.21) and (1.43). □

Note 1.11. A consequence of Note 1.10 for $V_H = 1$ is:

$$S_{B_{1/2},\,\mathrm{RL}}(t,\omega) = S_{B_{1/2},\,\mathrm{Weyl}}(t,\omega) = \frac{1}{\omega^2}(1 - \cos 2\omega t), \tag{1.48}$$

which is the PSD of the standard Brownian motion [72]. □

Note 1.12. The fractal dimension of two fBms is given by

$$D_{\mathrm{fBm}} = 2 - H_{\mathrm{fBm}}. \tag{1.49}$$

1.4.2 Generalized Fractional Brownian Motion with Holder Function

In the subsequent text $B_{H,\,\mathrm{Weyl}}(t)$ is simplified by $B_H(t)$ without causing any confusion. Recall that the fractal dimension of a sample path represents its self-similarity. For fBm, however, D_{fBm} is linearly related to H_{fBm}, see (1.49). Because (1.47) holds for all time scales, it represents a global self-similarity of fBm. This is a mono-fractal character, which may be too restrictive for many practical applications. Lim and Muniandy [76] replaced the Hurst parameter H in (1.43) by a continuously deterministic function $H(t)$ to obtain a form of the generalized fBm. The function $H(t)$ satisfies $H : [0,\infty) \to (0,1)$. Denote the generalized fBm by $X(t)$, instead of $B_H(t)$, so as to distinguish it from the standard one. Then,

$$X(t) = \frac{1}{\Gamma(H(t)+1/2)}\left\{\int_{-\infty}^{0}\left[(t-u)^{H-0.5} - (-u)^{H-0.5}\right]dB(u) + \int_{0}^{t}(t-u)^{H(t)-0.5}\,dB(u)\right\}. \tag{1.50}$$

By using $H(t)$, one has a tool to characterize local properties of fBm. The following ACF holds for $\tau \to 0$:

$$E[X(t)X(t+\tau)] = \frac{V_{H(t)}}{(H(t)+1/2)\Gamma(H(t)+1/2)}\left[|t|^{2H(t)} + |t+\tau|^{2H(t)} - |\tau|^{2H(t)}\right]. \tag{1.51}$$

The self-similarity expressed below is in the local sense, as $H(t)$ is time-varying

$$X(at) \equiv a^{H(t)}X(t), \quad a > 0. \tag{1.52}$$

Assume $H(t)$ is a β Holder function. Then, $0 < \inf[H(t)] \le \sup[H(t)] < \min(1,\beta)$. Therefore, one has the following local Hausdorff dimension of $X(t)$ for $[a, b] \subset \mathbf{R}_+$:

$$\dim\{X(t), t \in [a,b]\} = 2 - \min\{H(t), t \in [a,b]\}. \tag{1.53}$$

The above expression also exhibits the local self-similarity of $X(t)$.

Based on the local growth of the increment process, one may write a sequence expressed by

$$S_k(j) = \frac{m}{N-1} \sum_{j=0}^{j+k} |X(i+1) - X(i)|, \quad 1 < k < N, \tag{1.54}$$

where m is the largest integer not exceeding N/k. Then, $H(t)$ at point $t = j/(N-1)$ is given by

$$H(t) = -\frac{\log(\sqrt{\pi/2} S_k(j))}{\log(N-1)}, \tag{1.55}$$

see Peltier and Levy-Vehel [78, 79], and Li [80]. Muniandy et al. [81] demonstrates an application of this type of fBm to financial engineering.

1.4.3 Fractional Gaussian Noise

The continuous fGn is the derivative of the smoothed fBm that is in the domain of generalized functions. Its ACF, denoted by $C_H(\tau; \varepsilon)$, is given by

$$C_H(\tau; \varepsilon) = \frac{V_H \varepsilon^{2H-2}}{2} \left[\left(\frac{|\tau|}{\varepsilon} + 1 \right)^{2H} + \left| \frac{|\tau|}{\varepsilon} - 1 \right|^{2H} - 2 \left| \frac{\tau}{\varepsilon} \right|^{2H} \right], \tag{1.56}$$

where $0 < H < 1$ is the Hurst parameter and $\varepsilon > 0$ is used by smoothing fBm so that the smoothed fBm is differentiable [25].

FGn includes three classes of time series. When $0.5 < H < 1$, $C_H(\tau; \varepsilon)$ is positive and finite for all τ. It is non-integrable and the corresponding series is LRD. For $0 < H < 0.5$, the integral of $C_H(\tau; \varepsilon)$ is zero and $C_H(0; \varepsilon)$ diverges when $\varepsilon \to 0$. In addition, $C_H(\tau; \varepsilon)$ changes its sign and becomes negative for some τ proportional to ε in this parameter domain [25, p. 434]. FGn reduces to the white noise when $H = 0.5$.

Let $V_H = \sigma^2$. The PSD of fGn is given by (Li and Lim [36])

$$S_{fGn}(\omega) = \sigma^2 \sin(H\pi) \Gamma(2H+1) |\omega|^{1-2H}. \tag{1.57}$$

Denote the discrete fGn by dfGn. Then, the ACF of dfGn is given by

$$r_{dfGn}(k) = \frac{\sigma^2}{2} \left[(|k|+1)^{2H} + ||k|-1|^{2H} - 2|k|^{2H} \right]. \tag{1.58}$$

Its PSD (see Sinai [71]) is given by

$$S_{dfGn}(\omega) = 2C_f(1 - \cos\omega) \sum_{n=-\infty}^{\infty} |2\pi n + \omega|^{-2H-1}, \quad \omega \in [-\pi, \pi], \tag{1.59}$$

where $C_f = \sigma^2 (2\pi)^{-1} \sin(\pi H) \Gamma(2H+1)$.

Note that the expression $0.5[(k + 1)^{2H} - 2k^{2H} + (k - 1)^{2H}]$ is the finite second-order difference of $0.5(k)^{2H}$. Approximating it with the second-order differential of $0.5(k)^{2H}$ yields

$$0.5\left[(k+1)^{2H} - 2k^{2H} + (k-1)^{2H}\right] \approx H(2H-1)(k)^{2H-2}.$$

The above approximation is quite accurate for $k > 5$. Hence, taking into account (1.37) and (1.38), the following immediately appears (Li and Lim [42])

$$D_{fGn} = 2 - H_{fGn}. \tag{1.60}$$

Hence, the following notes are derived.

Note 1.13. The fGn as the increment process of $B_{H,\,Weyl}(t)$ is stationary. It is exactly self-similar with the global self-similarity described by (1.60). □

Note 1.14. The PSD of the fGn is divergent at $\omega = 0$ when $0.5 < H < 1$. □

Again, we remark that the fGn may be too strict for modeling a real series in practice. Hence, generalized versions of fGn are expected. One of the generalizations of fGn is to replace H by $H(t)$ in (1.58) ([76, 80]) so that

$$r_{dfGn}(k; H(t)) = \frac{\sigma^2}{2}\left[(|k|+1)^{2H(t)} + \big||k|-1\big|^{2H(t)} - 2|k|^{2H(t)}\right]. \tag{1.61}$$

Another generalization by Li [49, 50] is given by

$$r_{dfGn}(k; H, a) = \frac{\sigma^2}{2}\left(\big||k|^a + 1\big|^{2H} - 2\big||k|^a\big|^{2H} + \big||k|^a - 1\big|^{2H}\right), \quad 0 < a \le 1. \tag{1.62}$$

In (1.61), if $H(t) = $ const, the ACF reduces to that of the standard fGn. On the other side, $r_{dfGn}(k; H, a)$ in (1.62) becomes the ACF of the standard fGn if $a = 1$.

1.4.4 Generalized Cauchy (GC) Process

As discussed in Section 1.2, we use two parameters, namely, D and H, to respectively measure the local behavior and the global one of fractal time series instead of variance and mean. More precisely, the former measures a local property, namely, local irregularity, of a sample path while the latter characterizes a global property, namely, LRD. The parameters $1 < D < 2$ and $0 < H < 1$ are two different measures. They may be linearly related, see (1.60), or coupled (Li [49]), or independent of each other (Li and Lim [42]). By using a single-parameter model, such as fGn and fBm, D and H happen to be linearly related. Hence, a single-parameter model fails to separately capture the local irregularity and LRD. To release such relationship, two-parameter model is needed. The generalized Cauchy (GC) process is one of such models.

A series $X(t)$ is called the GC process if it is a stationary Gaussian centered process with the normalized ACF given by

$$C_{GC}(\tau) = E[X(t+\tau)X(t)] = (1+|\tau|^{\alpha})^{\frac{\beta}{\alpha}},$$

(1.63)

where $0 < \alpha \leq 2$ and $\beta > 0$. The ACF $C_{GC}(\tau)$ is positive-definite for the above ranges of α and β and it is completely monotone for $0 < \alpha \leq 1$, $\beta > 0$. When $\alpha = \beta = 2$, one gets the usual Cauchy process that is modeled by its ACF expressed by

$$C(\tau) = (1+|\tau|^{2})^{-1},$$

(1.64)

which has been applied in geostatistics (see, e.g., Chiles and Delfiner [82]).

The function $C_{GC}(\tau)$ has the asymptotic expressions of (1.36) and (1.39). More precisely, we have

$$C_{GC}(\tau) \sim |\tau|^{\alpha}, \quad \tau \to 0,$$

(1.65)

$$C_{GC}(\tau) \sim |\tau|^{-\beta}, \quad \tau \to \infty.$$

(1.66)

According to (1.37) and (1.38), therefore, one has

$$D_{GC} = 2 - \frac{\alpha}{2},$$

(1.67)

$$H_{GC} = 1 - \frac{\beta}{2}.$$

(1.68)

When considering the multiscale property of a series, one may utilize the time-varying D_{GC} and H_{GC} on an interval-by-interval basis (Li [49]). Denote the fractal dimension and the Hurst parameter in the nth interval by $D_{GC}(n)$ and $H_{GC}(n)$, respectively. Then, we have the ACF in the Ith interval given by

$$C_{GC}(\tau; n) = \left[1 + \tau^{\alpha(n)}\right]^{-\frac{\beta(n)}{\alpha(n)}}, \quad \tau \geq 0.$$

(1.69)

Consequently, we have

$$D_{GC}(n) = 2 - \frac{\alpha(n)}{2},$$

(1.70)

$$H_{GC}(n) = 1 - \frac{\beta(n)}{2}.$$

(1.71)

Let $\mathrm{Sa}(\omega) = (\sin\omega)/\omega$. Then, the PSD of the GC process is given by (Li and Lim [37])

$$S_{\mathrm{GC}}(\omega) = \sum_{k=0}^{\infty} \frac{(-1)^k \Gamma[(\beta/\alpha)+k]}{\pi \Gamma(\beta/\alpha)\Gamma(1+k)} I_1(\omega) * \mathrm{Sa}(\omega)$$

$$+ \sum_{k=0}^{\infty} \frac{(-1)^k \Gamma[(\beta/\alpha)+k]}{\pi \Gamma(\beta/\alpha)\Gamma(1+k)} [\pi I_2(\omega) - I_2(\omega) * \mathrm{Sa}(\omega)], \tag{1.72}$$

where

$$I_1(\omega) = -2 \sin(\alpha k \pi/2)\Gamma(\alpha k+1)|\omega|^{-\alpha k-1}, \tag{1.73}$$

$$I_2(\omega) = 2 \sin[(\beta+\alpha k)\pi/2]\Gamma[1-(\beta+\alpha k)]|\omega|^{(\beta+\alpha k)-1}. \tag{1.74}$$

In practice, the asymptotic expressions of $S_{\mathrm{GC}}(\omega)$ for small frequency and large one may be useful. The PSD of the GC process for $\omega \to 0$ is given by

$$S_{\mathrm{GC}}(\omega) \sim \frac{1}{\Gamma(\beta)\cos(\beta\pi/2)}|\omega|^{\beta-1}, \quad \omega \to 0, \tag{1.75}$$

which is actually the inverse Fourier transform of $C_{\mathrm{GC}}(\tau)$ for $\tau \to \infty$. On the other hand, $S_{\mathrm{GC}}(\omega)$ for $\omega \to \infty$ is given by

$$S_{\mathrm{GC}}(\omega) \sim \frac{\beta \Gamma(1+\alpha)\sin(\alpha\pi/2)}{\pi\alpha}|\omega|^{-(1+\alpha)}, \quad \omega \to \infty, \tag{1.76}$$

see ref [47] for details. As shown in (1.75) and (1.76), one may easily observe the power law that $S_{\mathrm{GC}}(\omega)$ obeys.

Note 1.15. The GC process is LRD if $0 < \beta < 1$. It is SRD if $1 < \beta$. Its statistical dependence is measured by H, see (1.68). □

Note 1.16. The GC process has the local self-similarity measured by D_{GC} expressed by (1.67). □

Note 1.17. The GC process is non-Markovian since $C_{\mathrm{GC}}(t_1, t_2)$ does not satisfy the triangular relation given by

$$C_{\mathrm{GC}}(t_1, t_3) = \frac{C_{\mathrm{GC}}(t_1, t_2)C_{\mathrm{GC}}(t_2, t_3)}{C_{\mathrm{GC}}(t_2, t_2)}, \quad t_1 < t_2 < t_3, \tag{1.77}$$

which is a necessary condition for a Gaussian process to be Markovian (Todorovic [83]). In fact, up to a multiplicative constant, the Ornstein–Uhlenbeck process is the only stationary Gaussian Markov process (Lim and Muniandy [84], Wolperta and Taqqu [85]). □

The above discussions exhibit that the GC model can be used to decouple the local behavior and the global one of fractal time series, and for flexibly better agreement with the real data for both short-term and long-term lags. Li gave an analysis of the modeling performance of the GC model in Hilbert space [49]. The application of the GC process to network traffic modeling refers to ref [42], Li [49], Li and Wang [86]. Lim and Teo [87] extended the GC model to describe the Gaussian fields and Gaussian sheets. Vengadesh et al. [88] applied it to the analysis of bacteriorhodopsin in material science.

1.4.5 Alpha-Stable Processes

As previously mentioned, two-parameter models are useful as they can separately characterize the local irregularity and global persistence. The CG process is one of such models and it is Gaussian. In some applications, for example, network traffic at small scales, a series is non-Gaussian (see, e.g., Scherrer et al. [89]). One type of model that is two-parameter and non-Gaussian in general is α-stable process.

Stable distributions imply a family of distributions. They are defined by their characteristic functions given by ([69, p. 5]), for a random variable Y,

$$\Phi(\theta) = E(e^{j\theta Y}) = \begin{cases} \exp\{j\mu\theta - |\sigma\theta|^{\alpha}[1 + j\beta\text{sign}(\theta)\tan(\pi\alpha/2)]\}, \alpha \neq 1, \\ \exp\{j\mu\theta - |\sigma\theta|[1 + j\beta\text{sign}(\theta)\ln|\theta|]\}, \alpha = 1. \end{cases} \quad (1.78)$$

The expression $Y \sim S_{\sigma,\beta,\mu}^{(\alpha)}$ implies that Y follows $\Phi(\theta)$.

The parameters in $\Phi(\theta)$ are explained as follows.

- The parameter $0 < \alpha \leq 2$ is characteristic exponent. It specifies the level of local roughness in the distribution, that is, the weight of the distribution tail.

- The parameter $-1 \leq \beta \leq 1$ specifies the skewness. Its positive values correspond to the right tail while negative ones to the left.

- The parameter $\sigma \geq 0$ is a scale factor, implying the dispersion of the distribution.

- $\mu \in \mathbf{R}$ is the location parameter, expressing the mean or median of the distribution.

Note 1.18. The family of α-stable distributions does not have a closed form of expressions in general. A few exceptions are the Cauchy distribution and the Levy one. □

Note 1.19. The property of heavy tail is described as follows. $E[|Y|^p] < \infty$ for $p \in (0, \alpha)$. $E[|Y|^p] = \infty$ for $p \geq \alpha$. □

When $\alpha = 2$, the characteristic function (1.78) reduces to that of the Gaussian distribution with the mean denoted by μ and the variance denoted by $2\sigma^2$. That is,

$$\Phi(\theta) = E(e^{j\theta Y}) = \exp[j\mu\theta - (\sigma\theta)^2]. \quad (1.79)$$

In this case, the PDF of Y is symmetric about the mean.

Alpha-stable processes are in general non-Gaussian. They include two types of noise. One is linear fractional stable noise (LFSN) and the other log-fractional stable noise (Log-FSN).

The model of LFSM is defined by the following stochastic integral ([69, p. 366]). Denote by $L_{\alpha,H}(t)$ the LFSM. Then,

$$L_{\alpha,H}(t) = \int_{-\infty}^{\infty} \left\{ a\left[(t-u)_+^{H-1/\alpha} - (-u)_+^{H(t)-1/\alpha} \right] + b\left[(t-u)_-^{H-1/\alpha} - (-u)_-^{H(t)-1/\alpha} \right] \right\} M du, \quad (1.80)$$

where a and b are arbitrary constants, $M \in \mathbf{R}$ is a random measure, and H the Hurst parameter. The range of H is given by

$$H = \begin{cases} (0,1], & \alpha \geq 1, \\ (0,1/\alpha], & \alpha < 1. \end{cases} \quad (1.81)$$

Denote by $Lo_{\alpha,H}(t)$ the Log-FSM. Then,

$$Lo_\alpha(t) = \int_{-\infty}^{\infty} \left(\ln|t - Y| - \ln|Y| \right) M du. \quad (1.82)$$

LSFN is the increments process of LSFM while Log-FSN is the increment process of Log-FSM. Denote the LSFN and Log-FSN by $N_{\alpha,H}(i)$ and $NLo_{\alpha,H}(i)$., respectively Then,

$$N_{\alpha,H}(i) = L_{\alpha,H}(i+1) - L_{\alpha,H}(i), \ i \in \mathbf{Z}, \quad (1.83)$$

$$NLo_{\alpha,H}(i) = Lo_\alpha(i+1) - Lo_\alpha(i), \ i \in \mathbf{Z}. \quad (1.84)$$

LSFN is non-Gaussian except $\alpha = 2$. It is stationary self-similar with the self-similarity measured by H and the local roughness characterized by α, see ref [69]. However, two parameters are not independent because the LRD condition ([69], Karasaridis and Hatzinakos [90]) relates them by

$$\alpha H > 1. \quad (1.85)$$

1.4.6 Ornstein–Uhlenbeck (OU) Processes and Their Generalizations

In the above subsections, the focus of the series considered is LRD. We now turn to a type of SRD fractal time series called OU processes.

1.4.6.1 Ordinary OU Process

Following the idea addressed by Uhlenbeck and Ornstein [91], the ordinary OU process is regarded as the solution to the Langevin equation ([84, 85], Lu [92], Valdivieso et al. [93]), which is a stochastic differential equation given by

$$\begin{cases} \left(\dfrac{d}{dt} + \lambda \right) X(t) = w(t), \\ X(0) = X_0, \end{cases} \quad (1.86)$$

where λ is a positive parameter, $w(t)$ is the white noise with zero mean, X_0 is a random variable independent of the standard Brownian motion $B(t)$. The stationary solution to the above equation is given by

$$X(t) = X_0 e^{-\lambda t} + \int_{-\infty}^{t} e^{\lambda u} w(u) du. \tag{1.87}$$

Denote the Fourier transforms of $w(t)$ and $X(t)$ by $W(\omega)$ and $X(\omega)$, respectively. Note that the system function of (1.86) in the frequency domain is given by

$$G_{OU}(\omega) = \frac{1}{\lambda + j\omega}. \tag{1.88}$$

Then, according to the convolution theorem, one has

$$X(\omega) = G_{OU}(\omega) W(\omega). \tag{1.89}$$

Since the PSD of the normalized $w(t)$ equals to 1, that is, $|W(\omega)|^2 = 1$, we immediately obtain the PSD of the OU process given by

$$S_{OU}(\omega) = \frac{1}{\lambda^2 + \omega^2}. \tag{1.90}$$

Consequently, the ACF of the OU process is given by

$$E[X(t)X(t+\tau)] = F^{-1}[S_{OU}(\omega)] = \frac{e^{-\lambda|\tau|}}{2\lambda}, \tag{1.91}$$

where F^{-1} is the operator of the inverse Fourier transform.

The ordinary OU process is obviously SRD. It is a one-dimensional process. What interests people in the field of fractal time series is the generalized OU processes described below.

1.4.6.2 Generalized Version I of the OU Process

Consider the following fractional Langevin equation with a single parameter $\beta > 0$

$$\left(\frac{d}{dt} + \lambda\right)^{\beta} X_1(t) = w(t). \tag{1.92}$$

Denote the impulse response function of the above system by $g_{X_1}(t)$. Then, it is the solution to the following equation

$$\left(\frac{d}{dt} + \lambda\right)^{\beta} g_{X_1}(t) = \delta(t), \tag{1.93}$$

where $\delta(t)$ is the δ function. Doing the Fourier transforms on the both sides on the above equation yields

$$G_{X_1}(\omega) = \frac{1}{(\lambda - j\omega)^\beta}, \tag{1.94}$$

where $G_{X_1}(\omega)$ is the Fourier transform of $g_{X_1}(t)$.

Note that the PSD of $X_1(t)$ is equal to

$$S_{X_1}(\omega) = G_{X_1}(\omega)[G_{X_1}(\omega)]^*, \tag{1.95}$$

where $[G_{X_1}(\omega)]^*$ is the complex conjugate of $G_{X_1}(\omega)$. Then,

$$S_{X_1}(\omega) = \frac{1}{(\lambda^2 + \omega^2)^\beta}, \tag{1.96}$$

which is the solution to (1.92) in the frequency domain. The solution to (1.92) in the time domain, therefore, is given by

$$C_{X_1}(\tau) = \mathrm{F}^{-1}[S_{X_1}(\omega)] = \frac{\lambda^{-2v}}{2^v \sqrt{\pi}\Gamma(v+1/2)}|\lambda\tau|^v K_v(|\lambda\tau|), \tag{1.97}$$

where $v = \beta - 1/2$ and K_v is the modified Bessel function of the second kind of order v [27, 84].

Let $v = H \in (0, 1)$. Then, one has

$$S_{X_1}(\omega) = \frac{1}{(\lambda^2 + \omega^2)^{H+1/2}}, \tag{1.98}$$

which exhibits that $X_1(t)$ is SRD because its PSD is convergent for $\omega \to 0$.

Keep in mind that the Langevin equation is in the sense of generalized functions since we take $w(t)$ as the differential of the standard Brownian motion $B(t)$, which is differentiable if it is regarded as a generalized function only. In the domain of generalized functions and following [16, p. 278], there is a generalized limit given by

$$\lim_{\omega \to \infty} \cos \omega t = 0. \tag{1.99}$$

Therefore, the PSD of the fBm of the Weyl type (see (1.44)) has the following asymptotic property:

$$\lim_{\omega \to \infty} S_{B_H, \text{Weyl}}(t, \omega) \sim \frac{1}{|\omega|^{2H+1}}. \tag{1.100}$$

On the other hand, from (1.98), for when $\omega \gg \lambda$, we see that the PSD of $X_1(t)$ has the asymptotic expression given by

$$S_{X_1}(\omega) \sim \frac{1}{|\omega|^{2H+1}}. \tag{1.101}$$

Therefore, we see that $S_{X_1}(\omega)$ has the approximation given by

$$S_{X_1}(\omega) \sim S_{B_H,\text{Weyl}}(t,\omega) \quad \text{for } \omega \to \infty. \tag{1.102}$$

Hence, comes Note 1.20.

Note 1.20. The generalized OU process governed by (1.92) can be taken as the locally stationary counterpart of the fBm. □

According to (1.29), we have

$$\mathrm{F}^{-1}\left(\frac{1}{|\omega|^{2H+1}}\right) \sim |\tau|^{2H}. \tag{1.103}$$

Therefore, we obtain

$$C_{X_1}(\tau) \sim c_{X_1}|\tau|^{2H} \quad \text{for } \tau \to 0, \tag{1.104}$$

where c_{X_1} is a constant. Following (1.36) and (1.37), we have the fractal dimension of $X_1(t)$ given by

$$D_{X_1} = 2 - H. \tag{1.105}$$

1.4.6.3 Generalized Version II of the OU Process (Lim et al. [35])

We now further extend the Langevin equation by indexing it with two fractions $\alpha, \beta > 0$ so that

$$\left(_{-\infty}D_t^\alpha + \lambda\right)^\beta X_2(t) = w(t), \tag{1.106}$$

where $_{-\infty}D_t^\alpha$ is the operator of the Weyl fractional derivative. Denote the impulse response function of the above system by $g_{X_2}(t)$. Then,

$$\left(_{-\infty}D_t^\alpha + \lambda\right)^\beta g_{X_2}(t) = \delta(t). \tag{1.107}$$

The Fourier transform of $g_{X_2}(t)$, which is denoted by $G_{X_2}(\omega)$, is given by

$$G_{X_2}(\omega) = \frac{1}{\left[\lambda + (-j\omega)^\alpha\right]^\beta}. \tag{1.108}$$

Therefore, the PSD of $X_2(t)$ is

$$S_{X_2}(\omega) = G_{X_2}(\omega)[G_{X_2}(\omega)]^* = \frac{1}{\left|\lambda + (j\omega)^\alpha\right|^{2\beta}}. \tag{1.109}$$

Note that

$$S_{X_2}(\omega) \sim \frac{1}{\omega^{2\alpha\beta}} \quad \text{for } \omega \to \infty. \tag{1.110}$$

In addition,

$$F^{-1}\left(\frac{1}{|\omega|^{2\alpha\beta}}\right) \sim |\tau|^{2\alpha\beta-1}. \tag{1.111}$$

Thus, the ACF of $X_2(t)$ has the asymptotic expression given by

$$C_{X_2}(\tau) \sim c_{X_2} |\tau|^{2\alpha\beta-1} \quad \text{for } \tau \to 0, \tag{1.112}$$

where c_{X_2} is a constant. Hence, the fractal dimension of $X_2(t)$ is given by

$$D_{X_2} = \frac{5}{2} - \alpha\beta. \tag{1.113}$$

In the above, $1/2 < \alpha\beta < 3/2$, which is a condition to assure $1 < D_{X_2} < 2$.

Note 1.21. The local irregularity of series relies on the fractal dimension instead of the statistical dependence. The local irregularity of an SRD series may be strong if its fractal dimension is large. □

1.5 SUMMARY

The concepts such as $1/f$ noise, power laws in PDF, ACF, and PSD in fractal time series have been discussed. The topics of $1/f$ noise and power laws will be further addressed in Chapters 2 and 3, respectively.

Both LRD and SRD series have been explained in this chapter. Several models, fBms, fGns, the GC process, alpha-stable processes, and generalized OU processes have been interpreted. Note that several models that are revisited above are a few in the family of fractal time series. There are other models as well (see, e.g., [72, 94–106]). As a matter of fact, the family of fractal time series is affluent but those discussed might yet be adequate to describe the fundamentals of fractal time series in traffic engineering. Two types of multifractal time series will be addressed in Chapters 9 and 10. Other important properties of LRD traffic, namely, ergodicity and predictability, will be individually introduced in Chapters 4 and 5. The convergence issue of sample ACF of LRD traffic will be discussed in Appendix.

REFERENCES

1. D. H. Griffel, *Applied Functional Analysis*, John Wiley & Sons, New York, 1981.
2. C. K. Liu, *Applied Functional Analysis*, Defence Industry Press, Beijing, China, 1986. In Chinese.
3. B. B. Mandelbrot, *The Fractal Geometry of Nature*, W. H. Freeman, New York, 1982.
4. G. Korvin, *Fractal Models in the Earth Science*, Elsevier, Amsterdam, The Netherlands, 1992.
5. E. E. Peters, *Fractal Market Analysis: Applying Chaos Theory to Investment and Economics*, John Wiley & Sons, New York, 1994.
6. J. B. Bassingthwaighte, L. S. Liebovitch, and B. J. West, *Fractal Physiology*, Oxford University Press, Oxford, 1994.
7. W. A. Fuller, *Introduction to Statistical Time Series*, 2nd ed., John Wiley & Sons, New York, 1996.
8. G. E. P. Box, G. M. Jenkins, and G. C. Reinsel, *Time Series Analysis: Forecasting and Control*, Prentice Hall, New York, 1994.
9. S. K. Mitra, and J. F. Kaiser, *Handbook for Digital Signal Processing*, John Wiley & Sons, New York, 1993.
10. J. S. Bendat, and A. G. Piersol, *Random Data: Analysis and Measurement Procedure*, 3rd ed., John Wiley & Sons, New York, 2000.
11. B. B. Mandelbrot, *Gaussian Self-Affinity and Fractals*, Springer, New York, 2001.
12. J. Beran, *Statistics for Long-Memory Processes*, Chapman & Hall, New York, 1994.
13. J. Levy-Vehel, E. Lutton, and C. Tricot (eds.), *Fractals in Engineering*, Springer, Berlin, 1997.
14. W. H. Press, S. A. Teukolsky, W. T. Vetterling, and B. P. Flannery, *Numerical Recipes in C_{++} The Art of Scientific Computing*, 2nd ed., Cambridge University Press, Cambridge, 2002.
15. H. Y.-F. Lam, *Analog and Digital Filters: Design and Realization*, Prentice-Hall, New York, 1979.
16. A. Papoulis, *The Fourier Integral and Its Applications*, McGraw-Hill, New York, 1962.
17. R. O. Harger, *An Introduction to Digital Signal Processing with MATHCAD*, PWS Publishing Company, New York, 1999.
18. J. Van de Vegte, *Fundamentals of Digital Signal Processing*, Prentice Hall, New York, 2003.
19. K. S. Miller, and B. Ross, *An Introduction to the Fractional Calculus and Fractional Differential Equations*, John Wiley & Sons, New York, 1993.
20. M. D. Ortigueira, Introduction to fractional linear systems, part I: Continuous-time systems, *IEE Proc. Vis. Image Sig Proc.*, 147(1) 2000, 62–70.
21. M. D. Ortigueira, Introduction to fractional linear systems, part ii: Discrete-time systems, *IEE Proceedings: Vision, Image and Signal Processing*, 147(1) 2000, 71–78.
22. M. D. Ortigueira, An introduction to the fractional continuous-time linear systems: The 21st century systems, *IEEE Circuits and Systems Magazine*, 8(3) 2008, 19–26.
23. Y. Q. Chen, and K. L. Moore, Discretization schemes for fractional order differentiators and integrators, *IEEE Transactions on Circuits and Systems I: Fundamental Theory and Applications*, 49(3) 2002, 363–367.
24. B. M. Vinagre, Y. Q. Chen, and I. Petras, Two direct Tustin discretization methods for fractional-order differentiator/integrator, *Journal of The Franklin Institute*, 340(5) 2003, 349–362.
25. B. B. Mandelbrot, and J. W. van Ness, Fractional Brownian motions, fractional noises and applications, *SIAM Review*, 10(4) 1968, 422–437.
26. T. Hida, *Brownian Motion*, Springer, Berlin, 1980.
27. I. M. Gelfand, and K. Vilenkin, *Generalized Functions*, vol. 1, Academic Press, New York, 1964.
28. J. Mikusinski, *Operational Calculus*, Pergamon Press, Oxford, 1959.

29. M. Li, and C.-H. Chi, A correlation-based computational method for simulating long-range dependent data, *Journal of the Franklin Institute*, 340(6–7) 2003, 503–514.
30. L. Daw, A uniform result for the dimension of fractional Brownian motion level sets, *Statistics & Probability Letters*, 169, 2021, 108984.
31. B. Buchmann, and C. Klüppelberg, Maxima of stochastic processes driven by fractional Brownian motion, *Advances in Applied Probability*, 37(3) 2005, 743–764
32. N. Zeinali, and A. Pourdarvish, An entropy-based estimator of the Hurst exponent in fractional Brownian motion, *Physica A*, 591, 2022, 126690.
33. I. Podlubny, *Fractional Differential Equations*, Academic Press, San Diego, 1999.
34. S. C. Lim, M. Li, and L. P. Teo, Locally self-similar fractional oscillator processes, *Fluctuation and Noise Letters*, 7(2) 2007, L169–179.
35. S. C. Lim, M. Li, and L. P. Teo, Langevin equation with two fractional orders, *Physics Letters A*, 372(42) 2008, 6309–6320.
36. M. Li, and S. C. Lim, A rigorous derivation of power spectrum of fractional Gaussian noise, *Fluctuation and Noise Letters*, 6(4) 2006, C33–36.
37. M. Li, and S. C. Lim, Power spectrum of generalized Cauchy process, *Telecommunication Systems*, 43(3–4) 2010, 219–222.
38. B. B. Mandelbrot, *Multifractals and 1/f Noise*, Springer, New York, 1998.
39. M. Li, Fractional Gaussian noise and network traffic modeling, *Proc., the 8th WSEAS Int. Conf. on Applied Computer and Applied Computational Science*, pp. 34–39, Hangzhou, China, May 2009.
40. M. Li, Self-similarity and long-range dependence in teletraffic, *Proc., the 9th WSEAS Int. Conf. on Multimedia Systems and Signal Processing*, pp. 19–24, Hangzhou, China, May 2009.
41. P. Abry, P. Borgnat, F. Ricciato, A. Scherrer, and D. Veitch, Revisiting an old friend: On the observability of the relation between long range dependence and heavy tail, *Telecommunication Systems*, 43(3–4) 2010, 147–165.
42. M. Li, and S. C. Lim, Modeling network traffic using generalized Cauchy process, *Physica A*, 387(11) 2008, 2584–2594.
43. J. T. Kent, and A. T. Wood, Estimating the fractal dimension of a locally self-similar Gaussian process by using increments, *Journal of Royal Statistical Society Series B*, 59(3) 1997, 679–699.
44. P. Hall, and R. Roy, On the relationship between fractal dimension and fractal index for stationary stochastic processes, *The Annals of Applied Probability*, 4(1) 1994, 241–253.
45. A. J. Adler, *The Geometry of Random Fields*, John Wiley & Sons, New York, 1981.
46. T. Gneiting, and M. Schlather, Stochastic models that separate fractal dimension and Hurst effect, *SIAM Rev.*, 46(2) 2004, 269–282.
47. S. C. Lim, and M. Li, Generalized Cauchy process and its application to relaxation phenomena, *Journal of Physics A: Mathematical and General*, 39(2), 2006, 2935–2951.
48. M. S. Taqqu, V. Teverovsky, and W. Willinger, Estimators for long-range dependence: An empirical study, *Fractals*, 3(4) 1995, 785–798.
49. M. Li, *Fractal Teletraffic Modeling and Delay Bounds in Computer Communications*, 1st ed., CRC Press, Boca Raton, 2022.
50. M. Li, Modeling autocorrelation functions of long-range dependent teletraffic series based on optimal approximation in Hilbert space: a further study, *Applied Mathematical Modelling*, 31(3) 2007, 625–631.
51. M. Li, Change trend of averaged Hurst parameter of traffic under DDOS flood attacks, *Computers & Security*, 25(3) 2006, 213–220.
52. M. Li, An approach to reliably identifying signs of DDOS flood attacks based on LRD traffic pattern recognition, *Computers & Security*, 23(7) 2004, 549–558.
53. M. Li, Long-range dependence and self-similarity of teletraffic with different protocols at the large time scale of day in the duration of 12 years: Autocorrelation modeling, *Physica Scripta*, 95(4) 2020, 065222 (15 pp).

54. G. M. Raymond, D. B. Percival, and J. B. Bassingthwaighte, The spectra and periodograms of anti-correlated discrete fractional Gaussian noise, *Physica A*, 322, 2003, 169–179.
55. J. Beran, Fitting long-memory models by generalized linear regression, *Biometrika*, 80(4) 1993, 817–822.
56. J. Beran, On parameter estimation for locally stationary long-memory processes, *Journal of Statistical Planning and Inference*, 139(3) 2009, 900–915.
57. M. J. Cannon, D. B. Percival, D. C. Caccia, G. M. Raymond, and J. B. Bassingthwaighte, Evaluating scaled windowed variance methods for estimating the Hurst coefficient of time series, *Physica A*, 241(3–4) 1997, 606–626.
58. D. C. Caccia, D. B. Percival, M. J. Cannon, G. M. Raymond, and J. B. Bassingthwaighte, Analyzing exact fractal time series: Evaluating scaled windowed variance methods for estimating the Hurst coefficient of time series, *Physica A*, 246(3–4) 1997, 609–632.
59. J. B. Bassingthwaighte, and G. M. Raymond, Evaluating rescaled range analysis for time series, *Annals of Biomedical Engineering*, 22(4) 1994, 432–444.
60. G. M. Raymond, and J. B. Bassingthwaighte, Deriving dispersional and scaled windowed variance analyses of the correlation of discrete fractional Gaussian noise, *Physica A*, 265(1–2) 1999, 85–96.
61. D. C. Caccia, D. Percival, M. J. Cannon, G. Raymond, and J. B. Bassingthwaighte, Evaluating maximum likelihood estimation methods to determine the Hurst coefficient, *Physica A*, 273(3–4) 1999, 439–451.
62. A. Guerrero, and L. A. Smith, A maximum likelihood estimator for long-range persistence, *Physica A*, 355(2–4) 2005, 619–632.
63. D. Veitch, and P. Abry, Wavelet-based joint estimate of the long-range dependence parameters, *IEEE Trans. Information Theory*, 45(3) 1999, 878–897.
64. Y. Avraham, and M. Pinchas, A novel clock skew estimator and its performance for the IEEE 1588v2 (PTP) case in fractional Gaussian noise/generalized fractional Gaussian noise environment, *Frontiers in Physics*, 9, 2021, 796811.
65. Y. Avraham, and M. Pinchas, Two novel one-way delay clock skew estimators and their performances for the fractional Gaussian noise/generalized fractional Gaussian noise environment applicable for the IEEE 1588v2 (PTP) case, *Frontiers in Physics*, 10, 2022, 867861.
66. G. W. Wornell, Wavelet-based representations for the 1 over f family of fractal processes, *Proc. the IEEE*, 81(10) 1993, 1428–1450.
67. P. Abry, D. Veitch, and P. Flandrin, Long-range dependence: Revisiting aggregation with wavelets, *Journal of Time Series Analysis*, 19(3) 1998, 253–266.
68. Y.-Q. Chen, R. Sun, and A. Zhou, An improved Hurst parameter estimator based on fractional Fourier transform, *Telecommunication Systems*, 43(3–4) 2010, 197–206.
69. G. Samorodnitsky, and M. S. Taqqu, *Stable Non-Gaussian Random Processes: Stochastic Models with Infinite Variance*, Chapman and Hall, New York, 1994.
70. T. G. Sinai, Distribution of the maximum of a fractional Brownian motion, *Russian Mathematical Surveys*, 52(2) 1997, 119–138.
71. T. G. Sinai, Self-similar probability distributions, *Theory of Probability & Its Applications*, 21(1) 1976, 64–80.
72. V. M. Sithi, and S. C. Lim, On the spectra of Riemann-Liouville fractional Brownian motion, *Journal of Physics A: Mathematical and General*, 28(11) 2003, 2995–3003.
73. S. V. Muniandy, and S. C. Lim, Modeling of locally self-similar processes using multifractional Brownian motion of Riemann-Liouville type, *Physical Review E*, 63(4) 2001, 046104.
74. D. Feyel, and A. D. L. Pradelle, On fractional Brownian motion, *Potential Analysis*, 10(3) 1999, 273–288.
75. G. A. Korn, and T. M. Korn, *Mathematical Handbook for Scientists and Engineers*, McGraw-Hill, New York, 1961.

76. S. C. Lim, and S. V. Muniandy, On some possible generalizations of fractional Brownian motion, *Physics Letters A*, 226(2-3) 2000, 140-145.

77. P. Flandrin, On the spectrum of fractional Brownian motion, *IEEE Transactions on Information Theory*, 35(1) 1989, 197-199.

78. R. F. Peltier, and J. Levy-Vehel, *Multifractional Brownian Motion: Definition and Preliminaries Results*, INRIA TR 2645, France, 1995.

79. R. F. Peltier, and J. Levy-Vehel, *A New Method for Estimating the Parameter of Fractional Brownian Motion*, INRIA TR 2696, France, 1994.

80. M. Li, Multi-fractional generalized Cauchy process and its application to teletraffic, *Physica A*, 550, 2020, 123982.

81. S. V. Muniandy, S. C. Lim, and R. Murugan, Inhomogeneous scaling behaviors in Malaysian foreign currency exchange rates, *Physica A*, 301(1-4) 2001, 407-428.

82. J.-P. Chiles, and P. Delfiner, *Geostatistics, Modeling Spatial Uncertainty*, John Wiley & Sons, New York, 1999.

83. P. Todorovic, *An Introduction to Stochastic Processes and Their Applications*, Springer, New York, 1992.

84. S. C. Lim, and S. V. Muniandy, Generalized Ornstein-Uhlenbeck processes and associated self-similar processes, *Journal of Physics A: Mathematical and General*, 36(14) 2003, 3961-3982.

85. R. L. Wolperta, and M. S. Taqqu, Fractional Ornstein-Uhlenbeck Lévy processes and the telecom process upstairs and downstairs, *Signal Processing*, 85(8) 2005, 1523-1545.

86. M. Li, and A. Wang, Fractal teletraffic delay bounds in computer networks, *Physica A*, 557, 2020, 124903 (13 pp).

87. S. C. Lim, and L. P. Teo, Gaussian fields and Gaussian sheets with generalized Cauchy covariance structure, *Stochastic Processes and Their Applications*, 119(4) 2009, 1325-1356.

88. P. Vengadesh, S. V. Muniandy, and W. H. Abd. Majid, Fractal morphological analysis of bacteriorhodopsin (bR) layers deposited onto indium tin oxide (ITO) electrodes, *Materials Science and Engineering: C*, 29(5) 2009, 1621-1626.

89. A. Scherrer, N. Larrieu, P. Owezarski, P. Borgnat, and P. Abry, Non-Gaussian and long memory statistical characterizations for internet traffic with anomalies, *IEEE Transactions on Dependable and Secure Computing*, 4(1) 2007, 56-70.

90. A. Karasaridis, and D. Hatzinakos, Network heavy traffic modeling using α-stable self-similar processes, *IEEE Transactions on Communications*, 49(7) 2001, 1203-1214.

91. G. E. Uhlenbeck, and L. S. Ornstein, On the theory of the Brownian motion, *Physical Review*, 36(5) 1930, 823-841.

92. D.-X. Lu, *Stochastic Processes and Their Applications*, The Press of Tsinghua University, Beijing, China, 2006. In Chinese.

93. L. Valdivieso, W. Schoutens, and F. Tuerlinckx, Maximum likelihood estimation in processes of Ornstein-Uhlenbeck type, *Statistical Inference for Stochastic Processes*, 12(1) 2009, 1-19.

94. M. Shlesinger, G. M. Zaslavsky, and U. Frisch, (Eds.), *Lévy Flights and Related Topics in Physics*, Springer, New York, 1995.

95. R. N. Mantegna, and H. E. Stanley, Stochastic process with ultra-slow convergence to a Gaussian: The truncated Levy flight, *Physical Review Letters*, 73(22) 1994, 2946-2949.

96. J.-P. Bouchaud, and A. Georges, Anomalous diffusion in disordered media: Statistical mechanisms, models and physical applications, *Physics Reports*, 195(4-5) 1990, 127-293.

97. D. Applebaum, Lévy processes: From probability to finance and quantum groups, *Notices of the American Mathematical Society*, 51(11) 2004, 1336-1347.

98. R. J. Martin, and A. M. Walker, A power-law model and other models for long-range dependence, *Journal of Applied Probability*, 34(3) 1997, 657-670.

99. R. J. Martin, and J. A. Eccleston, A new model for slowly-decaying correlations, *Statistics & Probability Letters*, 13(2) 1992, 139-145.

100. M. Li, W. Jia, and W. Zhao, Correlation form of timestamp increment sequences of self-similar traffic on ethernet, *Electronics Letters*, 36(19) 2000, 1668–1669.

101. F. Chapeau-Blondeau, (Max, +) dynamic systems for modeling traffic with long-range dependence, *Fractals*, 6(4) 1998, 305–311.

102. B. Minasny, and A. B. McBratney, The Matérn function as a general model for soil variograms, *Geoderma*, 128(3–4) 2005, 192–207.

103. W. Z. Daoud, J. D. W. Kahl, and J. K. Ghorai, On the synoptic-scale Lagrangian autocorrelation function, *Journal of Applied Meteorology*, 42(2) 2003, 318–324.

104. T. von Karman, Progress in the statistical theory of turbulence, *Proc. N. A. S.*, 34(11) 1948, 530–539.

105. M. Li, Modified multifractional Gaussian noise and its application, *Physica Scripta*, 96(12) 2021, 125002.

106. M. Li, Generalized fractional Gaussian noise and its application to traffic modeling, *Physica A*, 579, 2021, 1236137 (22 pp).

CHAPTER 2

On 1/f Noise

In the frequency domain, traffic is of 1/f noise owing to its property of long-range dependence (LRD) no matter what a concrete LRD model we consider. We previously often mentioned the term "1/f noise." Due to this, we present the elementary theory with the particularities of 1/f noise in comparison with conventional random processes in this chapter. The theory consists of three theorems, namely, Theorems 2.1–2.3. They are given for highlighting the particularities of 1/f noise. Theorem 2.1 says that a random function with LRD is a 1/f noise and vice versa. Theorem 2.2 indicates that a heavy-tailed random function is a 1/f noise and vice versa. Theorem 2.3 provides a type of stochastic differential equations that produce 1/f noise.

2.1 INTRODUCTION

We mentioned the concept of 1/f noise in Chapter 1. Since it describes processes with LRD specifically in the frequency domain, we discuss it in this individual chapter.

The pioneering work of 1/f noise may refer to W. Schottky [1], where he introduced the concept of two classes of noise. One class is thermal noise, such as the random motion of molecules in the conductors. The other is the shot noise that may be caused by randomness of the emission from the cathode and the randomness of the velocity of the emitted electrons [1, 2]. J. B. Johnson described the latter using the term "Schottky effect" [3], and his paper might be the first one expressing such type of processes by the term "1/f noise."

Let $x(t)$ ($-\infty < t < \infty$) be a random function. Let $S_{xx}(f)$ be its power spectrum density (PSD) function, where $f = \omega/2\pi$ is frequency and ω is radian frequency. Then, by 1/f noise, one means that $S_{xx}(f) \to \infty$ for $f \to 0$. Note that the PSD of a conventional random function, such as $\frac{1}{a^2+(2\pi f)^2}(a \neq 0)$, is convergent at $f = 0$. In the field, the term "1/f noise" is a collective noun, which implies in fact $\frac{1}{f^\beta}$ noise for $\beta > 0$. In the general case, 1/f noise has the meaning of $\frac{1}{f^\alpha}$ noise for $\alpha \in \mathbf{R}$, where $\mathbf{R}$ is the set of real numbers (see Mandelbrot [4]). However, since $\lim_{f \to 0}\frac{1}{f^\alpha}=0$ for $\alpha < 0$, one may usually not be interested in the case of $\alpha < 0$. In the subsequent text, we discuss the noise of $\frac{1}{f^\beta}$ type for $\beta > 0$ unless otherwise stated. As a matter of fact, when talking about 1/f noise, it is defaulted that $\beta > 0$ for $\frac{1}{f^\beta}$. In most cases, the term "1/f noise" includes the case of $\frac{1}{f^\beta}$.

DOI: 10.1201/9781003354987-3

Since the notion of $1/f$ noise appeared in ref [3], it has gained increasing interests of scientists in various fields, ranging from bioengineering to computer networks to mention a few (see refs [4–54]). This fact gives rise to a question: what is $1/f$ noise? The question may be roughly answered in a way that a $1/f$ noise $x(t)$ is such that its PSD $S_{xx}(f)$ is divergent at $f = 0$, as previously mentioned. Nonetheless, this answer may never be enough to describe the full picture of $1/f$ noise. By full picture we mean that we should describe a set of main properties of $1/f$ noise in addition to the property $\lim_{f \to 0} \frac{1}{f^\beta} = \infty$ in frequency domain. When regarding $\lim_{f \to 0} \frac{1}{f^\beta} = \infty$ as the first property of $1/f$ noise, denoted by P1, we would like to list other three properties as follows based on following three questions: (1) What is the qualitative structure of an autocorrelation function (ACF) of $1/f$ noise? (2) What is the main property of its probability density function (PDF)? (3) What is the possible structure of the differential equation to synthesize $1/f$ noise?

- P2: Regarding P2, we shall discuss the statistical dependence based on the hyperbolically decayed ACF structure.

- P3: With P3, we will explain the heavy-tailed property of $1/f$ noise.

- P4: In P4, we shall explain the structure of differential equations to synthesize $1/f$ noise.

For facilitating the description of the full picture of $1/f$ noise, we shall brief the preliminaries in Section 2.2. Then, three theorems (Theorems 2.1–2.3) will be discussed in Sections 2.3–2.5, respectively. After that we shall summarize the chapter in Section 2.6.

2.2 PRELIMINARIES

2.2.1 Dependence of Random Variables

A time series $x(t)$ may also be called a random function (Yaglom [55]). The term random function apparently exhibits that $x(t_i)$ $(i = 0, 1, \ldots)$ is a random variable, implying

$$x(t_i) \neq x(t_j) \text{ for } i \neq j (i, j = 0, 1, \ldots). \tag{2.1}$$

We would like to discuss the dependence of $x(t_i)$ and $x(t_j)$ for $i \neq j$, as well as $x(t)$.

2.2.1.1 Dependence Description of Random Variables with Probability
Let $P\{x(t_i) < x_1\}$ be the probability of the event $\{x(t_i) < x_1\}$. Denote the probability of the event $\{x(t_j) < x_2\}$ by $P\{x(t_j) < x_2\}$. Then according to Papoulis and Pillai [56], $\{x(t_i) < x_1\}$ and $\{x(t_j) < x_2\}$ are said to be independent events if

$$P[\{x(t_i) < x_1\}\{x(t_j) < x_2\}] = P\{x(t_i) < x_1\}P\{x(t_j) < x_2\}. \tag{2.2}$$

If $\{x(t_i) < x_1\}$ and $\{x(t_j) < x_2\}$ are dependent, on the other side,

$$P[\{x(t_i) < x_1\}\{x(t_j) < x_2\}] = P\{x(t_j) < x_2\}P[\{x(t_i) < x_1\} | \{x(t_j) < x_2\}] \tag{2.3}$$

where $P[\{x(t_i) < x_1\}|\{x(t_j) < x_2\}]$ is the conditional probability, implying the probability of the event $\{x(t_i) < x_1\}$, provided that event $\{x(t_j) < x_2\}$ has occurred.

The dependence of $\{x(t_i) < x_1\}$ and $\{x(t_j) < x_2\}$ is reflected in the conditional probability $P[\{x(t_i) < x_1\}|\{x(t_j) < x_2\}]$. If $\{x(t_i) < x_1\}$ and $\{x(t_j) < x_2\}$ are independent, $P[\{x(t_i) < x_1\}|\{x(t_j) < x_2\}] = P\{x(t_i) < x_1\}$.

2.2.1.2 Dependence Description of Gaussian Random Variables with Correlation

Let $x(t)$ be a Gaussian random function. Denote the correlation coefficient between $\{x(t_i) < x_1\}$ and $\{x(t_j) < x_2\}$ by $corr[\{x(t_i) < x_1\}, \{x(t_j) < x_2\}]$. Then according to Lindgren and McElrath [57] and Doob [58], $\{x(t_i) < x_1\}$ and $\{x(t_j) < x_2\}$ are independent if

$$corr[\{x(t_i) < x_1\}, \{x(t_j) < x_2\}] = a = 0. \tag{2.4}$$

On the other hand, $\{x(t_i) < x_1\}$ and $\{x(t_j) < x_2\}$ are dependent if

$$corr[\{x(t_i) < x_1\}, \{x(t_j) < x_2\}] = b \neq 0. \tag{2.5}$$

The condition (2.4) or (2.5) expressed by the correlation coefficient regarding the independence or dependence may not be enough to identify the independence or dependence of a Gaussian random function completely. For example, when $t_j = t_i + \tau$, that (2.4) holds for large τ may not imply that it is valid for small τ. In other words, one may encounter the situations expressed by (2.6) and (2.7),

$$corr[\{x(t_i) < x_1\}, \{x(t_i + \tau) < x_2\}] \approx 0 \text{ for large } \tau, \tag{2.6}$$

and

$$corr[\{x(t_i) < x_1\}, \{x(t_i + \tau) < x_2\}] \neq 0 \text{ for small } \tau. \tag{2.7}$$

By using the concept of probability, (2.6) corresponds to

$$P[\{x(t_i) < x_1\}\{x(t_i + \tau) < x_2\}] = P\{x(t_i) < x_1\}P\{x(t_i + \tau) < x_2\} \text{ for large } \tau. \tag{2.8}$$

Similarly, (2.7) corresponds to (2.9)

$$\begin{aligned} P[\{x(t_i) < x_1\}\{x(t_i + \tau) < x_2\}] = P\{x(t_i + \tau) < x_2\} \\ P[\{x(t_i) < x_1\}|\{x(t_i + \tau) < x_2\}] \text{ for small } \tau. \end{aligned} \tag{2.9}$$

The notion of the dependence or independence of a set of random variables plays a role in the axiomatic approach of probability theory and stochastic processes (see Kolmogorov [59]). The above explanation exhibits an interesting fact that the dependence or independence relies on the observation scale or observation range. In conventional time series, we do not usually consider the observation scale, that is, (2.2) and (2.3) or (2.4) and (2.5) hold for all observation ranges no matter whether τ is small or large (see Box et al. [60], Bendat and Piersol [61]). Statistical properties that depend on observation ranges are mentioned by Papoulis and Pillai [56] and Fuller [62] but are detailed in Beran [63, 64].

The Kolmogorov's work on axiomatic approach of probability theory and stochastic processes needs the assumption that $\{x(t_i) < x_1\}$ and $\{x(t_j) < x_2\}$ are independent in most cases, likely for the completeness of his theory. Nevertheless, he contributed a lot in random functions that are range-dependent (see Monin and Yaglom [65], Kolmogorov [66]).

2.2.1.3 Dependence Description of Gaussian Random Variables with ACF

Denote the ACF of $x(t)$ by $r_{xx}(\tau, t)$. Denote the operator of mean by E. Then,

$$r_{xx}(\tau, t) = E[x(t)x(t+\tau)]. \tag{2.10}$$

It represents the correlation between the one point $x(t)$ and the other τ apart, i.e., $x(t + \tau)$. For facilitating the discussions, we assume that $x(t)$ is stationary in the wide sense (stationary for short). With this assumption, $r_{xx}(\tau, t)$ only replies on the lag τ. Therefore,

$$r_{xx}(\tau, t) = r_{xx}(\tau) = E[x(t)x(t+\tau)]. \tag{2.11}$$

In the normalized case,

$$0 \le |r_{xx}(\tau)| \le 1. \tag{2.12}$$

ACF is a convenient tool for describing the dependence of a Gaussian random function $x(t)$. For instance, on the one hand, we say that any two different points of $x(t)$ are uncorrelated, accordingly independent as $x(t)$ is Gaussian, if $r_{xx}(\tau) = 0$. That is the case of Gaussian white noise. On the other hand, any two different points of $x(t)$ are completely dependent if $r_{xx}(\tau) = 1$. This is the case of the strongest long-range dependence. In the case of $0 < |r_{xx}(\tau)| < 1$, the value of $|r_{xx}(\tau)|$ varies with the lag $\tau = (t + \tau) - t$.

A useful measure called correlation time, which is denoted by τ_c (Nigam [67, p. 74]), is defined by

$$\tau_c = \frac{1}{\lim_{T \to \infty} \int_0^T r_{xx}(\tau)d\tau} \lim_{T \to \infty} \int_0^T \tau r_{xx}(\tau)d\tau. \tag{2.13}$$

By correlation time we say that the correlation can be neglected if $\tau_c \le \tau$, where τ is the time scale of interest.

For a conventional Gaussian random function $x(t)$, its correlation can be neglected if $\tau_c \le \tau$. This implies that the statistical dependence or independence of $x(t)$ relies on its correlation time τ_c. However, we shall show later, in the end of Section 2.3, that correlation time fails regarding the statistical dependence if $x(t)$ is a $1/f$ noise.

2.2.2 ACF and PSD

The PSD of $x(t)$ is the Fourier transform of its ACF. That is,

$$S_{xx}(f) = \int_{-\infty}^{\infty} r_{xx}(t)e^{-j2\pi ft}dt, \quad j = \sqrt{-1}. \tag{2.14}$$

Equivalently,

$$r_{xx}(\tau) = \int_{-\infty}^{\infty} S_{xx}(f)e^{j2\pi f\tau}df. \tag{2.15}$$

Letting $f = 0$ in (2.14) yields

$$S_{xx}(0) = \int_{-\infty}^{\infty} r_{xx}(\tau)d\tau. \tag{2.16}$$

Similarly, letting $\tau = 0$ in (2.15) produces

$$r_{xx}(0) = \int_{-\infty}^{\infty} S_{xx}(f)df. \tag{2.17}$$

Eq. (2.16) implies that an ACF $r_{xx}(\tau)$ is integrable if $S_{xx}(0) < \infty$. On the other side, a PSD $S_{xx}(f)$ is integrable when $r_{xx}(0) < \infty$ as can be seen from (2.17). Both are usual cases in conventional random functions. The noise of $1/f$ type has the property $S_{xx}(0) = \infty$, which makes $1/f$ noise substantially different from conventional random functions.

2.2.3 Mean and Variance

Denote the PDF of $x(t)$ by $p(x)$. Then, the mean denoted by $\mu(t)$ is given by

$$\mu(t) = \mathrm{E}[x(t)] = \int_{-\infty}^{\infty} xp(x)dx. \tag{2.18}$$

The variance denoted by $\sigma^2(t)$ is in the form

$$\sigma^2(t) = \mathrm{E}\{[x(t)-\mu(t)]^2\} = \int_{-\infty}^{\infty} [x(t)-\mu(t)]^2 p(x)dx. \tag{2.19}$$

When $x(t)$ is stationary, $\mu(t)$ and $\sigma^2(t)$ do not rely on time t. In this case, they are expressed by

$$\mu(t) = \mu,$$

$$\sigma^2(t) = \sigma^2. \tag{2.20}$$

Without the generality losing, we always assume that $x(t)$ is stationary in what follows unless otherwise stated.

Denote the auto-covariance of $x(t)$ by $C_{xx}(\tau)$. Then,

$$C_{xx}(\tau) = \mathrm{E}\{[x(t) - \mu][x(t + \tau) - \mu]\} = \int\limits_{-\infty}^{\infty} [x(t) - \mu][x(t + \tau) - \mu]p(x)dx. \tag{2.21}$$

Taking into account (2.19) and (2.21), one has

$$C_{xx}(0) = \sigma^2. \tag{2.22}$$

Denote the mean square value of $x(t)$ by Ψ. Then,

$$\Psi = \mathrm{E}\left[x^2(t)\right] = \int\limits_{-\infty}^{\infty} x^2 p(x)dx. \tag{2.23}$$

Considering $r_{xx}(\tau)$ given by

$$r_{xx}(\tau) = \mathrm{E}[x(t)x(t + \tau)] = \int\limits_{-\infty}^{\infty} x(t)x(t + \tau)p(x)dx, \tag{2.24}$$

one has

$$\Psi = r_{xx}(0). \tag{2.25}$$

Therefore,

$$\sigma^2 = \Psi - \mu^2. \tag{2.26}$$

It is worth noting that the above number characteristics are crucial to the analysis of $x(t)$ in practice (see refs [60–87] just to cite a few). However, things turn complicated if $x(t)$ is $1/f$ noise. In Section 2.4, we shall show that μ and or σ^2 may not exist for specific types of $1/f$ noise.

2.3 HYPERBOLICALLY DECAYED ACFS AND 1/F NOISE

The qualitative structure of $1/f$ noise is in the form $\frac{K}{f^\beta}$, where K is a constant and $K > 0$. Since $\lim\limits_{f \to 0} \frac{K}{f^\beta} = \infty$, as we previously mentioned several times, its ACF is nonintegrable over $(-\infty, \infty)$. That is,

$$S_{xx}(0) = \int\limits_{-\infty}^{\infty} r_{xx}(\tau)d\tau = \infty. \tag{2.27}$$

Note that the above may be taken as a definition of LRD property of a random function (see refs [4, 88–92]). Thus, the above text exhibits that 1/*f* noise is LRD. Consequently, $r_{xx}(\tau)$ of a 1/*f* noise $x(t)$ may have the asymptotic property given by

$$r_{xx}(\tau) \sim c|\tau|^{-\gamma}, \quad c > 0, 0 < \gamma < 1, \tau \to \infty. \tag{2.28}$$

Following Gelfand and Vilenkin [91], Li and Lim [92, 93], we have the Fourier transform of $|\tau|^{-\gamma}$ in the form

$$F\left(|\tau|^{-\gamma}\right) = \int_{-\infty}^{\infty} |\tau|^{-\gamma} e^{-j2\pi ft} d\tau = 2\sin(\pi\gamma/2)\Gamma(1-\gamma)|2\pi f|^{\gamma-1}. \tag{2.29}$$

From the above, we have

$$\gamma = 1 - \beta. \tag{2.30}$$

Qualitatively, the ACF of 1/*f* noise is in the structure of power function. It follows power law. This is the explanation of P2 stated in the introduction section.

Example 2.1: Fractional Gaussian noise (fGn) of the Weyl type is a 1/*f* noise when 0.5 < *H* < 1.

Proof. The ACF of the fGn of the Weyl type is in the form

$$r_{\text{fGn}}(\tau) = \frac{V_H}{2}\left[\left(|\tau|+1\right)^{2H} + \left\||\tau|-1\right\|^{2H} - 2|\tau|^{2H}\right], \tag{2.31}$$

where $0 < H < 1$ is the Hurst parameter, $V_H = (H\pi)^{-1}\Gamma(1-2H)\cos(H\pi)$ is the strength of fGn [93, 94]. Its PSD is in the form (Li and Lim [93])

$$S_{\text{fGn}}(f) = V_H\sin(H\pi)\Gamma(2H+1)|2\pi f|^{1-2H}. \tag{2.32}$$

The above implies that $S_{\text{fGn}}(f)$ is 1/*f* noise when 0.5 < *H* < 1. □

Example 2.2: Two fractional Brownian motions are of 1/*f* noise when 0 < *H* < 1.

Proof. The PSD of the fractional Brownian motion (fBm) of the Weyl type is in the form for $V_H = 1$ [95, 96]

$$S_{\text{fBm,Weyl}}(t, f) = \frac{1}{|2\pi f|^{2H+1}}\left(1 - 2^{1-2H}\cos 4\pi ft\right), \quad 0 < H < 1. \tag{2.33}$$

The PSD of the fBm of the Riemann–Liouville type is given by ([97–99])

$$S_{\text{fBm,Riemann-Liouville}}(t,f)$$

$$= \frac{\pi \omega t}{(2\pi f)^{2H+1}} [J_H(4\pi ft)\mathbb{H}_{H-1}(4\pi ft) - J_{H-1}(4\pi ft)\mathbb{H}_H(4\pi ft)], \qquad (2.34)$$

where J_H is the Bessel function of order H and $\mathbb{H}_H$ is the Struve function of order H. The above exhibit that $S_{\text{fBm, Weyl}}(f) \sim 1/f$ and $S_{\text{fBm, Riemann-Liouville}}(f) \sim 1/f$. $\square$

The above implies that two fBms are LRD when $0 < H < 0.5$ and $0.5 < H < 1$. $\square$

Example 2.3: The Cauchy-class process with LRD discussed in refs [100, 101] is a 1/f noise.

Proof. The normalized ACF of the Cauchy-class process with LRD is in the form

$$r_{\text{Cauchy-class}}(\tau) = \left(1 + |\tau|^2\right)^{-b/2}, \quad 0 < b < 1. \qquad (2.35)$$

Its PSD is given by

$$S_{\text{Cauchy-class}}(f) = \int_{-\infty}^{\infty} \left(1 + |\tau|^2\right)^{-b/2} e^{-j2\pi f\tau} d\tau = \frac{2^{(1-b)/2}}{\sqrt{\pi}\Gamma(b/2)}|2\pi f|^{1/2(b-1)} K_{1/2(b-1)}\left(|2\pi f|\right), \quad (2.36)$$

where $K_v()$ is the modified Bessel function of the second kind. It has the asymptotic expression for $f \to 0$ in the form being $1/f$ noise

$$S_{\text{Cauchy-class}}(f) \sim \frac{\Gamma\left[\frac{1}{2}(1-b)\right]}{2^b \sqrt{\pi}\Gamma(b/2)}|2\pi f|^{b-1}. \qquad (2.37)$$

Thus, the Cauchy-class process with LRD is a $1/f$ noise. $\square$

Example 2.4: The generalized Cauchy process with LRD reported in refs [102–105] is a 1/f noise.

Proof. The normalized ACF of generalized Cauchy process with LRD is given by

$$r_{\text{GC}}(\tau) = \left(1 + |\tau|^\alpha\right)^{-\beta/\alpha}, \qquad (2.38)$$

where $0 < \alpha \le 2$ and $0 < \beta < 1$. Its PSD in the complete form refers to Li and Lim [92]. The following may suffice to exhibit its $1/f$ noise behavior (Lim and Li [102]),

$$S_{\text{GC}}(f) \sim \frac{1}{\Gamma(\beta)\cos(\beta\pi/2)}|2\pi f|^{\beta-1}, \quad f \to 0. \qquad (2.39)$$

Therefore, the generalized Cauchy process with LRD is a $1/f$ noise. $\square$

It may be worthwhile for me to write a theorem and a note when this section will soon finish.

Theorem 2.1. If a random function $x(t)$ is LRD, it is 1/*f* noise and vice versa.

Proof. LRD implies that the right side of (2.16) is divergent when $f = 0$. It means that $x(t)$ is a 1/*f* noise. On the other side, when $x(t)$ is a 1/*f* noise, $S_{xx}(0) = \infty$, which means that its ACF is nonintegrable. Hence, LRD. □

Since any random function with LRD is 1/*f* noise, one may observe other types of random functions that belong to 1/*f* noise from a view of LRD processes, for example, those described in refs [106–112].

As 1/*f* noise is of LRD, its ACF is nonintegrable over $(-\infty, \infty)$. Thus, in (2.13), its correlation time $\tau_c \to \infty$. This implies that correlations at any time scales cannot be neglected. Consequently, the measure of correlation time fails to characterize the statistical dependence of 1/*f* noise.

2.4 HEAVY-TAILED PDFS AND 1/*F* NOISE

Heavy-tailed PDFs are widely observed in various fields of sciences and technologies, including life science and computer science (see refs [113–146]). Typical heavy-tailed PDFs are the Pareto distribution, the log-Weibull distribution, the stretched exponential distribution, the Zipfian distribution, the Lévy distribution, and the Cauchy distribution to cite a few (see refs [51, 96, 125, 143–169]).

By heavy tails, we mean that the tail of a PDF $p(x)$ decays slower than the tails of PDFs in the form of exponential functions. More precisely, the term "heavy tail" implies that $p(x)$ of a random function $x(t)$ decays slowly such that $C_{xx}(\tau)$ in (2.21), or $r_{xx}(\tau)$ in (2.24), decays hyperbolically such that it is of LRD. Thus, we may have the theorem below.

**Theorem 2.2. Let $x(t)$ be a heavy-tailed random function.
Then, it is 1/*f* noise and vice versa.**

Proof. Considering that $x(t)$ is heavy-tailed, we may assume its ACF decays hyperbolically, that is, $r_{xx}(\tau) \sim c|\tau|^{-\gamma}$ for $\tau \to \infty$ and $0 < \gamma < 1$. According to (2.29), therefore, we see that $x(t)$ is 1/*f* noise. On the other side, if $x(t)$ is a 1/*f* noise, it is LRD and accordingly heavy-tailed [54, 146]. This completes the proof. □

Theorem 2.2 may be taken as an explanation of P3 stated in the introduction section. Since γ in Theorem 2.2 is restricted to (0, 1), Theorem 2.1 is consistent with the result of the Taqqiu's law. Refer [53, 54, 89] for the details of the Taqqiu's law.

The tail of $p(x)$ may be so heavy that the mean or variance of $x(t)$ may not exist. A commonly used instance to clarify this point is the Pareto distribution (see refs [96, 134, 147, 148, 161]). To clarify it further, we would like to explain more. Denote the Cauchy distribution by $p_{Cauchy}(x)$. Then,

$$p_{Cauchy}(x) = \frac{b}{\pi\left[(x-m)^2 + b^2\right]}, \tag{2.40}$$

where b is the half width at half maximum and m is the statistical median [170]. Its nth moment denoted by m_n is computed by

$$m_n = \mathrm{E}\left[x^n(t)\right] = \int_{-\infty}^{\infty} x^n p_{\mathrm{Cauchy}}(x)dx = \frac{b}{2\pi}\int_{-\infty}^{\infty}\frac{x^n dx}{(x-m)^2 + (0.5b)^2}. \qquad (2.41)$$

Since the above integral is divergent for $n \geq 2$, the variance of $x(t)$ obeying $p_{\mathrm{Cauchy}}(x)$ does not exist.

Application of the Cauchy distribution to network traffic modeling refers to Field et al. [169]. The generalized Cauchy distribution is reported by Konno and Tamura [171], Konno and Watanabe [172], and Lubashevsky [173].

Another type of heavy-tailed random function without mean and variance is the Lévy distribution, (see refs [162–166, 174–177]). Suppose that $x(t)$ follows the Lévy distribution. Denote it by $p_{\mathrm{Levy}}(x)$. Then, for $x \geq \mu$, it is given by

$$p_{\mathrm{Levy}}(x) = p_{\mathrm{Levy}}(x;\mu,c) = \sqrt{\frac{c}{2\pi}}\,\frac{\exp\left[-\dfrac{c}{2(x-\mu)}\right]}{(x-\mu)^{3/2}}, \qquad (2.42)$$

where μ is the location parameter and c the scale parameter. The nth moment of a Lévy distributed random function, for the simplicity by letting $\mu = 0$, is given by

$$m_n = \sqrt{\frac{c}{2\pi}}\int_0^{\infty} x^n x^{-3/2} e^{-c/(2x)}dx. \qquad (2.43)$$

The integral in (2.43) is divergent for $n > 0$. Thus, its mean and variance do not exist as they approach ∞.

Application of the Lévy distribution to network traffic modeling is discussed by Terdik and Gyires [174]. Tools in Matlab for simulating Lévy distributed random data are described by Liang and Chen [177].

The previously discussed heavy-tailed distributions, such as the Cauchy distribution and the Lévy distribution, are special cases of stable distributions, which are detailed in refs [89, 120, 178–181]. Denote the PDF of a stable-distributed random function x by $p_{\mathrm{Stable}}(x)$. Then, it is indirectly defined by its characteristic function denoted by $\varphi_{\mathrm{Stable}}(t;\mu,c,\alpha,\beta)$. It is in the form:

$$\varphi_{\mathrm{Stable}}(t;\mu,c,\alpha,\beta) = \int_{-\infty}^{\infty} p_{\mathrm{Stable}}(x)e^{j2\pi xt}dx \qquad (2.44)$$

$$= \exp\left[jt\mu - |ct|^{\alpha}\left(1 - j\beta\,\mathrm{sgn}(t)\Phi\right)\right],$$

where $\alpha \in (0,2]$ is the stability parameter, $\beta \in [-1,1]$ is the skewness parameter, $c \in (0, \infty)$ is the scale parameter, $\mu \in (-\infty, \infty)$ the location parameter, and

$$\Phi = \begin{cases} \tan\dfrac{\pi\alpha}{2}, & \text{if } \alpha \neq 1 \\[2mm] -\dfrac{2}{\pi}\log|t|, & \text{if } \alpha = 1 \end{cases}. \tag{2.45}$$

It may be easy to see that x has mean μ when $\alpha > 1$. However, its mean is undefined otherwise. In addition, its variance equals to $2c^2$ if $\alpha = 2$. Otherwise, its variance is infinite.

A stable distribution is characterized by four parameters. Generally, its analytical expression is unavailable except for some specific values of parameters, due to the difficulties in performing the inverse Fourier transform of the right side of (2.44).

A stable distribution is generally non-Gaussian except for some specific values of parameters. When $\alpha = 2$, it is Gaussian with the mean μ and the variance of $2c^2$. Generally speaking, $p_{\text{Stable}}(x)$ is heavy-tailed. It reduces to the Landau distribution when $\alpha = \beta = 1$. Denote the Landau distribution by Landau(μ, c) (Landau [182]). Its characteristic function is given by

$$\varphi_{\text{Landau}}(t;\mu,c) = \int_{-\infty}^{\infty} p_{\text{Landau}}(x)e^{j2\pi xt}\,dx = \exp\left[jt\mu - |ct|\left(1 - \frac{2j}{\pi}\log|t|\right)\right], \tag{2.46}$$

where μ is the location parameter and c the scaled parameter. Its PDF is given by

$$p_{\text{Landau}}(x) = \frac{1}{\pi}\int_{0}^{\infty} e^{-t\log t - xt}\sin(\pi t)\,dt. \tag{2.47}$$

Applications of the Landau distribution can be found in nuclear physics (Wilkinson [183], Tabata and Ito [184]). It reduces to the Holtsmark distribution if $\alpha = 3/2$ and $\beta = 0$. Accordingly,

$$\varphi_{\text{Holtsmark}}(t;\mu,c) = \int_{-\infty}^{\infty} p_{\text{Holtsmark}}(x)e^{j2\pi xt}\,dx = \exp\left(jt\mu - |ct|^{3/2}\right). \tag{2.48}$$

Its PDF is given by

$$p_{\text{Holtsmark}}(x) = \frac{1}{2\pi}\int_{-\infty}^{\infty} \exp\left(jt\mu - |ct|^{3/2}\right)e^{-j2\pi xt}\,dt. \tag{2.49}$$

This type of random functions has mean μ but its variance is infinite (Holtsmark [185], Pittel et al. [186], Hummer [187]).

The literature regarding applications of stable distributions is rich. Their applications to network traffic modeling can also be found in Garroppo et al. [188], Gallardo et al. [189], Karasaridis and Hatzinakos [190].

It is worth noting that the observation of random functions without mean and variance may be traced back to the work of the famous statistician Daniel Bernoulli's cousin, Nicolas Bernoulli in 1713 (Montmort [191, 192]). Nicolas Bernoulli studied a casino game as follows. A player bets on how many tosses of a coin will be needed before it first turns up heads. If it falls heads on the first toss, the player wins \$2; if it falls tails, heads, the player wins \$4; if it falls tails, tails, heads, the player wins \$8, ... According to this game rule, if the probability of an outcome is 2^{-k}, the player wins \$$2^k$. Thus, the mean for $k \to \infty$ is given by

$$\sum_{k=1}^{\infty} 2^{-k}(\$2^k) = \infty. \tag{2.50}$$

The above may be used to express the game that is now termed "Petersburg Paradox." This paradox is now named after Daniel Bernoulli due to his presentation of the problem and his solution in the Commentaries of the Imperial Academy of Science of Saint Petersburg (Bernoulli [193]).

2.5 FRACTIONALLY GENERALIZED LANGEVIN EQUATION AND 1/F NOISE

The standard Langevin equation is in the form

$$\begin{cases} \left(\dfrac{d}{dt} + \lambda\right) X(t) = w(t), \\ \qquad X(0) = X_0, \end{cases} \tag{2.51}$$

where $\lambda > 0$ ([56, 194]) and $w(t)$ is a standard white noise. By standard white noise, we mean that its PSD $S_{ww}(f)$ is in the form

$$S_{ww}(f) = 1 \text{ for } f \in (-\infty, \infty). \tag{2.52}$$

The solution to (2.51) in frequency domain is given by

$$S_{XX}(f) = \frac{1}{\lambda^2 + (2\pi f)^2}. \tag{2.53}$$

The standard Langevin equation may not attract people much in the field of $1/f$ noise. People are usually interested in fractionally generalized Langevin equations. There are two types of fractionally generalized Langevin equations. One is in the form [195–210]

$$\left(\frac{d^{1+\varepsilon}}{dt^{1+\varepsilon}} + \lambda\right) X_1(t) = w(t), \quad \varepsilon > -1. \tag{2.54}$$

The other is expressed by [211–217]

$$\left(\frac{d}{dt}+\lambda\right)^{\alpha} X_1(t)=w(t), \quad \alpha>0. \tag{2.55}$$

Two are consistent when $\lambda = 0$. We now adopt the one expressed by (2.55).

Theorem 2.3. A solution to the stochastically fractional differential equation below belongs to 1/*f* noise

$$\frac{d^{\alpha}}{dt^{\alpha}}X_1(t)=w(t), \quad \alpha>0. \tag{2.56}$$

Proof. Denote the impulse response function of (2.55) by $g_{X_1}(t)$. That is,

$$\left(\frac{d}{dt}+\lambda\right)^{\alpha} g_{X_1}(t)=\delta(t), \tag{2.57}$$

where $\delta(t)$ is the δ function. When $f(t)$ is continuous at $t = 0$, it is given by

$$\begin{cases} \int\limits_{-\infty}^{\infty}\delta(t)f(t)dt = f(0), \\[2mm] \delta(t)=0, \qquad t\neq 0. \end{cases} \tag{2.58}$$

The function $\delta(t)$ is called the impulse function in linear systems [61, 218]. According to the theory of linear systems, $g_{X_1}(t)$ is the solution to (2.57) under the zero initial condition, which is usually called the impulse response function in linear systems [61, 68, 69, 76, 77, 218–224]. Denote the Fourier transform of $g_{X_1}(t)$ by $G_{X_1X_1}(f)$. Then, with the techniques in fractional calculus ([215, 217, 225]), doing the Fourier transforms on (2.57) yields

$$G_{X_1X_1}(f)=\frac{1}{(\lambda+j2\pi f)^{\alpha}}. \tag{2.59}$$

By taking into account (2.52), we have

$$S_{X_1X_1}(f)=|G_{X_1X_1}(f)|^2 S_{ww}(f)=|G_{X_1X_1}(f)|^2=\frac{1}{\left[\lambda^2+(2\pi f^2)\right]^{\alpha}}. \tag{2.60}$$

If $\lambda = 0$ in (2.60), we have 1/*f* noise expressed by

$$S_{X_1X_1}(\omega)=\frac{1}{(2\pi f)^{2\alpha}}. \tag{2.61}$$

This finishes the proof. □

From the above, we see that $X_1(t)$ belongs to $1/f$ noise. As a matter of fact, by using fractional integral in (2.56), we have

$$X_1(t) = \frac{d^{-\alpha}}{dt^{-\alpha}} w(t), \quad \alpha > 0. \tag{2.62}$$

The above expression implies that a $1/f$ noise may be taken as a solution to a stochastically fractional differential equation, being an explanation to P4 described in the introduction section. The following example will soon refine this point of view.

Let $B(t)$ be the Wiener Brownian motion for $t \in (0, \infty)$, see ref [226] for the details of the Brownian motion. Then, it is nondifferentiable in the domain of ordinary functions. It is differentiable, however, in the domain of generalized functions over the Schwartz space of test functions [91, 227]. Therefore, in the domain of generalized functions, we write the stationary Gaussian white noise by

$$w(t) = \frac{dB(t)}{dt}. \tag{2.63}$$

Based on the definitions of the fractional integrals of the Riemann–Liouville's and the Weyl's (Miller and Ross [225]), on the one hand, when using the Riemann–Liouville integral operator, we express the fBm of the Riemann–Liouville type, which is denoted by $B_{H,\mathrm{RL}}(t)$, in the form

$$B_{H,\mathrm{RL}}(t) = {}_0 D_t^{-(H+1/2)} \frac{dB(t)}{dt} = \frac{1}{\Gamma(H+1/2)} \int_0^t (t-u)^{H-1/2} dB(u), \tag{2.64}$$

where ${}_0 D_t^{-(H+1/2)}$ is the Riemann–Liouville integral operator of order $(0.5 + H)$ for $0 < H < 1$. On the other hand, the fBm of the Weyl type by using the Weyl fractional integral, denoted by $B_{H,\mathrm{Weyl}}(t)$, is given by

$$W^{-(H+1/2)} \frac{dB(t)}{dt} = B_{H,\mathrm{Weyl}}(t) - B_{H,\mathrm{Weyl}}(0) =$$

$$\frac{1}{\Gamma(H+1/2)} \left\{ \begin{array}{l} \displaystyle\int_{-\infty}^{0} \left[(t-u)^{H-0.5} - (-u)^{H-0.5} \right] dB(u) \\[2em] + \displaystyle\int_0^t (t-u)^{H-0.5} dB(u) \end{array} \right\}, \tag{2.65}$$

where $W^{-(H+1/2)}$ is the Weyl integral operator of order $(0.5 + H)$. Expressions (2.64) and (2.65) are the fBms introduced by Mandelbrot and van Ness in ref [94] but we provide a new outlook of describing them from the point of view of the fractional generalized Langevin equation with the topic of $1/f$ noise.

2.6 SUMMARY

We have explained the main properties of 1/f noise as follows: First, 1/f noise is LRD and its ACF is hyperbolically decayed such that the ACF is nonintegrable. Second, its PDF obeys power laws and it is heavy-tailed. Finally, it may be taken as a solution to a stochastic differential equation. Fractal time series, such as fGn, fBm, the generalized Cauchy process, the Lévy flights, and α-stable processes, are generally in the class of 1/f noise. One thing worth noting is that fBm is of 1/f noise for $0 < H < 1$. Consequently, it is always LRD when $0 < H < 1$.

REFERENCES

1. W. Schottky, Über spontane Stromschwankungen in verschiedenen Elektrizitätsleitern, *Annalen der Physik*, 362(23) 1918, 541–567.
2. W. Schottky, Zur Berechnung und Beurteilung des Schroteffektes, *Annalen der Physik*, 373(10) 1922, 157–176.
3. J. B. Johnson, The Schottky effect in low frequency circuits, *Physical Review*, 26(1) 1925, 71–85.
4. B. B. Mandelbrot, *Multifractals and 1/f Noise*, Springer, New York, 1998.
5. K. Fraedrich, U. Luksch, and R. Blender, 1/f model for long-time memory of the ocean surface temperature, *Physical Review E*, 70(3) 2004, 037301 (4 pp).
6. E.-J. Wagenmakers, S. Farrell, and R. Ratcliff, Estimation and interpretation of $\frac{1}{f^\alpha}$ noise in human cognition, *Psychonomic Bulletin & Review*, 11(4) 2004, 579–615.
7. F. Principato, and G. Ferrante, 1/f noise decomposition in random telegraph signals using the wavelet transform, *Physica A*, 380, 2007, 75–97.
8. V. P. Koverda, and V. N. Skokov, Maximum entropy in a nonlinear system with a 1/f power spectrum, *Physica A*, 391(1–2) 2012, 21–28.
9. Y. Nemirovsky, D. Corcos, I. Brouk, A. Nemirovsky, and S. Chaudhry, 1/f noise in advanced CMOS transistors, *Instrumentation & Measurement Magazine*, 14(1) 2011, 14–22.
10. O. Miramontes, and P. Rohani, Estimating $\frac{1}{f^\alpha}$ scaling exponents from short time-series, *Physica D*, 166(3–4) 2002, 147–154.
11. C. M. Van Vliet, Random walk and 1/f noise, *Physica A*, 303(3–4) 2002, 421–426.
12. J. S. Kim, Y. S. Kim, H. S. Min, and Y. J. Park, Theory of 1/f noise currents in semiconductor devices with one-dimensional geometry and its application to Si Schottky barrier diodes, *IEEE Transactions on Electron Devices*, 48(12) 2001, 2875–2883.
13. T. Antal, M. Droz, G. Györgyi, and Z. Rácz, 1/f noise and extreme value statistics, *Physical Review Letters*, 87(24) 2001, 240–601 (4 pp).
14. B. Pilgram, and D. T. Kaplan, A comparison of estimators for 1/f noise, *Physica D*, 114(1–2) 1998, 108–122.
15. H. J. Jensen, Lattic gas as a model of 1/f noise, *Physical Review Letters*, 64(26) 1990, 3103–3106.
16. E. Marinari, G. Parisi, D. Ruelle, and P. Windey, Random walk in random environment and 1/f noise, *Physical Review Letters*, 50(17) 1983, 1223–1225.
17. F. N. Hooge, 1/f noise, *Physica B*, 83(1) 1976, 14–23.
18. M. B. Weissman, Simple model for 1/f noise, *Physical Review Letters*, 35(11) 1975, 689–692.
19. F. N. Hooge, Discussion of recent experiments on 1/f noise, *Physica*, 60(1) 1972, 130–144.
20. Y. Avraham, and M. Pinchas, Two novel one-way delay clock skew estimators and their performances for the fractional Gaussian noise/generalized fractional Gaussian noise environment applicable for the IEEE 1588v2 (PTP) case, *Frontiers in Physics*, 10, 2022, 867861.
21. Y. Avraham, and M. Pinchas, A novel clock skew estimator and its performance for the IEEE 1588v2 (PTP) case in fractional Gaussian noise/generalized fractional Gaussian noise environment, *Frontiers in Physics*, 9, 2021, 796811.
22. M. Pinchas, Cooperative multi PTP slaves for timing improvement in an fGn environment, *IEEE Communications Letters*, 22(7), 2018, 1366–1369.

23. M. Pinchas, Residual ISI obtained by blind adaptive equalizers and fractional noise, *Mathematical Problems in Engineering*, 2013, Article ID 972174, 11 pages, 2013.
24. X. Xiao, C. Xu, Y. Yu, J. He, M. Li, and C. Cattani, Fat tail in the phytoplankton movement patterns and swimming behavior: New insights into the prey-predator interactions, *Fractal and Fractional*, 5(2) 2021, 49.
25. M. S. Keshner, 1/f noise, *Proceedings of the IEEE*, 70(3) 1982, 212–218.
26. B. Ninness, Estimation of 1/f noise, *IEEE Transactions on Information Theory*, 44(1) 1998, 32–46.
27. B. Yazici, and R. L. Kashyap, A class of second-order stationary self-similar processes for 1/f phenomena, *IEEE Transactions on Signal Processing*, 45(2) 1997, 396–410.
28. G. W. Wornell, Wavelet-based representations for the 1/f family of fractal processes, *Proceedings of the IEEE*, 81(10) 1993, 1428–1450.
29. B. B. Mandelbrot, Some noises with 1/f spectrum, a bridge between direct current and white noise, *IEEE Transactions on Information Theory* 13(2) 1967, 289–298.
30. N. J. Kasdin, Discrete simulation of colored noise and stochastic processes and $\frac{1}{f^\alpha}$ generation, *Proceedings of the IEEE*, 83(5) 1995, 802–827.
31. G. Corsini, and R. Saletti, A $\frac{1}{f^\gamma}$ power spectrum noise sequence generator, *IEEE Transactions on Instrumentation and Measurement*, 37(4) 1988, 615–619.
32. W.-T. Li, and D. Holste, Universal 1/f noise, crossovers of scaling exponents, and chromosome-specific patterns of guanine-cytosine content in DNA sequences of the human genome, *Physical Review E*, 71(4) 2005, 041910 (9 pp).
33. W.-T. Li, G. Stolovitzky, P. Bernaola-Galvan, and J. L. Oliver, Compositional heterogeneity within, and uniformity between, DNA sequences of yeast chromosomes, *Genome Research*, 8(9) 1998, 916–928.
34. W.-T. Li, and K. Kaneko, Long-range correlation and partial $\frac{1}{f^\alpha}$ spectrum in a noncoding DNA sequence, *Europhysics Letters*, 17(7) 1992, 655–660.
35. P. C. Ivanov, L. Amaral, A. L. Goldberger, S. Havlin, M. G. Rosenblum, H. E. Stanley, and Z. R. Struzik, From 1/f noise to multifractal cascades in heartbeat dynamics, *Chaos*, 11(3) 2001, 641–652.
36. N. M. Rosario, and H. E. Stanley, *An Introduction to Econophysics: Correlations and Complexity in Finance*, Cambridge University Press, Cambridge, 2000.
37. W.-Q. Duan, and H. E. Stanley, Cross-correlation and the predictability of financial return series, *Physica A*, 390(2) 2011, 290–296.
38. B. Podobnik, D. Horvatic, A. Lam Ng, H. E. Stanley, and P. Ivanov Ch., Modeling long-range cross-correlations in two-component ARFIMA and FIARCH processes, *Physica A*, 387(15) 2008, 3954–3959.
39. G. Aquino, M. Bologna, P. Grigolini, and B. J. West, Beyond the death of linear response: 1/f optimal information transport, *Physical Review Letters*, 105(4) 2010, 040601.
40. B. J. West, and P. Grigolini, Chipping away at memory, *Biological Cybernetics*, 103(2) 2010, 167–174.
41. B. J. West, and M. F. Shlesinger, On the ubiquity of 1/f noise, *International Journal of Modern Physics B*, 3(6) 1989, 795–819.
42. A. L. Goldberger, V. Bhargava, B. J. West, and A. J. Mandell, On a mechanism of cardiac electrical stability, *Biophysical Journal*, 48(3) 1985, 525–528.
43. T. Musha, H. Takeuchi, and T. Inoue, 1/f fluctuations in the spontaneous spike discharge intervals of a giant snail neuron, *IEEE Transactions on Biomedical Engineering*, 30(3) 1983, 194–197.
44. M. Kobayashi, and T. Musha, 1/f fluctuation of heartbeat period, *IEEE Transactions on Biomedical Engineering*, 29(6) 1982, 456–457.
45. B. Neumcke, 1/f noise in membranes, *Biophysics of Structure and Mechanism*, 4(3) 1978, 79–199.

46. J. R. Clay, and M. F. Shlesinger, Unified theory of 1/*f* and conductance noise in nerve membrane, *Journal of Theoretical Biology*, 66(4) 1977, 763–773.

47. E. Frehland, Diffusion as a source of 1/*f* noise, *The Journal of Membrane Biology*, 32(1) 1977, 195–196.

48. M. E. Green, Diffusion and 1/*f* noise, *Journal of Membrane Biology*, 28(1) 1976, 181–186.

49. I. Csabai, 1/*f* noise in computer network traffic, *Journal of Physics A. Mathematical and General*, 27(2) 1994, L417–421.

50. M. Takayasu, H. Takayasu, and T. Sato, Critical behaviors and 1/*f* noise in information traffic, *Physica A*, 233(3–4) 1996, 824–834.

51. V. Paxson, and S. Floyd, Wide area traffic: The failure of Poisson modeling, *IEEE/ACM Transactions on Networking*, 3(3) 1995, 226–244.

52. W. Willinger, R. Govindan, S. Jamin, V. Paxson, and S. Shenker, Scaling phenomena in the internet critically, *Proceedings of the National Academy of Sciences of the United States of America*, 99 (Suppl 1) 2002, 2573–2580.

53. P. Loiseau, P. Gonçalves, G. Dewaele, P. Borgnat, P. V.-B. Abry, and Primet, Investigating self-similarity and heavy-tailed distributions on a large-scale experimental facility, *IEEE/ACM Transactions on Networking*, 18(4), 2010, 1261–1274.

54. P. Abry, P. Borgnat, F. Ricciato, A. Scherrer, and D. Veitch, Revisiting an old friend: On the observability of the relation between long range dependence and heavy tail, *Telecommunication Systems*, 43(3–4) 2010, 147–165.

55. A. M. Yaglom, *Correlation Theory of Stationary and Related Random Functions*, Vol. I, Springer, Berlin, 1987.

56. A. Papoulis, and S. U. Pillai, *Probability, Random Variables and Stochastic Processes*, McGraw-Hill, New York, 1997.

57. B. W. Lindgren, and G. W. McElrath, *Introduction to Probability and Statistics*, The Macmillan Company, New York, 1959.

58. J. L. Doob, The elementary Gaussian processes, *The Annals of Mathematical Statistics*, 15(3) 1944, 229–281.

59. А. Н. КолМогороВ, *Основные лонятния теории вероятностей*, translated by S.-T. Ding, Shangwu Yinshuguan, Shanghai, 1954, in Chinese.

60. G. E. P. Box, G. M. Jenkins, and G. C. Reinsel, *Time Series Analysis: Forecasting and Control*, Prentice Hall, New York, 1994.

61. J. S. Bendat, and A. G. Piersol, *Random Data: Analysis and Measurement Procedure*, 3rd ed., John Wiley & Sons, New York, 2000.

62. W. A. Fuller, *Introduction to Statistical Time Series*, 2nd ed., John Wiley & Sons, New York, 1996.

63. J. Beran, *Statistics for Long-Memory Processes*, Chapman & Hall, New York, 1994.

64. J. Beran, Statistical methods for data with long-range dependence, *Statistical Science*, 7(4) 1992, 404–416.

65. A. S. Monin, and A. M. Yaglom, *Statistical Fluid Mechanics: Mechanics of Turbulence*, Vol. 2, The MIT Press, New York, 1971.

66. A. N. Kolmogorov, Local structure of turbulence in an incompressible viscous fluid at very high Reynolds numbers, *Soviet Physics Uspekhi*, 10(6) 1968, 734–736.

67. N. C. Nigam, *Introduction to Random Vibrations*, The MIT Press, New York, 1983.

68. T. T. Song, and M. Grigoriu, *Random Vibration of Mechanical and Structural Systems*, Prentice Hall, New York, 1993.

69. C. M. Harris, *Shock and Vibration Ha*, 4th ed., McGraw-Hill, New York, 1995.

70. H. Czichos, T. Saito, and L. Smith, *Springer Handbook of Metrology and Testing*, Springer, Berlin, 2011.

71. W. N. Sharpe Jr., *Springer Handbook of Experimental Solid Mechanics*, Springer, Berlin, 2008.

72. C. Tropea, A. L. Yarin, and J. F. Foss (eds.), *Springer Handbook of Experimental Fluid Mechanics*, Springer, Berlin, 2007.

73. K.-H. Grote, and E. K. Antonsson (eds.), *Springer Handbook of Mechanical Engineering*, Springer, Berlin, 2009.

74. R. Kramme, K.-P. Hoffmann, and R. S. Pozos, *Springer Handbook of Medical Technology*, Springer, Berlin, 2012.

75. W. Kresse, and D. M. Danko, *Springer Handbook of Geographic Information*, Springer, Berlin, 2012.

76. S. S. Bhattacharyya, F. Deprettere, R. Leupers, and J. Takala (eds.), *Handbook of Signal Processing Systems*, Springer, Berlin, 2010.

77. S. K. Mitra, and J. F. Kaiser, *Handbook for Digital Signal Processing*, John Wiley & Sons, New York, 1993.

78. W. A. Woyczy'nski, *A First Course in Statistics for Signal Analysis*, Birkhäuser, Boston, 2006.

79. R. A. Bailey, *Design of Comparative Experiments*, Cambridge University Press, Cambridge, 2008.

80. ASME, *Measurement Uncertainty Part 1, Instruments and Apparatus, Supplement to ASME, Performance Test Codes*, ASME, New York, 1986.

81. D. Sheskin, *Statistical Tests and Experimental Design: A Guidebook*, Gardner Press, New York, 1984.

82. T. W. MacFarland, *Two-Way Analysis of Variance*, Springer, Berlin, 2012.

83. A. K. Gupta, W.-B. Zeng, and Y. Wu, *Probability and Statistical Models: Foundations for Problems in Reliability and Financial Mathematics*, Birkhäuser, Boston, 2010.

84. P. Fieguth, *Statistical Image Processing and Multidimensional Modeling*, Springer, Berlin, 2011.

85. J. Nauta, *Statistics in Clinical Vaccine Trials*, Springer, Berlin, 2011.

86. A. Gelman, Analysis of variance? Why it is more important than ever, *The Annals of Statistics*, 33(1) 2005, 1–53.

87. C. G. Pendse, A note on mathematical expectation, *The Mathematical Gazette*, 22(251) 1938, 399–402.

88. B. B. Mandelbrot, *Gaussian Self-Affinity and Fractals*, Springer, New York, 2001.

89. G. Samorodnitsky, and M. S. Taqqu, *Stable Non-Gaussian Random Processes*, Chapman & Hall, New York, 1994.

90. J. Beran, R. Shernan, M. S. Taqqu, and W. Willinger, Long-range dependence in variable bit-rate video traffic, *IEEE Transactions on Communications*, 43(2/3/4) 1995, 1566–1579.

91. I. M. Gelfand, and K. Vilenkin, *Generalized Functions*, Vol. 1, Academic Press, New York, 1964.

92. M. Li, and S. C. Lim, Power spectrum of generalized Cauchy process, *Telecommunication Systems*, 43(3–4) 2010, 219–222.

93. M. Li, and S. C. Lim, A rigorous derivation of power spectrum of fractional Gaussian noise, *Fluctuation and Noise Letters*, 6(4) 2006, C33–C36.

94. B. B. Mandelbrot, and J. W. van Ness, Fractional Brownian motions, fractional noises and applications, *SIAM Review*, 10(4) 1968, 422–437.

95. P. Flandrin, On the spectrum of fractional Brownian motion, *IEEE Transactions on Information Theory*, 35(1) 1989, 197–199.

96. M. Li, Fractal time series: A tutorial review, *Mathematical Problems in Engineering*, 2010, Article ID 157264, 26 pages, 2010.

97. S. V. Muniandy, and S. C. Lim, Modeling of locally self-similar processes using multi-fractional Brownian motion of Riemann-Liouville type, *Physical Review E*, 63(4) 2001, 046104.

98. V. M. Sithi, and S. C. Lim, On the spectra of Riemann-Liouville fractional Brownian motion, *Journal of Physics A: Mathematical and General*, 28(11) 2003, 2995–3003.

99. S. C. Lim, and S. V. Muniandy, On some possible generalizations of fractional Brownian motion, *Physics Letters A*, 226(2–3) 2000, 140–145.

100. J.-P. Chiles, and P. Delfiner, *Geostatistics, Modeling Spatial Uncertainty*, John Wiley & Sons, New York, 1999.

101. M. Li, and J.-Y. Li, Generalized Cauchy model of sea level fluctuations with long-range dependence, *Physica A*, 484, 2017, 309–335.

102. S. C. Lim, and M. Li, Generalized Cauchy process and its application to relaxation phenomena, *Journal of Physics A: Mathematical and General*, 39(12) 2006, 2935–2951.

103. S. C. Lim, and L. P. Teo, Gaussian fields and Gaussian sheets with generalized Cauchy covariance structure, *Stochastic Processes and Their Applications*, 119(4) 2009, 1325–1356.

104. M. Li, and S. C. Lim, Modeling network traffic using generalized Cauchy process, *Physica A*, 387(11) 2008, 2584–2594.

105. P. Vengadesh, S. V. Muniandy, and W. H. A. Majid, Fractal morphological analysis of bacteriorhodopsin (bR) layers deposited onto indium tin oxide (ITO) electrodes, *Materials Science and Engineering: C*, 29(5) 2009, 1621–1626.

106. R. J. Martin, and A. M. Walker, A power-law model and other models for long-range dependence, *Journal of Applied Probability*, 34(3) 1997, 657–670.

107. R. J. Martin, and J. A. Eccleston, A new model for slowly-decaying correlations, *Statistics & Probability Letters*, 13(2) 1992, 139–145.

108. W. A. Woodward, Q. C. Cheng, and H. L. Gray, A k-factor GARMA long-memory model, *Journal of Time Series Analysis*, 19(4) 1998, 485–504.

109. C. Ma, Power-law correlations and other models with long-range dependence on a lattice, *Journal of Applied Probability*, 40(3) 2003, 690–703.

110. C. Ma, A class of stationary random fields with a simple correlation structure, *Journal of Multivariate Analysis*, 94(2) 2005, 317–327.

111. M. Li, W. Jia, and W. Zhao, Correlation form of timestamp increment sequences of self-similar traffic on ethernet, *Electronics Letters*, 36(19) 2000, 1668–1669.

112. M. Li, and W. Zhao, Quantitatively investigating locally weak stationarity of modified multifractional Gaussian noise, *Physica A*, 391(24) 2012, 6268–6278.

113. E. G. Tsionas, Estimating multivariate heavy tails and principal directions easily, with an application to international exchange rates, *Statistics & Probability Letters*, 82(11) 2012, 1986–1989.

114. J. Lin, Second order asymptotics for ruin probabilities in a renewal risk model with heavy-tailed claims, *Insurance: Mathematics and Economics*, 51(2) 2012, 422–429.

115. K. Yu, M.-L. Huang, and P. H. Brill, An algorithm for fitting heavy-tailed distributions via generalized hyperexponentials, *INFORMS Journal on Computing*, 24(1) 2012, 42–52.

116. R. Luger, Finite-sample bootstrap inference in GARCH models with heavy-tailed innovations, *Computational Statistics & Data Analysis*, 56(11) 2012, 3198–3211.

117. T. Ishihara, and Y. Omori, Efficient Bayesian estimation of a multivariate stochastic volatility model with cross leverage and heavy-tailed errors, *Computational Statistics & Data Analysis*, 56(11) 2012, 3674–3689.

118. J. E. Methni, L. Gardes, S. Girard, and A. Guillou, Estimation of extreme quantiles from heavy and light tailed distributions, *Journal of Statistical Planning and Inference*, 142(10) 2012, 2735–2747.

119. J. Beran, B. Das, and D. Schell, On robust tail index estimation for linear long-memory processes, *Journal of Time Series Analysis*, 33(3) 2012, 406–423.

120. P. Barbe, and W. P. McCormick, Heavy-traffic approximations for fractionally integrated random walks in the domain of attraction of a non-Gaussian stable distribution, *Stochastic Processes and Their Applications*, 122(4) 2012, 1276–1303.

121. C. Weng, and Y. Zhang, Characterization of multivariate heavy-tailed distribution families via copula, *Journal of Multivariate Analysis*, 106, 2012, 178–186.

122. C. B. García, J. García Pérez, and J. R. van Dorp, Modeling heavy-tailed, skewed and peaked uncertainty phenomena with bounded support, *Statistical Methods & Applications*, 20(4) 2011, 146–166.

123. V. H. Lachos, T. Angolini, and C. A. Abanto-Valle, On estimation and local influence analysis for measurement errors models under heavy-tailed distributions, *Statistical Papers*, 52(3) 2011, 567–590.

124. V. Ganti, K. M. Straub, E. Foufoula-Georgiou, and C. Paola, Space-time dynamics of depositional systems: Experimental evidence and theoretical modeling of heavy-tailed statistics, *Journal of Geophysical Research*, 116(F2) 2011, F02011 (17 pp).

125. U. J. Dixit, and M. J. Nooghabi, Efficient estimation in the Pareto distribution with the presence of outliers, *Statistical Methodology*, 8(4) 2011, 340–355.

126. P. Nándori, Recurrence properties of a special type of heavy-tailed random walk, *Journal of Statistical Physics*, 142(2) 2011, 342–355.

127. D. Ceresetti, G. Molinié, and J.-D. Creutin, Scaling properties of heavy rainfall at short duration: a regional analysis, *Water Resources Research*, 46, 2010, W09531 (12 pp).

128. A. Charpentier, and A. Oulidi, Beta kernel quantile estimators of heavy-tailed loss distributions, *Statistics and Computing*, 20(1) 2010, 35–55.

129. P. Embrechts, J. Nešlehová, and M. V. Wüthrich, Additivity properties for value-at-risk under Archimedean dependence and heavy-tailedness, *Insurance: Mathematics and Economics*, 44(2) 2009, 164–169.

130. I. F. Alves, L. de Haan, and C. Neves, A test procedure for detecting super-heavy tails, *Journal of Statistical Planning and Inference*, 139(2) 2009, 213–227.

131. J. Beirlant, E. Joossens, and J. Segers, Second-order refined peaks-over-threshold modelling for heavy-tailed distributions, *Journal of Statistical Planning and Inference*, 139(8) 2009, 2800–2815.

132. R. Ibragimov, Heavy-tailedness and threshold sex determination, *Statistics and Probability Letters*, 78(16) 2008, 2804–2810.

133. R. Delgado, A reflected fBm limit for fluid models with ON/OFF sources under heavy traffic, *Stochastic Processes and Their Applications*, 117(2) 2007, 188–201.

134. M. S. Taqqu, The modelling of ethernet data and of signals that are heavy-tailed with infinite variance, *Scandinavian Journal of Statistics*, 29(2) 2002, 273–295.

135. B. G. Lindsay, J. Kettenring, and D. O. Siegmund, A report on the future of statistics, *Statistical Science*, 19(3) 2004, 387–413.

136. S. Resnick, On the foundations of multivariate heavy-tail analysis, *Journal of Applied Probability*, 41, 2004, 191–212.

137. S. Resnick, and H. Rootzén, Self-similar communication models and very heavy tails, *The Annals of Applied Probability*, 10(3) 2000, 753–778.

138. J. Cai, and Q. Tang, On max-sum equivalence and convolution closure of heavy-tailed distributions and their applications, *Journal of Applied Probability*, 41(1) 2004, 117–130.

139. V. Limic, A LIFO queue in heavy traffic, *The Annals of Applied Probability*, 11(2) 2001, 301–331.

140. H. Le, and A. O'Hagan, A class of bivariate heavy-tailed distributions, *Sankhyā: The Indian Journal of Statistics, Series B*, 60(1) 1998, 82–100

141. M. C. Bryson, Heavy-tailed distributions: Properties and tests, *Technometrics*, 16(1) 1974, 61–68.

142. J. Beran, Discussion: Heavy tail modeling and teletraffic data, *The Annals of Statistics*, 25(5) 1997, 1852–1856.

143. S. Ahn, J. H. T. Kim, and V. Ramaswami, A new class of models for heavy tailed distributions in finance and insurance risk, *Insurance: Mathematics and Economics*, 51(1) 2012, 43–52.

144. V. Pisarenko, and M. Rodkin, *Heavy-Tailed Distributions in Disaster Analysis*, Springer, Berlin, 2010.

145. S. I. Resnick, *Heavy-Tail Phenomena Probabilistic and Statistical Modeling*, Springer, Berlin, 2007.

146. R. J. Adler, R. E. Feldman, and M. S. Taqqu (eds.), *A Practical Guide to Heavy Tails: Statistical Techniques and Applications*, Birkhäuser, Boston, 1998.

147. M. Li, and W. Zhao, Visiting power laws in cyber-physical networking systems, *Mathematical Problems in Engineering*, 2012, 2012, Article ID 302786 (13 pp).

148. L. Xu, P. Ivanov Ch., K. Hu, Z. Chen, A. Carbone, and H. E. Stanley, Quantifying signals with power-law correlations: A comparative study of detrending and moving average techniques, *Physical Review E*, 71(5) 2005, 051101 (14 pp).

149. M. Li, and J.-Y. Li, On the predictability of long-range dependent series, *Mathematical Problems in Engineering*, 2010, 2010, Article ID 397454 (9 pp).

150. W. Hürlimann, From the general affine transform family to a Pareto type IV model, *Journal of Probability and Statistics*, 2009, 2009, Article ID 364901 (10 pp).

151. A. Adler, Limit theorems for randomly selected adjacent order statistics from a Pareto distribution, *International Journal of Mathematics and Mathematical Sciences*, 2005(21) 2005, 3427–3441.

152. H. E. Stanley, Phase transitions: Power laws and universality, *Nature*, 378(6557) 1995, 554.

153. I. Eliazar, and J. Klafter, A probabilistic walk up power laws, *Physics Reports*, 511(3) 2012, 143–175.

154. A. R. Bansal, G. Gabriel, and V. P. Dimri, Power law distribution of susceptibility and density and its relation to seismic properties: An example from the German continental deep drilling program (KTB), *Journal of Applied Geophysics*, 72(2) 2010, 123–128.

155. S. Milojević, Power law distributions in information science: Making the case for logarithmic binning, *Journal of the American Society for Information Science and Technology*, 61(12) 2010, 2417–2425.

156. Y. Wu, Q. Ye, and J. Xiao, and L.-X. Li, Modeling and statistical properties of human view and reply behavior in on-line society, *Mathematical Problems in Engineering*, 2012, 2012, Article ID 969087 (7 pp).

157. A. Fujihara, M. Uchida, and H. Miwa, Universal power laws in the threshold network model: A theoretical analysis based on extreme value theory, *Physica A*, 389(5) 2010, 1124–1130.

158. A. Saiz, Boltzmann power laws, *Physica A*, 389(2) 2010, 225–236.

159. J. Behboodian, A. Jamalizadeh, and N. Balakrishnan, A new class of skew-Cauchy distributions, *Statistics and Probability Letters*, 76(14) 2006, 1488–1493.

160. X. Zhao, P.-J. Shang, and Y.-L. Pang, Power law and stretched exponential effects of extreme events in Chinese stock markets, *Fluctuation and Noise Letters*, 9(2) 2010, 203–217.

161. P. Kokoszka, and T. Mikosch, The integrated periodogram for long-memory processes with finite or infinite variance, *Stochastic Processes and Their Applications*, 66(1) 1997, 55–78.

162. D. Belomestny, Spectral estimation of the Lévy density in partially observed affine models, *Stochastic Processes and Their Applications*, 121(6) 2011, 1217–1244.

163. T. Simon, Fonctions de Mittag-Leffler et processus de Lévy stables sans sauts négatifs, *Expositiones Mathematicae*, 28(3) 2010, 290–298.

164. R. Lambiotte, and L. Brenig, Truncated Lévy distributions in an inelastic gas, *Physics Letters A*, 345(4–6) 2005, 309–313.

165. G. Terdik, W. A. Woyczynski, and A. Piryatinska, Fractional- and integer-order moments, and multiscaling for smoothly truncated Lévy flights, *Physics Letters A*, 348(3–6) 2006, 94–109.

166. I. Koponen, Analytic approach to the problem of convergence of truncated lévy flights towards the Gaussian stochastic process, *Physical Review E*, 52(1) 1995, 1197–1199.

167. B. Lacaze, A stochastic model for propagation through tissue, *IEEE Transactions on Ultrasonics, Ferroelectrics and Frequency Control*, 56(10) 2009, 2180–2186.

168. P. Garbaczewski, Cauchy flights in confining potentials, *Physica A*, 389(5) 2010, 936–944.

169. A. J. Field, U. Harder, and P. G. Harrison, Measurement and modelling of self-similar traffic in computer networks, *IEEE Proceedings on Communications*, 151(4) 2004, 355–363.
170. G. A. Korn, and T. M. Korn, *Mathematical Handbook for Scientists and Engineers*, McGraw-Hill, New York, 1961.
171. H. Konno, and Y. Tamura, A generalized Cauchy process having cubic nonlinearity, *Reports on Mathematical Physics*, 67(2) 2011, 179–195.
172. H. Konno, and F. Watanabe, Maximum likelihood estimators for generalized Cauchy processes, *Journal of Mathematical Physics*, 48(10) 2007, 103303 (19 pp).
173. I. A. Lubashevsky, Truncated Lévy flights and generalized Cauchy processes, *European Physical Journal B*, 82(2) 2011, 189–195.
174. G. Terdik, and T. Gyires, Levy flights and fractal modeling of internet traffic, *IEEE/ACM Transactions on Networking*, 17(1) 2009, 120–129.
175. E. Di Nardo, and I. Oliva, Multivariate Bernoulli and Euler polynomials via Lévy processes, *Applied Mathematics Letters*, 25(9) 2012, 1179–1184.
176. D. V. Vinogradov, Arbitrary truncated Levy flight: Asymmetrical truncation and high-order correlations, *Physica A*, 391(22) 2012, 5584–5597.
177. Y. Liang, and W. Chen, A survey on computing Lévy stable distributions and a new MATLAB toolbox, *Signal Processing*, 93(1) 2013, 244–251.
178. E. E. Kuruoglu, Density parameter estimation of skewed α-stable distributions, *IEEE Transactions on Signal Processing*, 49(10) 2001, 2091–2101.
179. A. P. Petropulu, J.-C. Pesquet, X. Yang, and J. J. Yin, Power-law shot noise and its relationship to long-memory α-stable processes, *IEEE Transactions on Signal Processing*, 48(7) 2000, 1883–1892.
180. S. Cohen, and G. Samorodnitsky, Random rewards, fractional Brownian local times and stable self-similar processes, *The Annals of Applied Probability*, 16(3) 2006, 1432–1461.
181. M. Shao, and C. L. Nikias, Signal processing with fractional lower order moment: Stable processes and their applications, *Proceedings of the IEEE*, 81(7) 1993, 986–1010.
182. L. Landau, On the energy loss of fast particles by ionization, *Journal of Physics*, 8, 1944, 201–205.
183. D. H. Wilkinson, Ionization energy loss by charged particles part I. The Landau distribution, *Nuclear Instruments and Methods in Physics Research Section A*, 383(2–3) 1996, 513–515.
184. T. Tabata, and R. Ito, Approximations to Landau's distribution functions for the ionization energy loss of fast electrons, *Nuclear Instruments and Methods*, 158, 1979, 521–523.
185. J. Holtsmark, Uber die Verbreiterung von Spektrallinien, *Annalen der Physik*, 363(7) 1919, 577–630.
186. B. Pittel, W. A. Woyczynski, and J. A. Mann, Random tree-type partitions as a model for acyclic polymerization: Holtsmark (3/2-stable) distribution of the supercritical gel, *The Annals of Probability*, 18(1) 1990, 319–341.
187. D. G. Hummer, Rational approximations for the Holtsmark distribution, its cumulative and derivative, *Journal of Quantitative Spectroscopy and Radiative Transfer*, 36(1) 1986, 1–5.
188. R. G. Garroppo, S. Giordano, M. Pagano, and G. Procissi, Testing α-stable processes in capturing the queuing behavior of broadband teletraffic, *Signal Processing*, 82(12) 2002, 1861–1872.
189. J. R. Gallardo, D. Makrakis, and L. Orozco-Barbosa, Use of α-stable self-similar stochastic processes for modeling traffic in broadband networks, *Performance Evaluation*, 40(1–3) 2000, 71–98.
190. A. Karasaridis, and D. Hatzinakos, Network heavy traffic modeling using α-stable self-similar processes, *IEEE Transactions on Communications*, 49(7) 2001, 1203–1214.
191. P. R. de Montmort, *Essay d'analyse sur les jeux de hazard [Essays on the analysis of games of chance]*, 1713.
192. P. R. de Montmort, *Essay d'analyse sur les jeux de hazard*, American Mathematical Society, New York, 1980.

193. D. Bernoulli, Originally published in 1738; Translated by L. Sommer, Exposition of a new theory on the measurement of risk, *Econometrica*, 22(1) 1954, 22–36.
194. W. T. Coffey, Y. P. Kalmykov, and J. T. Waldron, *The Langevin Equation*, 2nd ed., World Scientific, Singapore, 2004.
195. S. F. Kwok, Langevin equation with multiplicative white noise: Transformation of diffusion processes into the Wiener process in different prescriptions, *Annals of Physics*, 327(8) 2012, 1989–1997.
196. A. V. Medino, S. R. C. Lopes, R. Morgado, and C. C. Y. Dorea, Generalized Langevin equation driven by Lévy processes: A probabilistic, numerical and time series based approach, *Physica A*, 391(3) 2012, 572–581.
197. D. Panja, Generalized Langevin equation formulation for anomalous polymer dynamics, *Journal of Statistical Mechanics*, 2010, 2010, L02001.
198. A. Bazzani, G. Bassi, and G. Turchetti, Diffusion and memory effects for stochastic processes and fractional Langevin equations, *Physica A*, 324(3–4) 2003, 530–550.
199. E. Lutz, Fractional Langevin equation, *Physical Review E*, 64(5) 2001, 051106 (4 pp).
200. M. G. McPhie, P. J. Daivis, I. K. Snook, J. Ennis, and D. J. Evans, Generalized Langevin equation for nonequilibrium systems, *Physica A*, 299(3–4) 2001, 412–426.
201. K. S. Fa, Fractional Langevin equation and Riemann-Liouville fractional derivative, *European Physical Journal E*, 139(2) 2007, 139–143.
202. B. Ahmad, J. J. Nieto, A. Alsaedi, and M. El-Shahed, A study of nonlinear Langevin equation involving two fractional orders in different intervals, *Nonlinear Analysis: Real World Applications*, 13(2) 2012, 599–606.
203. B. Ahmad, and J. J. Nieto, Solvability of nonlinear Langevin equation involving two fractional orders with Dirichlet boundary conditions, *International Journal of Differential Equations*, 2010, Article ID 649486, 10 pages, 2010.
204. S. C. Kou, and X. S. Xie, Generalized Langevin equation with fractional Gaussian noise: Subdiffusion within a single protein molecule, *Physical Review Letters*, 93(18) 2004, 180603 (4 pp).
205. H. C. Fogedby, Langevin equations for continuous time Lévy flights, *Physical Review E*, 50(2) 1994, 1657–1660.
206. Y. Fukui, and T. Morita, Derivation of the stationary generalized Langevin equation, *Journal of Physics A: Mathematical and Theoretical*, 4(4) 1971, 477–490.
207. S. C. Kou, Stochastic modeling in nanoscale biophysics: Subdiffusion within proteins, *The Annals of Applied Statistics*, 2(2) 2008, 501–535.
208. V. V. Anh, C. C. Heyde, and N. N. Leonenko, Dynamic models of long-memory processes driven by Lévy noise, *Journal of Applied Probability*, 39(4) 2002, 730–747.
209. B. N. Narahari Achar, J. W. Hanneken, T. Enck, and T. Clarke, Damping characteristics of a fractional oscillator, *Physica A*, 399(3–4) 2004, 311–319.
210. B. N. N. Achar, J. W. Hanneken, T. Enck, and T. Clarke, Response characteristics of a fractional oscillator, *Physica A*, 309(3–4) 2002, 275–288.
211. C. H. Eab, and S. C. Lim, Fractional generalized Langevin equation approach to single-file diffusion, *Physica A*, 389(13) 2010, 2510–2521.
212. C. H. Eab, and S. C. Lim, Fractional Langevin equations of distributed order, *Physical Review E*, 83(3) 2011, 031136 (10 pp).
213. S. C. Lim, and L. P. Teo, Modeling single-file diffusion with step fractional Brownian motion and a generalized fractional Langevin equation, *Journal of Statistical Mechanics*, 2009(8) 2009, P08015.
214. S. C. Lim, M. Li, and L. P. Teo, Locally self-similar fractional oscillator processes, *Fluctuation and Noise Letters*, 7(2) 2007, L169–179.
215. M. Li, S. C. Lim, and S. Y. Chen, Exact solution of impulse response to a class of fractional oscillators and its stability, *Mathematical Problems in Engineering*, 2011, 2011, Article ID 657839 (9 pp).

216. S. C. Lim, C. H. Eab, K.-H. Mak, M. Li, and S. Chen, Solving linear coupled fractional differential equations and their applications, *Mathematical Problems in Engineering*, 2012, 2012, Article ID 653939 (28 pp).

217. S. C. Lim, M. Li, and L. P. Teo, Langevin equation with two fractional orders, *Physics Letters A*, 372(42) 2008, 6309–6320.

218. R. A. Gabel, and R. A. Roberts, *Signals and Linear Systems*, John Wiley & Sons, New York, 1973.

219. K. Torre, and E. J. Wagenmakers, Theories and models for $1/f^{beta}$ noise in human movement science, *Human Movement Science*, 28(3) 2009, 297–318.

220. B. Kaulakys, M. Alaburda, and J. Ruseckas, $1/f$ noise from the nonlinear transformations of the variables, *Modern Physics Letters B*, 29(34) 2015, 1550223.

221. T. S. Rekha, and C. K. Mitra, $1/f$ correlations in viral genomes: A fast-Fourier transformation (FFT) study, *Indian Journal of Biochemistry & Biophysics*, 43(3) 2006, 137–142.

222. D. Storch, K. J. Gaston, and J. Cepak, Pink landscapes: $1/f$ spectra of spatial environmental variability and bird community, *Proceedings of the Royal Society of London (B): Biological Sciences*, 269(1502) 2002, 1791–1796.

223. J. He, G. Christakos, J. Wu, M. Li, and J. Leng, Spatiotemporal BME characterization and mapping of sea surface chlorophyll in Chesapeake Bay (USA) using auxiliary sea surface temperature data, *Science of the Total Environment*, 794, 2021, 148670.

224. M. Li, Long-range dependence and self-similarity of teletraffic with different protocols at the large time scale of day in the duration of 12 years: Autocorrelation modeling, *Physica Scripta*, 95(4) 2020, 065222 (15 pp).

225. K. S. Miller, and B. Ross, *An Introduction to the Fractional Calculus and Fractional Differential Equations*, John Wiley & Sons, New York, 1993.

226. T. Hida, *Brownian Motion*, Springer, Berlin, 1980.

227. A. H. Zemanian, An introduction to generalized functions and the generalized Laplace and Legendre transformations, *SIAM Review*, 10(1) 1968, 1–24.

CHAPTER 3

Power Laws of Fractal Data in Cyber-Physical Networking Systems

Cyber-physical networking systems (CPNS) are made up of various physical systems, including the Internet, that are heterogeneous in nature. Therefore, exploring universalities in CPNS for either data or systems is desired in its fundamental theory. This chapter is in the aspect of fractal time series, network traffic in particular, and it aims to address that power laws may yet be a universality of data in CPNS. The chapter is divided into three main sections. First, we provide a short tutorial about power laws. Then, we address the power laws related to some physical systems. Finally, we discuss that power-law-type data may be governed by stochastically differential equations of fractional order. As a side product, we present the point of view that the upper bound of fractal traffic data flow at large-time scaling and the small one also follows power laws.

3.1 BACKGROUND

Previously in Chapter 1 we mentioned the power laws regarding fractal time series. In this chapter, we specifically discuss them from the point of view of CPNS.

CPNS consist of computational and physical elements integrated, including the Internet and other computer networks, toward specific tasks [1–3]. Generally, both data and systems in CPNS are heterogeneous. For instance, network traffic differs from transportation traffic, letting along other data in CPNS, such as those in physiology and so on. Therefore, one of fundamental questions is what possible general laws are to meet CPNS in theory? The answer to this question is in two folds. One is data. The other is systems that transmit data from sources to destinations within a predetermined restrict period of time according to a given quality of service.

In general, both data and systems in CPNS are multi-dimensional. For instance, data from sources to be transmitted may be from a set of sensors distributed in a certain area.

DOI: 10.1201/9781003354987-4

Destinations receiving data may be a set of actuators, for example, a set of cars distributed in a certain area. Systems to transmit data are generally distributed.

Denote the n-dimensional Euclidean space by $\mathbf{R}^n$. Denote data at sources and destinations by $X(t)$ that is supposed to be n-dimensional and $Y(t)$ that is supposed to be m-dimensional, respectively. They are given by

$$X(t) = (x_1(t), \ldots, x_n(t)), \tag{3.1}$$

$$Y(t) = (y_1(t), \ldots, y_m(t)). \tag{3.2}$$

A stochastic equation describing an abstract relationship between $X(t)$ and $Y(t)$ may be expressed by

$$Y^T(t) = S(t) \otimes X^T(t) \oplus B^T(t), \tag{3.3}$$

where T implies the transposition, $S(t)$ is a service matrix of $n \times m$ order of a system, and $B(t)$, which is a vector with the same dimension as that of $Y(t)$, may represent uncertainty for the operation of $S(t) \otimes X^T(t)$. The operations $\otimes$ and $\oplus$ are to be studied from a view of systems and they are out of the scope of this chapter.

Note that $X(t)$ is usually a random field (see Chiles and Delfiner [4] in geosciences, Granichin et al. [5] and Mao et al. [6] in telecommunications, Muniandy et al. [7] in medical images, Mason et al. [8] in wind engineering, simply citing a few). The statistics of $X(t)$ is obviously crucial for the performance analysis of physical systems. It should be noted that the physical meaning of $X(t)$ is diverse. For example, it may represent a two-dimensional aeromagnetic data (Spector and Grant [9]), a medical image (Fortin et al. [10]), vegetation data (Myrhaug et al. [11]), surface crack in material science (Tanaka et al. [12]), data in physiology (Werner [13], West [14]), traffic and its delay bounds (Li and Wang [15]), data in stock markets (Rosenow et al. [16]), just mentioning a few. Therefore, seeking for possible universalities of $X(t)$ in CPNS is desired.

Without the generality losing, we rewrite (3.1) as

$$X(t) = X(t_1, \ldots, t_n), \tag{3.4}$$

where $t = (t_1, \ldots, t_n)$. The norm of t is given by

$$\|t\| = \sqrt{t_1^2 + \cdots + t_n^2}. \tag{3.5}$$

The autocovariance function (ACF) of $X(t)$ is given, over the hyperrectangle $C = \prod_{i=1}^{n} [a_i, b_i]$ for $a_i, b_i \in \mathbf{R}$ (Adler [17], Li [18]), by

$$C(\tau) = \mathrm{E}[X(t)X(t+\tau)], \tag{3.6}$$

where E is the mean operator, $\tau = (\tau_1, \ldots, \tau_n)$ and

$$\|\tau\| = \sqrt{\tau_1^2 + \cdots + \tau_n^2}. \tag{3.7}$$

The ACF $C(\tau)$ measures how $X(t)$ correlates to $X(t + \tau)$.

From the point of view of applications of CPNS, we are interested in two asymptotic expressions of $C(\tau)$. One is $C(\tau)$ for $\|\tau\| \to 0$. The other is $C(\tau)$ for $\|\tau\| \to \infty$. The former characterizes the small scaling phenomenon of $X(t)$. The latter measures the large scaling phenomenon of $X(t)$. It is quite natural for us to investigate two types of scaling phenomena. As a matter of fact, one may be interested in small scaling in some applications, for example, admission control in computer communications or monitoring sudden disaster in geoscience. On the other side, one may be interested in large scaling in applications such as long-term performance analysis of systems. Exact expression of $C(\tau)$ is certainly useful but it may usually be application dependent. Consequently, we study possible generalities of $C(\tau)$ for $\|\tau\| \to 0$ and $\|\tau\| \to \infty$ instead of its exactly full expressions. The aim of this chapter is to explain that both the small scaling described by $C(\tau)$ for $\|\tau\| \to 0$ and the large scaling described by $C(\tau)$ for $\|\tau\| \to \infty$, in some fields related to CPNS, ranging from geoscience to computer communications, follow power laws.

The rest of chapter is organized as follows: Short tutorial about power laws is given in Section 3.2. Some cases of power laws relating to computational and physical systems in CPNS are described in Section 3.3. Stochastically differential equations to govern power-law-type data are discussed in Section 3.4, which is followed by summary.

3.2 BIREF OF POWER LAWS

Denote the probability space by (Ω, T, P). Then, $x(t, \varsigma)$ is said to be a stochastic process when the random variable x represents the value of the outcome of an experiment T for every time t, where Ω represents the sample space, T is the event space, and P is the probability measure.

As usual, $x(t, \varsigma)$ is simplified to be written as $x(t)$. That is, the event space is usually omitted. Denote the probability function of x by $P(x)$. Then, one can define the general nth order, time varying, joint distribution function $P(x_1, \ldots, x_n; t_1, \ldots, t_n)$ for the random variables $x(t_1), \ldots, x(t_n)$. The joint probability density function (PDF) is written by

$$p(x_1, \ldots, x_n; t_1, \ldots, t_n) = \frac{\partial^n P(x_1, \ldots, x_n; t_1, \ldots, t_n)}{\partial x_1 \ldots \partial x_n}. \tag{3.8}$$

For simplicity, we write $P(X) = P(x_1, \ldots, x_n; t_1, \ldots, t_n)$ and $p(X) = p(x_1, \ldots, x_n; t_1, \ldots, t_n)$. Then, the probability is given by

$$P(X_2) - P(X_1) = \text{Prob}[X_1 < \xi < X_2] = \int_{X_1}^{X_2} p(\xi) d\xi. \tag{3.9}$$

The mean and the ACF of X based on PDF are written as (3.10) and (3.11), respectively,

$$\mu_X = \int_{-\infty}^{\infty} X p(X) dX, \tag{3.10}$$

$$C_{XX}(\tau) = \int_{-\infty}^{\infty} X(t) X(t+\tau) p(X) dX. \tag{3.11}$$

Let V_X be the variance of X. Then,

$$V_X = \mathrm{E}\left[(X - \mu_X)^2\right] = \int_{-\infty}^{\infty} (X - \mu_X)^2 \, p(X) dX. \tag{3.12}$$

The above expressions imply that the integrals in (3.10) and (3.12) are convergent in the domain of ordinary functions if $p(X)$ is light-tailed, that is, exponentially-decayed. Light-tailed PDFs are not our interests. We are interested in heavy-tailed PDFs. By heavy tail, we mean that $p(X)$ decays so slowly that (3.10) or (3.12) may be divergent or very large such that μ_X or V_X may not be usable in practice. In the following subsections, we shall describe power laws in probability space, ACF, and power spectrum density (PSD) function.

3.2.1 Power Law in PDF

A typical heavy-tailed case is the Pareto distribution. Denote the PDF of the Pareto distribution by $p_{\mathrm{Pareto}}(X)$. Then,

$$p_{\mathrm{Pareto}}(X) = \frac{ab}{X^{a+1}}, \tag{3.13}$$

where a and b are parameters and $X \geq a$. The mean and variance of X that follows $p_{\mathrm{Pareto}}(X)$ are given by (3.14) and (3.15), respectively

$$\mu_{\mathrm{Pareto}} = \frac{ab}{a-1}, \tag{3.14}$$

$$\mathrm{Var}(X)_{\mathrm{Pareto}} = \frac{ab^2}{(a-1)^2(a-2)}. \tag{3.15}$$

It is easily seen that μ_{Pareto} and $\mathrm{Var}(X)_{\mathrm{Pareto}}$ do not exist if $a = 1$. Note that μ_X implies a global property of X while $\mathrm{Var}(X)$ represents a local property of X. Therefore, heavy-tailed PDFs imply that X is in wild randomness due to infinite or very large variance, see Mandelbrot [19] for the meaning of wild randomness.

Note 3.1. The Pareto distribution is an instance of power-law-type PDF. □

3.2.2 Power Law in ACF

A consequence of a heavy-tailed random variable in ACF is that $C_{XX}(\tau)$ is slowly decayed. By slowly decayed, we mean that, for $d \in \mathbf{R}_+$, $C_{XX}(\tau)$ decays hyperbolically in power law given by (Adler et al. [20])

$$C_{XX}(\tau) \sim \tau^{-d}. \tag{3.16}$$

The Taqqu's theorem describes the relationship between a heavy-tailed PDF and hyperbolically decayed ACF (Abry et al. [21]).

3.2.3 Power Law in PSD

Denote the PSD of X by $S_{XX}(\omega)$. Then,

$$S_{XX}(\omega) = \int\limits_{-\infty}^{\infty} C_{XX}(\tau)e^{-j\omega\tau}d\tau, \quad j = \sqrt{-1}. \tag{3.17}$$

According to the theory of generalized functions (Kanwal [22]), one has

$$S_{XX}(\omega) \sim |\omega|^{d-1}. \tag{3.18}$$

Therefore, power law in PSD, which is usually termed $1/f$ noise, see Wornell [23], Keshner [24], Ninness [25], Corsini and Saletti [26], Li [27].

3.2.4 Power Laws in Describing Scaling Phenomena

We now turn to scaling descriptions. Small scaling phenomenon may be investigated by $C_{XX}(\tau)$ for $\tau \to 0$ and large scaling one for $\tau \to \infty$, respectively (Li and Zhao [28]). On the one side, following Davies and Hall [29], if $C_{XX}(\tau)$ is sufficiently smooth on $(0, \infty)$ and if

$$C_{XX}(0) - C_{XX}(\tau) \sim c_1|\tau|^{\alpha} \text{ for } |\tau| \to 0, \tag{3.19}$$

where c_1 is a constant and α is the fractal index of X, the fractal dimension, denoted by D, of X is expressed by

$$D = 2 - \frac{\alpha}{2}. \tag{3.20}$$

Note 3.2. Fractal dimension is a parameter to characterize small scaling phenomenon (Mandelbrot [30], Gneiting and Schlather [31], Li [32]). □

On the other side, if

$$C_{XX}(\tau) \sim |\tau|^{-\beta} \; (\tau \to \infty), \tag{3.21}$$

the parameter β is used to measure the statistical dependence of X. If $\beta > 1$, $C_{XX}(\tau)$ is integrable and accordingly X is short-range-dependent (SRD). If $0 < \beta < 1$, $C_{XX}(\tau)$ is nonintegrable and X is of long-range dependence (LRD) (Beran [33]). Representing β by the Hurst parameter $H \in (0, 1)$ yields

$$H = 1 - \frac{\beta}{2}. \tag{3.22}$$

Note 3.3. Statistical dependence, either SRD or LRD, is a property for large scaling phenomenon. □

3.3 CASES OF POWER LAWS IN CPNS

We address some application cases of power laws in CPNS in this section.

3.3.1 Power Laws in the Internet

Let $x(t)$ be the teletraffic time series. It may represent the packet size of teletraffic at time t. Denote the ACF of $x(t)$ by

$$R(\tau) = E[X(t+\tau)X(t)]. \tag{3.23}$$

Then, we have (Li and Lim [34], Li [35])

$$R(\tau) \sim |\tau|^{\alpha}, \quad \tau \to 0, \tag{3.24}$$

$$R(\tau) \sim |\tau|^{-\beta}, \quad \tau \to \infty. \tag{3.25}$$

From (3.24) and (3.25), the fractal dimension D and the Hurst parameter H of teletraffic are respectively given by

$$D = 2 - \frac{\alpha}{2}, \tag{3.26}$$

$$H = 1 - \frac{\beta}{2}. \tag{3.27}$$

The above expressions exhibit that both the small scaling and the large one follow power laws.

It is worth noting that the upper bounds of teletraffic also follow power laws. In fact, the amount of teletraffic accumulated in the interval $[0, t]$ is upper bounded by

$$\int_0^t x(u)du \le \sigma + \rho t, \tag{3.28}$$

where σ and ρ are constants and $t > 0$ (Cruz [36]). Following Li and Zhao [28], we have the bounds of both the small time scaling and the large one, respectively, expressed by

$$\int_0^t x(u)du \le r^{2D-5}\sigma \text{ for small } t, \tag{3.29}$$

$$\int_0^t x(u)du \le a^{-H}\rho \text{ for large } t, \tag{3.30}$$

where $r > 0$ is a small-scale factor, $a > 0$ is a large-scale factor. Therefore, we have the following theorem.

Theorem 3.1. Both the small-scale factor and the large-scale one of teletraffic obey power laws, that is, r^{2D-5} and a^{-H}.

Proof. Two scaling factors follow r^{2D-5} and a^{-H}, respectively. Thus, they obey power laws. This completes the proof.

In addition to teletraffic, others with respect to the Internet also follow power laws. Some are listed below.

Note 3.4. Barabasi and Albert [37] studied several large databases in the World Wide Web (WWW), where they defined vertices by HyperText Markup Language (HTML) documents. They inferred that the probability $P(k)$ that a vertex in the network interacts with k other vertices decays hyperbolically as $P(k) \sim k^{-\gamma}$ for $\gamma > 0$. Hence, power law. □

Note 3.5. Let $P_{out}(k)$ and $P_{in}(k)$ be the probabilities of a document to have k outgoing and incoming links, respectively. Then, $P_{out}(k)$ and $P_{in}(k)$ obey power laws (Albert et al. [38]). □

Note 3.6. The probability of web pages among sites is of power law (Huberman and Adamic [39]). □

3.3.2 Power Laws in Geosciences

Let $(x, y, z) \in \mathbf{R}^3$ be a spatial point. The physical meaning of a random function $U(x, y, z)$ may be diverse in the field. For instance, it may represent prospected gold amount at (x, y, z) in a gold mine or a value of pollution index for pollution alert at (x, y, z) in a city.

For simplicity, denote a vector by $l = (x, y, z)$. Let

$$\rho = |l| = \sqrt{x^2 + y^2 + z^2}. \tag{3.31}$$

Then, one may be interested in the covariance function of $U(\rho)$. Denote the covariance function of $U(\rho)$ by $C(\tau)$. Then,

$$C(\tau) = E\{[U(\rho) - EU(\rho)][U(\rho + \tau) - EU(\rho)]\}. \tag{3.32}$$

One of the commonly used models of covariance functions in geosciences is given by

$$C(\tau) = \frac{1}{1+|\tau|^2}.$$ (3.33)

The above constant power is the case of the standard Cauchy process (Webster and Oliver [40]). It fits with some cases in geosciences (see Wackernagel [41], Li and Li [42]). We list some in the following notes.

Note 3.7. Let $C(s)$ be the covariance function between yield densities at any two points in a region, where s represents the distance difference between two points. Then, for large v,

$$C(s) \sim |s|^{-v} \quad (v > 0),$$ (3.34)

see Whittle [43]. □

Note 3.8. Sea level fluctuations, river flow, and flood height follow power laws (Li and Li [42], Lefebvre [44], Lawrance and Kottegoda [45]). □

Note 3.9. Urban growth obeys power laws (Makse et al. [46]). □

3.3.3 Power Laws in Wind Engineering

Wind engineering is an important field related to wind power generation and disaster preventions from a view of CPNS. In this field, studying fluctuations of wind speed is essential.

The PSD introduced by von Kármán [47], known as the von Kármán spectrum (VKS), is widely used in the diverse fields, ranging from turbulence to acoustic wave propagation in random media (see Goedecke et al. [48], Hui et al. [49]). For the VKS expressed in (3.35), we use the term VKSW for short,

$$S_{\mathrm{von}}(f) = \frac{4u_f^2 b_v w}{f\left(1+70.8w^2\right)^{5/6}}, \quad w = \frac{fL_u^x}{U},$$ (3.35)

where f is frequency (Hz), L_u^x is turbulence integral scale, U is mean speed, u_f is friction velocity (ms^{-1}), b_v is friction velocity coefficient such that the variance of wind speed is $\sigma_u^2 = b_v u_f^2$. The above equation implies that VKSW obeys power law for $f \in (0, \infty)$.

Another famous PSD in wind engineering is the one introduced by Davenport [50], which is expressed by

$$\frac{fS_{\mathrm{Dav}}(f)}{u_f^2} = \frac{4u^2}{\left(1+u^2\right)^{4/3}}, \quad u = \frac{1200n}{z},$$ (3.36)

where n is the normalized frequency ($fz/U(10\mathrm{m})$), $U(10\mathrm{m})$ is the mean wind speed (ms^{-1}) measured at height 10 m, $U(z)$ mean wind speed (ms^{-1}) measured at height z.

Davenport's PSD exhibits a power law of wind speed. Other forms of the PSDs of wind speed, such as those discussed by Kaimal [51], Antoniou et al. [52], Hiriart et al. [53], all follow power laws.

3.4 SOME EQUATIONS FOR POWER-LAW-TYPE DATA

The cases of power laws mentioned in the previous section are a few that people may be interested in from a view of CPNS. There are others that are essential in the field of CPNS, such as power laws in earthquake (see Pisarenko and Rodkin [54]). Now, we turn to the discussions about the generality about equations that may govern data of power law type.

Conventionally, a stationary random function $y(t)$ may be taken as a solution of a differential equation of integer order, which is driven by white noise $w(t)$. This equation may be written as

$$\sum_{i=0}^{p} a_i \frac{d^{p-i} y(t)}{dt^{p-i}} = w(t), \tag{3.37}$$

where p and i are integers.

Let $v > 0$ and $f(t)$ be a piecewise continuous on $(0, \infty)$ and integrable on any finite subinterval of $[0, \infty)$. For $t > 0$, denote the Riemann–Liouville integral operator of order v by $_0D_t^{-v}$ [55–58]. Then,

$$_0D_t^{-v} f(t) = \frac{1}{\Gamma(v)} \int_0^t (t-u)^{v-1} f(u) du, \tag{3.38}$$

where Γ is the Gamma function. For simplicity, we write $_0D_t^{-v}$ as D^{-v} below.

Let $v_p, v_{p-1}, \ldots, v_0$ be a strictly decreasing sequence of nonnegative numbers. Then, for the constants a_i, we have

$$\sum_{i=0}^{p} a_{p-i} D^{v_i} y(t) = w(t). \tag{3.39}$$

The above is a stochastically fractional differential equation with constant coefficients of order v_p. This class of equations yield random functions with power laws (Li [27] or Chapter 1 in this book). In the case of random fields, (3.39) is extended to be a partial differential equation of fractional order given by

$$\sum_{i=0}^{p} a_{p-i} D^{v_i} y = w, \tag{3.40}$$

where both w and y are multidimensional, and D is an operator of partial differentiation.

Another class of stochastically differential equations of fractional order is given by (Lim and Muniandy [59])

$$\left(\sum_{i=0}^{p} a_i \frac{d^{p-i} y(t)}{dt^{p-i}} \right)^{\beta_1} = w(t) \quad (\beta_1 > 0). \tag{3.41}$$

Note that (3.39) or (3.40) or (3.41) should not be taken as a simple extension of conventional equation (3.37) from integer order to fractional one. As a matter of fact, there are challenge issues with respect to differential equations of fractional order. Since data of power-law type may be with infinite variance (Samorodnitsky and Taqqu [60]), variance analysis, which is a powerful tool in the analysis of conventional random functions, fails to describe random data with infinite variance. Power-law-type data may be LRD, which makes the stationarity test of data a tough issue (see Mandelbrot [61], Abry et al. [62], Li et al. [63], Li and Zhao [64]). We shall address this issue in Chapter 7 later in the book. In addition, prediction of data with power laws considerably differs from that of conventional data (Li and Li [65], Hall and Yao [66]).

3.5 SUMMARY

We have discussed the basics of power laws from either a mathematical point of view or applications in a number of fields in CPNS. The purpose of this chapter is to exhibit that power laws may yet serve as a universality of fractal time series, including fractal traffic, in CPNS. This point of view may be useful for data modeling and analysis in CPNS, including computer communication networks.

REFERENCES

1. E. A. Lee, *Cyber Physical Systems: Design Challenges*, University of California, Berkeley Technical Report No. UCB/EECS-2008-8, 2008.
2. J. Stankovic, I. Lee, A. Mok, and R. Rajkumar, Opportunities and obligations for physical computing systems, *Computer*, 38(11) 2005, 23–31.
3. R. Alur, D. Thao, J. Esposito, H. Yerang, F. Ivancic, V. Kumar, P. Mishra, G. J. Pappas, and O. Sokolsky, Hierarchical modeling and analysis of embedded systems, *Proceedings of the IEEE*, 91(1) 2003, 11–28.
4. J.-P. Chiles, and P. Delfiner, *Geostatistics, Modeling Spatial Uncertainty*, John Wiley & Sons, New York, 1999.
5. O. Granichin, V. Erofeeva, Y. Ivanskiy, and Y. Jiang, Simultaneous perturbation stochastic approximation-based consensus for tracking under unknown-but-bounded disturbances, *IEEE Transactions on Automatic Control*, 66(8) 2021, 3710–3717.
6. Z. Mao, Y. Jiang, X. Di, and Y. T. Woldeyohannes, Joint head selection and airtime allocation for data dissemination in mobile social networks, *Computer Networks*, 166, 2020, 106990.
7. S. V. Muniandy, and J. Stanslas, Modelling of chromatin morphologies in breast cancer cells undergoing apoptosis using generalized Cauchy field, *Computerized Medical Imaging and Graphics*, 32(7) 2008, 631–637.
8. M. S. Mason, D. F. Fletcher, and G. S. Wood, Numerical simulation of idealised three-dimensional downburst wind fields, *Engineering Structures*, 32(11) 2010, 3558–3570.

9. A. Spector, and F. S. Grant, Statistical models for interpreting aeromagnetic data, *Geophysics*, 35(2) 1970, 293–302.

10. C. S. Fortin, R. Kumaresan, W. J. Ohley, and S. Hoefer, Fractal dimension in the analysis of medical images, *IEEE Engineering in Medicine and Biology Magazine*, 11(2) 1992, 65–71.

11. D. Myrhaug, L. E. Holmedal, and M. C. Ong, Nonlinear random wave-induced drag force on a vegetation field, *Coastal Engineering*, 56(3) 2009, 371–376.

12. M. Tanaka, R. Kato, Y. Kimura, and A. Kayama, Automated image processing and analysis of fracture surface patterns formed during creep crack growth in austenitic heat-resisting steels with different microstructures, *ISIJ International*, 42(12) 2002, 1412–1418.

13. G. Werner, Fractals in the nervous system: Conceptual implications for theoretical neuroscience, *Frontiers in Fractal Physiology*, 1, 2010, 00015.

14. B. J. West, Fractal physiology and the fractional calculus: A perspective, *Frontiers in Fractal Physiology*, 1, 2010, 00012.

15. M. Li, and A. Wang, Fractal teletraffic delay bounds in computer networks, *Physica A*, 557, 2020, 124903 (13 pp).

16. B. Rosenow, P. Gopikrishnan, V. Plerou, and H. E. Stanley, Dynamics of cross-correlations in the stock market, *Physica A*, 324(1–2) 2003, 241–246.

17. R. J. Adler, *The Geometry of Random Fields*, John Wiley & Sons, New York, 1981.

18. M. Li, Generalized fractional Gaussian noise and its application to traffic modeling, *Physica A*, 579, 2021, 1236137 (22 pp).

19. B. B. Mandelbrot, *Multifractals and 1/f Noise*, Springer, New York, 1998.

20. R. J. Adler, R. E. Feldman, and M. S. Taqqu (eds.), *A Practical Guide to Heavy Tails: Statistical Techniques and Applications*, Birkhäuser, Boston, 1998.

21. P. Abry, P. Borgnat, F. Ricciato, A. Scherrer, and D. Veitch, Revisiting an old friend: On the observability of the relation between long range dependence and heavy tail, *Telecommunication Systems*, 43(3–4) 2010, 147–165.

22. R. P. Kanwal, *Generalized Functions: Theory and Applications*, 3rd ed., Birkhäuser, Boston, 2004.

23. G. W. Wornell, Wavelet-based representations for the 1/f family of fractal processes, *Proceedings of the IEEE*, 81(10) 1993, 1428–1450.

24. M. S. Keshner, 1/f noise, *Proceedings of the IEEE*, 70(3) 1982, 212–218.

25. B. Ninness, Estimation of 1/f noise, *IEEE Transactions on Information Theory*, 44(1) 1998, 32–46.

26. G. Corsini, and R. Saletti, A $1/f^\gamma$ power spectrum noise sequence generator, *IEEE Transactions on Instrumentation and Measurement*, 37(4) 1988, 615–619.

27. M. Li, Fractal time series: A tutorial review, *Mathematical Problems in Engineering*, 2010, 2010, Article ID 157264 (26 pp).

28. M. Li, and W. Zhao, Representation of a stochastic traffic bound, *IEEE Transactions on Parallel and Distributed Systems*, 21(9) 2010, 1368–1372.

29. S. Davies, and P. Hall, Fractal analysis of surface roughness by using spatial data, *Journal of Royal Statistical Society Series B*, 61(1) 1999, 3–37.

30. B. B. Mandelbrot, *The Fractal Geometry of Nature*, W. H. Freeman, New York, 1982.

31. T. Gneiting, and M. Schlather, Stochastic models that separate fractal dimension and the Hurst effect, *SIAM Review*, 46(2) 2004, 269–282.

32. M. Li, A class of negatively fractal dimensional Gaussian random functions, *Mathematical Problems in Engineering*, 2011, 2011, Article ID 291028 (18 pp).

33. J. Beran, *Statistics for Long-Memory Processes*, Chapman & Hall, New York, 1994.

34. M. Li, and S. C. Lim, Modeling network traffic using generalized Cauchy process, *Physica A*, 387(11) 2008, 2584–2594.

35. M. Li, Generation of teletraffic of generalized Cauchy type, *Physica Scripta*, 81(2) 2010, 025007 (10pp).

36. R. L. Cruz, A calculus for network delay, part I: Network elements in isolation, part II: Network analysis, *IEEE Transactions on Information Theory*, 37(1) 1991, 114–141.

37. A.-L. Barabasi, and R. Albert, Emergence of scaling in random networks, *Science*, 286(5439) 1999, 509–512.

38. R. Albert, H. Jeong, and A.-L. Barabási, Internet: Diameter of the world-wide web, *Nature*, 401(6749) 1999, 130.

39. B. A. Huberman, and L. A. Adamic, Internet: Growth dynamics of the world-wide web, *Nature*, 401(6749) 1999, 131.

40. R. Webster, and M. A. Oliver, *Geostatistics for Environmental Scientists*, John Wiley and Sons, New York, 2007.

41. H. Wackernagel, *Multivariate Geostatistics: An Introduction with Applications*, Springer, New York, 2005.

42. M. Li, and J.-Y. Li, Generalized Cauchy model of sea level fluctuations with long-range dependence, *Physica A*, 484, 2017, 309–335.

43. P. Whittle, On the variation of yield variance with plot size, *Biometrika*, 43(3–4) 1962, 337–343.

44. M. Lefebvre, A one- and two-dimensional generalized Pareto model for a river flow, *Applied Mathematical Modelling*, 30(2) 2006, 155–163.

45. A. J. Lawrance, and N. T. Kottegoda, Stochastic modelling of riverflow time series, *Journal of Royal Statistical Society Series A*, 140(1) 1977, 1–47.

46. H. Makse, S. Havlin, and H. E. Stanley, Modeling urban growth patterns, *Nature*, 377, 1995, 608–612.

47. T. von Kármán, Progress in the statistical theory of turbulence, *Proceedings of the National Academy of Sciences of United States of America*, 34(11) 1948, 530–539.

48. G. H. Goedecke, V. E. Ostashev, D. K. Wilson, and H. J. Auvermann, Quasi-wavelet model of Von Kármán spectrum of turbulent velocity fluctuations, *Boundary-Layer Meteorology*, 112(1) 2004, 33–56.

49. M. C. H. Hui, A. Larsen, and H. F. Xiang, Wind turbulence characteristics study at the stonecutters bridge site: Part II wind power spectra, integral length scales and coherences, *Journal of Wind Engineering and Industrial Aerodynamics*, 97(1) 2009, 48–59.

50. A. G. Davenport, The spectrum of horizontal gustiness near the ground in high winds, *Quarterly Journal of the Royal Meteorological Society*, 87(372) 1961, 194–211.

51. J. C. Kaimal, J. C. Wyngaard, Y. Izumi, and O. R. Coté, Spectral characteristics of surface-layer turbulence, *Quarterly Journal of the Royal Meteorological Society*, 98(417) 1972, 563–589.

52. I. Antoniou, D. Asimakopoulos, A. Fragoulis, A. Kotronaros, D. P. Lalas, and I. Panourgias, Turbulence measurements on top of a steep hill, *Journal of Wind Engineering and Industrial Aerodynamics*, 39(1–3) 1992, 343–355.

53. D. Hiriart, J. L. Ochoa, and B. Garcia, Wind power spectrum measured at the San Pedro Martir Sierra, *Revista Mexicana de Astronomia y Astrofisica*, 37, 2001, 213–220.

54. V. Pisarenko, and M. Rodkin, *Heavy-Tailed Distributions in Disaster Analysis*, Springer, New York, 2010.

55. C. A. Monje, Y.-Q. Chen, B. M. Vinagre, D. Xue, and V. Feliu, *Fractional Order Systems and Controls: Fundamentals and Applications*, Springer, New York, 2010.

56. M. D. Ortigueira, An introduction to the fractional continuous-time linear systems: The 21st century systems, *IEEE Circuits and Systems Magazine*, 8(3) 2008, 19–26.

57. Y. Q. Chen, and K. L. Moore, Discretization schemes for fractional order differentiators and integrators, *IEEE Transactions on Circuits and Systems I: Fundamental Theory and Applications*, 49(3) 2002, 363–367.

58. B. M. Vinagre, Y. Q. Chen, and I. Petras, Two direct Tustin discretization methods for fractional-order differentiator/integrator, *Journal of the Franklin Institute*, 340(5) 2003, 349–362.

59. S. C. Lim, and S. V. Muniandy, Self-similar Gaussian processes for modelling anomalous diffusion, *Physical Review E*, 66(2) 2002, 0211–14.
60. G. Samorodnitsky, and M. S. Taqqu, *Stable Non-Gaussian Random Processes: Stochastic Models with Infinite Variance*, Chapman and Hall, New York, 1994.
61. B. B. Mandelbrot, Note on the definition and the stationarity of fractional Gaussian noise, *Journal of Hydrology*, 30(4) 1976, 407–409.
62. P. Abry, and D. Veitch, Wavelet analysis of long-range dependent traffic, *IEEE Transactions on Information Theory*, 44(1) 1998, 2–15.
63. M. Li, W.-S. Chen, and L. Han, Correlation matching method of the weak stationarity test of LRD traffic, *Telecommunication Systems*, 43(3–4) 2010, 181–195.
64. M. Li, and W. Zhao, Quantitatively investigating locally weak stationarity of modified multifractional Gaussian noise, *Physica A*, 391(24) 2012, 6268–6278.
65. M. Li, and J.-Y. Li, On the predictability of long-range dependent series, *Mathematical Problems in Engineering*, 2010, 2010, Article ID 397454 (9 pp).
66. P. Hall, and Q.-W. Yao, Inference in ARCH and GARCH models with heavy-tailed errors, *Econometrica*, 71(1) 2003, 285–317.

CHAPTER 4

Ergodicity of Long-Range-Dependent Traffic

There are two contributions discussed in this chapter. First, we show that long-range-dependent (LRD) Gaussian processes with the autocovariance $C_{xx}(\tau) \sim |\tau|^{2H-2}$ for the Hurst parameter H in the range of $0.5 < H < 1$ are ergodic. Second, as a consequence, we explain that such a class of LRD processes is stationary. The ergodicity and stationarity of two LRD processes, namely, the fractional Gaussian noise (fGn) and the generalized Cauchy (GC) process, are addressed.

4.1 BACKGROUND

Denote the nth sample function for $n \in \mathbf{N}$ and $t \in \mathbf{R}$ by $x_n(t)$, where $\mathbf{N}$ is the set of natural numbers, $\mathbf{R}$ is the set of real numbers. Let $\mathbf{P}$ be the probability set for $\text{Prob}[x_n(t) \le a]$. Then, the set $\{x_n(t)\}$ is called a random process, where $x_n(t) \in [\mathbf{N}, \mathbf{R}, \mathbf{P}]$. For facilitating the discussions and without the generality losing, people often denote $x_n(t)$ by $x(t)$ and simply call $x(t)$ a random process or random function in short (Yaglom [1], Li [2], Papoulis and Pillai [3]).

Stationarity is an essential property of a process. The literature regarding stationarity test is rich. For instance, Borgnat and Flandrin explained their stationarity test method using surrogates [4], Von Sachs and Neumann stated their stationarity test method via wavelets [5], Ling studied his method with double autoregressive models [6], Psaradakis discussed his method using blockwise bootstrap [7], Rodrigues and Rubia described their method using autoregressive processes with level-dependent conditional heteroskedasticity [8], and so on (see Al-Shoshan [9], Darné and Diebolt [10], Sheskin [11], simply mentioning a few). However, conventional methods of stationarity test cease for processes with long-range dependence (LRD) as stated by Abry and Veitch [12, right column, p. 7].

Processes with LRD have gained interests of researchers in various fields, ranging from physics to computer science (see Li [2], Abry and Veitch [12], Li [13–17], Loiseau et al. [18], Fontugne et al. [19], Sheng et al. [20], Sheng and Chen [21], Heydari et al. [22, 23], Pinchas [24, 25], Avraham and Pinchas [26, 27], Silva and Rocha [28], Nashat and Hussain

DOI: 10.1201/9781003354987-5

[29], Mitzenmacher [30], Tadaki [31], Ercan and Kavvas [32], Jiang et al. [33], Abolbashari et al. [34], Yang et al. [35], Gneiting [36], Beran [37], Mandelbrot [38], Hurst [39], just citing a few). Because how to test the stationarity of processes with LRD remains a problem unsolved (Abry and Veitch [12]), how to clarify the ergodicity of processes with LRD is also a tough problem open. In this chapter, we aim at giving the solution to the above problem by proving the ergodicity and stationarity of LRD processes with $C_{xx}(\tau) \sim |\tau|^{2H-2}$ for $0.5 < H < 1$.

The rest of the chapter is organized as follows: In Section 4.2, we brief the preliminaries. Problem statements are given in Section 4.3. We shall present our results in Section 4.4, which is followed by discussions and summary.

4.2 PRELIMINARIES

4.2.1 Processes with LRD

Let $C_{xx}(\tau)$ be the autocovariance function (ACF) of $x(t)$. If $C_{xx}(\tau)$ decays slowly such that

$$\int_{-\infty}^{\infty} C_{xx}(\tau)d\tau = \infty, \tag{4.1}$$

we say that $x(t)$ is of LRD (Gneiting [36], Beran [37]). If $\int_{-\infty}^{\infty} C_{xx}(\tau)d\tau < \infty$, $x(t)$ is said to be of short-range dependence (SRD). A common approximation of $C_{xx}(\tau)$ for large τ is given by a power function in the form

$$C_{xx}(\tau) \sim c|\tau|^{-\beta}, \tag{4.2}$$

where $c > 0$ is a constant and $\beta > 0$ is the index of statistical dependences. When $0 < \beta < 1$, $x(t)$ is of LRD because $\int_{-\infty}^{\infty} C_{xx}(\tau)d\tau = \infty$. On the other side, $x(t)$ is of SRD for $\beta > 1$. In the field, for the purpose of dedicating H. E. Hurst for his pioneering work on processes with LRD (Hurst [39]), people often utilize the Hurst parameter, H, to express β as

$$\beta = 2 - 2H. \tag{4.3}$$

Therefore, the LRD condition $0 < \beta < 1$ implies $0.5 < H < 1$. On the other hand, $0 < H < 0.5$ represents the SRD condition. In this case, (4.2) is rewritten as

$$C_{xx}(\tau) \sim c|\tau|^{2H-2}. \tag{4.4}$$

A commonly used traffic model with LRD is the fGn introduced by Mandelbrot and van Ness [40]. Its ACF is in the form:

$$C_{fGn}(\tau) = \frac{V_H}{2}\left[\left(|\tau|+1\right)^{2H} + \left\||\tau|-1\right\|^{2H} - 2|\tau|^{2H}\right], \tag{4.5}$$

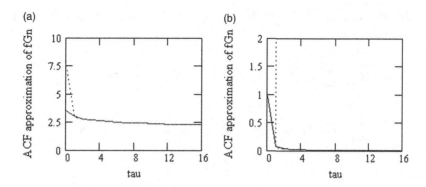

FIGURE 4.1 Illustrations of the ACF approximation of fGn. (a) $H = 0.95$. Solid: $r_{fGn}(\tau)$. Dot: Approximating $r_{fGn}(\tau)$ with (4.7). (b) $H = 0.55$. Solid: $r_{fGn}(\tau)$. Dot: Approximating $r_{fGn}(\tau)$ with (4.7).

where

$$V_H = \Gamma(1-2H)\frac{\cos \pi H}{\pi H}. \tag{4.6}$$

Because $r_{fGn}(\tau)$ is the finite second-order difference of $0.5V_H|\tau|^{2H}$, $r_{fGn}(\tau)$ is approximated by

$$C_{fGn}(\tau) \approx V_H H(2H-1)|\tau|^{2H-2}. \tag{4.7}$$

The above approximation is quite accurate for $\tau > 5$, as can be seen in Figure 4.1 (also see Li [2, 41]).

Let $S_{fGn}(\omega)$ be the power spectrum density (PSD) function of the fGn. Following our previous work (Li and Lim [42]), $S_{fGn}(\omega)$ is in the form

$$S_{fGn}(\omega) = V_H \sin(H\pi)\Gamma(2H+1)|\omega|^{1-2H}. \tag{4.8}$$

Another LRD model of traffic is the generalized Cauchy (GC) process. The GC process is Gaussian. Its ACF is in the form:

$$C_{GC}(\tau) = \left(1+|\tau|^{4-2D}\right)^{\frac{1-H}{2-D}}, \tag{4.9}$$

where $1 < D < 2$ is the fractal dimension and H the Hurst parameter (Li [2, 15–17], Gneiting [36], Li [43, 44], Li and Li [45], Li and Lim [46, 47], Lim and Li [48], Gneiting and Schlather [49]). When τ is extremely large, for $\tau \to \infty$, we have

$$C_{GC}(\tau) \sim |\tau|^{2H-2}. \tag{4.10}$$

Both the fGn and the GC models have the LRD property when $0.5 < H < 1$. More LRD models refer to Li [2, 14, 15] and Gneiting [36].

Denote the PSD of the GC process by $S_{GC}(\omega)$. Based on our previous work (Li and Lim [46]), it is expressed by

$$S_{GC}(\omega) = \sum_{k=0}^{\infty} A_1(D, H, k) |\omega|^{-(4-2D)k-1} * Sa(\omega)$$

$$+ \sum_{k=0}^{\infty} A_2(D, H, k) \left[\pi |\omega|^{2[0.5-H+(2-D)k]} - |\omega|^{2[0.5-H+(2-D)k]} * Sa(\omega) \right], \tag{4.11}$$

where $Sa(\omega) = \frac{\sin\omega}{\omega}$,

$$A_1(D, H, k) = \frac{(-1)^{k+1} 2\Gamma\{[(1-H)/(2-D)]+k\} \sin[(2-D)k\pi] \Gamma[(4-2D)k+1]}{\pi \Gamma[(1-H)/(2-D)]\Gamma(1+k)},$$

$$A_2(D, H, k) = \frac{(-1)^k 2\Gamma\{[(1-H)/(2-D)]+k\} \sin\{[1-H+(2-D)k]\pi\}}{\Gamma\{2[H-0.5-(2-D)k]\}}{\pi \Gamma[(1-H)/(2-D)]\Gamma(1+k)}. \tag{4.12}$$

In this research, we consider a class of LRD processes with the ACF form (4.4), that is, the mono-fractal traffic model with LRD.

4.2.2 Ergodicity of Mean

Let $\mu = E[x(t)]$ be the ensemble mean of $x(t)$. By ensemble mean, we imply that

$$\mu = E[x(t)] = \mu(t) = \lim_{N \to \infty} \frac{1}{N} \sum_{n=1}^{N} x_n(t). \tag{4.13}$$

The above $\mu(t)$ is the point estimate of $E[x(t)]$ at t. It requires infinite sample functions $x_n(t)$ for $n = 1, \ldots, \infty$. In practical applications, however, a random function $x(t)$ is often with single history (Bendat and Piersol [50]). For instance, it is impossible to measure infinite sample functions of a traffic trace at a network point. Thus, it greatly desires other estimation method of the ensemble mean.

Consider a single history of $x(t)$ with the finite record for $t \in (-T, T)$. Denote the mean of $x(t)$ by μ_T (time average) for $t \in (-T, T)$. It is given by

$$\mu_T = \frac{1}{2T} \int_{-T}^{T} x(t)dt. \tag{4.14}$$

For the random variable μ_T, one has

$$E(\mu_T) = \frac{1}{2T} \int_{-T}^{T} E[x(t)]dt = \mu. \tag{4.15}$$

Thus, μ_T as an estimate of μ is unbiased.

Denote the variance of μ_T by σ_T^2. When $\lim_{T\to\infty}\sigma_T^2=0$, one has

$$\lim_{T\to\infty}\mu_T = \frac{1}{2T}\int_{-T}^{T}x(t)dt = \mu. \tag{4.16}$$

When the above holds, we say that $x(t)$ is mean-ergodic. When $x(t)$ is mean-ergodic, the ensemble mean can be replaced by the time average, significantly simplifying the computation of the mean of $\{x_n(t)\}$. Obviously, the stationarity of $x(t)$ is a premise of its mean ergodicity.

4.2.3 Ergodicity of Autocovariance

Denote the ensemble ACF of $x(t)$ by $C_{xx}(t, t+\tau) = E[x(t)x(t+\tau)]$. By ensemble ACF, we mean

$$C_{xx}(t,t+\tau)= \lim_{N\to\infty}\frac{1}{N}\sum_{n=1}^{N}\{x_n(t)-E[x_n(t)]\}\{x_n(t+\tau)-E[x_n(t)]\}. \tag{4.17}$$

When $x(t)$ is stationary, $C_{xx}(t, t+\tau)$ only replies on the lag τ so that

$$C_{xx}(t,t+\tau)=C_{xx}(\tau). \tag{4.18}$$

The computation of an ensemble ACF needs infinite sample functions $x_n(t)$ for $n \in \mathbf{N}$. Thus, it is impossible to be used in practical applications when $x(t)$ is only a single history. However, when $x(t)$ is ACF-ergodic, the ensemble ACF equals to the one by time average, with probability one, for any sample function with $n \in \mathbf{N}$, in the form

$$C_{xx}(\tau)= \lim_{T\to\infty}\frac{1}{2T}\int_{-T}^{T}[x(t)-\mu][x(t+\tau)-\mu]dt. \tag{4.19}$$

If the above holds, we say that $x(t)$ is ACF-ergodic. The premise of $x(t)$ being ACF-ergodic is that $x(t)$ is stationary. A consequence of $x(t)$ being ACF-ergodic is that it is autocovariance-ergodic.

Denote the ACF of $x(t)$ by $C_{xx,T}(\tau)$ recorded in the finite interval $(-T, T)$. It means that

$$C_{xx,T}(\tau)=\frac{1}{2T}\int_{-T}^{T}[x(t)-\mu][x(t+\tau)-\mu]dt. \tag{4.20}$$

Then, $C_{xx,T}(\tau)$ is a random variable. Let σ_{CT}^2 be the variance of $C_{xx,T}(\tau)$. Then, $x(t)$ is ACF-ergodic if $\lim_{T\to\infty}\sigma_{CT}^2=0$.

4.2.4 Ergodicity of Variance

The variance of $x(t)$ is given by

$$\text{Var}\big[x(t)\big] = \text{E}\big\{x(t) - \text{E}\big[x(t)\big]\big\}^2. \tag{4.21}$$

By the ensemble average, one has

$$\text{Var}[x(t)] = \lim_{N \to \infty} \frac{1}{N} \sum_{n=1}^{N} \big\{x_n(t) - \text{E}[x(t)]\big\}\big\{x_n(t) - \text{E}[x(t)]\big\}. \tag{4.22}$$

The quantity $\text{Var}[x(t)]$ based on the ensemble average desires infinite sample functions $x_n(t)$ for $n \in \text{N}$. Therefore, it is very difficult, if not impossible, to be used in practical applications when $x(t)$ is only a single history. Nevertheless, if $x(t)$ is variance-ergodic, the ensemble variance equals to the one by time average with probability one. It is computed by

$$\text{Var}[x(t)] = \lim_{T \to \infty} \frac{1}{2T} \int_{-T}^{T} \big\{x(t) - \text{E}[x(t)]\big\}\big\{x(t) - \text{E}[x(t)]\big\}\, dt. \tag{4.23}$$

Denote the variance for $x(t)$ by $\text{Var}[x(t)]_T$ with the finite record in $(-T, T)$. Then,

$$\text{Var}[x(t)]_T = \frac{1}{2T} \int_{-T}^{T} \big\{x(t) - \text{E}[x(t)]\big\}^2\, dt. \tag{4.24}$$

Let $\text{Var}\{\text{Var}[x(t)]_T\}$ be the variance of random variable $\text{Var}[x(t)]_T$. Then, if $\text{Var}\{\text{Var}[x(t)]_T\} \to 0$ for $T \to \infty$, the ensemble variance equals to the one based on time average. Accordingly, $x(t)$ is variance-ergodic.

When $x(t)$ is mean-ergodic, ACF-ergodic, and variance-ergodic, we say that it is ergodic in the wide sense and ergodic in short.

4.3 PROBLEM STATEMENTS

Let $w(t)$ be a white noise. Denote the impulse response function of a linear filter by $h(t)$. Then, random function $x(t)$ is given by

$$x(t) = w(t) * h(t), \tag{4.25}$$

where $*$ implies the convolution operation. The impulse response function $h(t)$ is expressed by

$$h(t) = \text{F}^{-1}\left\{\sqrt{\text{F}\big[C_{xx}(t)\big]}\right\}, \tag{4.26}$$

where F and F^{-1} are the operators of the Fourier transform and its inverse, respectively, see my previous work (Li [43]).

From the point of view of linear filters, Box et al. [51] states that $x(t)$ is stationary if

$$\int_0^\infty h(t)dt < \infty. \tag{4.27}$$

On the other hand, when

$$\int_0^\infty h(t)dt = \infty, \tag{4.28}$$

$x(t)$ is non-stationary (Box et al. [51, Chap. 1]). However, such a test method ceases to the processes with LRD. We utilize the LRD model of fGn to explain this point.

Denote the impulse response function of a linear filter with the output of fGn under the excitation of $w(t)$ by $h_{fGn}(t)$. Following our previous work (Li et al. [52]), $h_{fGn}(t)$ is given by

$$h_{fGn}(t) = \frac{\sqrt{V_H \sin(H\pi)\Gamma(2H+1)}}{-2\sin\left(\frac{(H-3/2)\pi}{2}\right)\Gamma\left(H - \frac{1}{2}\right)} \frac{1}{|t|^{\frac{3}{2}-H}}. \tag{4.29}$$

From the above, for the fGn with LRD, we have

$$\int_0^\infty h_{fGn}(t)dt = \infty. \tag{4.30}$$

Can we say that the fGn with LRD is non-stationary based on the Box method described by (4.28)? The answer is negative. Precisely, we can only say that the conventional test methods of stationarity, including the one by Box et al. [51], in general, may be unavailable for the processes with LRD. Specifically for the fGn with LRD, Mandelbrot wrote a paper for clarifying its stationarity (Mandelbrot [53]). However, how to test the stationarity of a class of LRD processes with $C_{xx}(\tau) \sim |\tau|^{2H-2}$ for $0.5 < H < 1$ in general remains an unsolved problem. Consequently, how to clarify whether that class of LRD processes is ergodic is also a problem open. In this chapter, we aim at proving that LRD processes with $C_{xx}(\tau) \sim |\tau|^{2H-2}$ for $0.5 < H < 1$ are ergodic. As a consequence, we conclude that such a class of LRD processes are stationary.

4.4 RESULTS

For the variance-ergodic processes, we present a theorem to describe it.

Theorem 4.1. **If $x(t)$ is ACF-ergodic and mean-ergodic, it is also variance-ergodic.**

Proof. Note that

$$\text{Var}\big[x(t)\big] = r_{xx}(0) - \mu^2, \tag{4.31}$$

where $r_{xx}(\tau)$ is the autocorrelation function of $x(t)$. Because $x(t)$ is ACF-ergodic, it is also autocorrelation-ergodic. Thus, when $x(t)$ is ACF-ergodic and mean-ergodic, the left side in the above equation is also ergodic. That is, $x(t)$ is variance-ergodic. $\square$

Lemma 4.1. (Slutsky) (Papoulis and Pillai [3, Eq. (12-7), Chap. 12]).
A process is mean-ergodic if and only if

$$\lim_{T \to \infty} \frac{1}{T} \int_0^T C_{xx}(\tau) d\tau = 0. \tag{4.32}$$

Proof. Consider the cross-covariance of $x(0)$ and μ_T. It is given by

$$\text{Cov}[\mu_T, x(0)] = \text{E}\{[\mu_T - \mu][x(0) - \mu]\}. \tag{4.33}$$

The above can be expressed by

$$\text{Cov}[\mu_T, x(0)] = \frac{1}{2T} \int_{-T}^T C_{xx}(\tau) d\tau. \tag{4.34}$$

Because

$$\text{Cov}[\mu_T, x(0)] = \text{E}[\mu_T x(0)] - \text{E}(\mu_T)\text{E}\big[x(0)\big], \tag{4.35}$$

one has

$$\text{Cov}^2[\mu_T, x(0)] \leq \text{Var}(\mu_T)\text{Var}\big[x(0)\big] = \sigma_T^2 C_{xx}(0). \tag{4.36}$$

Thus, if $\lim_{T \to \infty} \sigma_T^2 = 0$, we have (4.32).
    For any $\varepsilon > 0$, there is a constant c_0 such that the following holds when $c > c_0$

$$\frac{1}{T} \int_c^T C_{xx}(\tau) d\tau. \tag{4.37}$$

Therefore, one has

$$\sigma_T^2 = \frac{1}{T} \int_0^{2T_0} C_{xx}(\tau)\left(1 - \frac{\tau}{2T}\right)d\tau + \frac{1}{T} \int_{2T_0}^{2T} C_{xx}(\tau)\left(1 - \frac{\tau}{2T}\right)d\tau. \tag{4.38}$$

Note that $|C_{xx}(\tau)| \le C_{xx}(0)$ implies that the integral from 0 to $2T_0$ is less than $\frac{2T_0 C_{xx}(0)}{T}$. That is,

$$\frac{1}{T} \int_0^{2T_0} C_{xx}(\tau)\left(1 - \frac{\tau}{2T}\right)d\tau < \frac{2T_0 C_{xx}(0)}{T}. \tag{4.39}$$

Thus, (4.38) becomes

$$\sigma_T^2 < \frac{2T_0 C_{xx}(0)}{T} + \frac{1}{T} \int_{2T_0}^{2T} C_{xx}(\tau)\left(1 - \frac{\tau}{2T}\right)d\tau. \tag{4.40}$$

Note that

$$\int_{2T_0}^{2T} C_{xx}(\tau)(2T - \tau)d\tau = \int_{2T_0}^{2T} C_{xx}(\tau)\int_{\tau}^{2T} dt\, d\tau = \int_{2T_0}^{2T}\int_{2T_0}^{\tau} C_{xx}(\tau)d\tau\, dt. \tag{4.41}$$

Thus,

$$\int_{2T_0}^{\tau} C_{xx}(\tau)d\tau < \varepsilon t. \tag{4.42}$$

Hence,

$$\lim_{T \to \infty} \left[\frac{2T_0 C_{xx}(0)}{T} + \frac{1}{T} \int_{2T_0}^{2T} C_{xx}(\tau)\left(1 - \frac{\tau}{2T}\right)d\tau \right] \to 2\varepsilon. \tag{4.43}$$

Consequently, $\lim_{T \to \infty} \sigma_T^2 = 0$. This finishes the proof. $\square$

Lemma 4.2. A process is mean-ergodic if

$$\lim_{\tau \to \infty} C_{xx}(\tau) = 0. \tag{4.44}$$

Proof. For any $\varepsilon > 0$, there is a constant T_0 such that $|C_{xx}(\tau)| < \varepsilon$ when $\tau > T_0$. In fact,

$$\frac{1}{T} \int_0^T C_{xx}(\tau)d\tau = \frac{1}{T} \int_0^{T_0} C_{xx}(\tau)d\tau + \frac{1}{T} \int_{T_0}^T C_{xx}(\tau)d\tau < \frac{T_0}{T} C_{xx}(0) + \varepsilon \frac{T - T_0}{T}.$$

Thus,

$$\lim_{T\to\infty}\left[\frac{T_0}{T}C_{xx}(0)+\varepsilon\frac{T-T_0}{T}\right]=\varepsilon.$$

Hence, Lemma 4.1 holds if (4.32) is satisfied. □

Lemma 4.2 can be equivalently represented by Lemma 4.3 below.

Lemma 4.3. A process is mean-ergodic if

$$\lim_{\tau\to\infty}r_{xx}(\tau)=\mu^2. \tag{4.45}$$

Proof. Since $C_{xx}(\tau) = r_{xx}(\tau) - \mu^2$, $\lim_{\tau\to\infty}C_{xx}(\tau)=0$ produces the above. □

From the above, we propose the following theorem to describe the mean ergodicity of processes with LRD.

Theorem 4.2. Processes with LRD with $C_{xx}(\tau)\sim c|\tau|^{2H-2}$ are mean-ergodic.

Proof. For $0.5 < H < 1$, we have

$$\lim_{\tau\to\infty}|\tau|^{2H-2}=0. \tag{4.46}$$

According to Lemma 4.2, Theorem 4.2 is true. □

In order to prove the following Theorem 4.3, we need Lemma 4.4 first.

Lemma 4.4. A process $x(t)$ is ACF-ergodic if and only if

$$\lim_{T\to\infty}\frac{1}{T}\int_0^T r_{yy}(\tau)d\tau = 0, \tag{4.47}$$

where $y(t) = [x(t + \tau) - \mu][x(t) - \mu]$.

Proof. For a process $y(t) = [x(t + \tau) - \mu][x(t) - \mu]$, when we prove that it is mean-ergodic, we actually prove that $x(t)$ is ACF-ergodic. □

Theorem 4.3. Let $y(t) = [x(t + \tau) - \mu][x(t) - \mu]$. Denote the ACF of $y(t)$ by $C_{yy}(\tau)$. Then, a Gaussian process $x(t)$ is ACF-ergodic if and only if

$$\lim_{T\to\infty}\frac{1}{T}\int_0^T C_{yy}(\tau)d\tau = 0. \tag{4.48}$$

Proof. Considering the time average of $y(t)$ over $(-T, T)$ yields

$$C_{yy,T}(\tau) = \frac{1}{2T} \int_{-T}^{T} [x(t) - \mu][x(t+\tau) - \mu] dt. \tag{4.49}$$

The above $r_{yy,T}(\tau)$ is an estimate of the mean in the form

$$C_{yy}(\tau) = \lim_{T \to \infty} \frac{1}{2T} \int_{-T}^{T} [x(t) - \mu][x(t+\tau) - \mu] dt. \tag{4.50}$$

It is unbiased. Considering the ACF of $y(t)$, we have

$$C_{yy}(\tau) = E\{[x(t+\tau+\tau) - \mu][x(t+\tau) - \mu][x(t+\tau) - \mu][x(t) - \mu]\} - C_{xx}^2(\tau). \tag{4.51}$$

In order to prove that $y(t)$ is mean-ergodic, following Lemma 4.2, we may prove that

$$\lim_{\tau \to \infty} C_{yy}(\tau) = 0. \tag{4.52}$$

For facilitating discussions, we denote by $t_1 = t$, $t_2 = t + \tau$. Then, (4.51) is rewritten as

$$C_{yy}(\tau) = E\{[x(t_1) - \mu][x(t_2) - \mu][x(t_1+\tau) - \mu][x(t_2+\tau) - \mu]\} - C_{xx}^2(\tau). \tag{4.53}$$

Let

$$X_1 = x(t_1) - \mu, X_2 = x(t_2) - \mu, X_3 = x(t_1+\tau) - \mu, X_4 = x(t_2+\tau) - \mu. \tag{4.54}$$

Then,

$$E\{[x(t_1) - \mu][x(t_2) - \mu][x(t_1+\tau) - \mu][x(t_2+\tau) - \mu]\} = E(X_1 X_2 X_3 X_4). \tag{4.55}$$

Since x is Gaussian, X_1, X_2, X_3, and X_4 have a joint-normal distribution. Thus,

$$E(X_1 X_2 X_3 X_4) = m_{12} m_{34} + m_{13} m_{24} + m_{14} m_{23}, \tag{4.56}$$

where

$$\begin{cases} m_{12} = E\{[x(t_1) - \mu][x(t_2) - \mu]\} = C_{xx}(t_2 - t_1), \\ m_{13} = E\{[x(t_1) - \mu][x(t_1+\tau) - \mu]\} = C_{xx}(\tau), \\ m_{14} = E\{[x(t_1) - \mu][x(t_2+\tau) - \mu]\} = C_{xx}(t_2 - t_1 + \tau), \\ m_{23} = E\{[x(t_2) - \mu][x(t_1+\tau) - \mu]\} = C_{xx}(t_1 - t_2 + \tau), \\ m_{24} = E\{[x(t_2) - \mu][x(t_2+\tau) - \mu]\} = C_{xx}(\tau), \\ m_{34} = E\{[x(t_1) - \mu][x(t_2+\tau) - \mu]\} = C_{xx}(t_2 - t_1). \end{cases} \tag{4.57}$$

Therefore, following Li [17, 54] and Laning and Battin [55], one has

$$E(X_1X_2X_3X_4)=C_{xx}^2(t_2-t_1)+C_{xx}^2(\tau)+C_{xx}(t_2-t_1+\tau)C_{xx}(t_1-t_2+\tau). \qquad (4.58)$$

Thus,

$$C_{yy}(\tau)=C_{xx}^2(t_2-t_1)+C_{xx}(t_2-t_1+\tau)C_{xx}(t_1-t_2+\tau). \qquad (4.59)$$

Since $t_2 - t_1 = \tau$, the above becomes

$$C_{yy}(\tau)=C_{xx}^2(\tau)+C_{xx}(0)C_{xx}(2\tau). \qquad (4.60)$$

Because $\lim_{\tau \to \infty} C_{xx}(\tau)=0$, we have $\lim_{\tau \to \infty} C_{yy}(\tau)=0$. □

Theorem 4.3 can be equivalently expressed by the following corollary.

Corollary 4.1. Let $y(t) = [x(t + \tau) - \mu][x(t) - \mu]$. Denote the ACF of $y(t)$ by $C_{yy}(\tau)$. Then, a Gaussian process $x(t)$ is ACF-ergodic if

$$\lim_{\tau \to \infty} C_{yy}(\tau)=0. \qquad (4.61)$$

Proof. For any $\varepsilon > 0$, there is a constant T_0 such that $|r_{yy}(\tau)| < \varepsilon$ when $\tau > T_0$. As a fact of fact, because

$$\frac{1}{T}\int_0^T C_{yy}(\tau)d\tau = \frac{1}{T}\int_0^{T_0} C_{yy}(\tau)d\tau + \frac{1}{T}\int_{T_0}^T C_{yy}(\tau)d\tau < \frac{T_0}{T}C_{yy}(0)+\varepsilon\frac{T-T_0}{T}, \qquad (4.62)$$

we have

$$\lim_{T \to \infty}\left[\frac{T_0}{T}C_{yy}(0)+\varepsilon\frac{T-T_0}{T}\right]=\varepsilon.$$

Hence, Corollary 4.1 is true. □

Since $\lim_{\tau \to \infty} C_{yy}(\tau)=0$ implies that $\lim_{\tau \to \infty} C_{xx}(\tau)=0$, a process $x(t)$ is mean-ergodic if it is ACF-ergodic. However, the inverse may not be true. In fact, in Lemma 4.2 for the mean ergodicity, we do not assume that $x(t)$ is Gaussian but it requires that $x(t)$ is Gaussian in Theorem 4.3 for the ACF ergodicity.

Theorem 4.4. Let $x(t)$ be a Gaussian process with LRD with the ACF $C_{xx}(\tau) \sim |\tau|^{2H-2}$ for $0.5 < H < 1$. Then, $x(t)$ is ergodic.

Proof. Let $y(t) = [x(t + \tau) - \mu][x(t) - \mu]$. Since $C_{xx}(\tau) \to 0$ for $\tau \to \infty$, $\lim_{\tau \to \infty} C_{yy}(\tau)=0$. According to Corollary 4.1, $x(t)$ is ACF-ergodic. In addition, following Lemma 4.2, $x(t)$ is mean-ergodic due to $C_{xx}(\tau) \to 0$ for $\tau \to \infty$. Based on Theorem 4.1, therefore, when $x(t)$ is mean-ergodic and ACF-ergodic, it is variance-ergodic. Thus, $x(t)$ is ergodic. $\square$

Example 4.1: Let $x(t)$ be the fGn. Then, $x(t)$ is ergodic.

Proof. Since $C_{fGn}(\tau) \to 0$ for $\tau \to \infty$, $x(t)$ is mean-ergodic according to Lemma 4.2. Let $y(t) = [x(t+\tau) - \mu][x(t) - \mu]$. Then, $C_{yy}(\tau) = C_{fGn}^2(\tau) + C_{fGn}(0)C_{fGn}(2\tau)$. Because $C_{fGn}(2\tau) \to 0$ and $C_{fGn}^2(\tau) \to 0$ for $\tau \to \infty$, according to Corollary 4.1, the fGn is ACF-ergodic. As the fGn is mean-ergodic and ACF-ergodic, based on Theorem 4.1, it is variance-ergodic. Thus, the fGn is ergodic. $\square$

Example 4.2: The GC process is ergodic.

Proof. Since $C_{GC}(\tau) \to 0$ for $\tau \to \infty$, the GC process is mean-ergodic according to Lemma 4.2. Denote the GC process by $x(t)$. Let $y(t) = [x(t+\tau) - \mu][x(t) - \mu]$. Note that the GC process is Gaussian. Thus, $C_{yy}(\tau) = C_{GC}^2(\tau) + C_{GC}(0)C_{GC}(2\tau)$. Since $C_{GC}(2\tau) \to 0$ and $C_{GC}^2(\tau) \to 0$ when $\tau \to \infty$, based on Corollary 4.1, the GC process is ACF-ergodic. Because the GC process is both mean-ergodic and ACF-ergodic, it is variance-ergodic. Hence, it is ergodic. $\square$

Finally, we discuss the stationarity of processes with LRD.

Lemma 4.5. A process is stationary if it is ergodic (Khinchin [56]). $\square$

Example 4.3: Let $x(t)$ be the fGn. Then, $x(t)$ is stationary.

Proof. In Example 4.1, we see that the fGn is ergodic. Thus, it is stationary. $\square$

Example 4.4: The GC process is stationary.

Proof. Since the GC process is ergodic (see Example 4.2), it is stationary. $\square$

4.5 DISCUSSIONS AND SUMMARY

We have given the proofs that LRD traffic series with $C_{xx}(\tau) \sim |\tau|^{2H-2}$ for $0.5 < H < 1$ are ergodic, and consequently they are stationary. However, in practice, a measured LRD traffic series is of finite length. Thus, stationarity test of a measured LRD traffic series with finite record length is still desired. We shall address that in Chapter 7 and Chapter 10. Note that the convergence of sample ACF of LRD traffic is a consequence of the ergodicity. Nevertheless, Resnick et al. questioned the convergence of sample ACF of LRD series [57]. For that reason, we arrange an Appendix in the end of the monograph for showing the convergence of sample ACF of LRD traffic.

As an application of Theorem 4.2, we have explained the ergodicity and stationarity of two LRD traffic models, namely, the fGn and the GC process.

REFERENCES

1. A. M. Yaglom, *An Introduction to the Theory of Stationary Random Functions*, Prentice-Hall, New York, 1962.
2. M. Li, *Fractal Teletraffic Modeling and Delay Bounds in Computer Communications*, CRC Press, Boca Raton, 2022.
3. A. Papoulis, and S. U. Pillai, *Probability, Random Variables, and Stochastic Processes*, 4th ed., McGraw-Hill, New York, 2002, Chap. 12.
4. P. Borgnat, and P. Flandrin, Stationarization via surrogates, *Journal of Statistical Mechanics: Theory and Experiment*, 29, 2009, 1–14.
5. R. Von Sachs, and M. H. Neumann, A wavelet-based test for stationarity, *Journal of Time Series Analysis*, 21(5) 2000, 597–613.
6. S. Ling, Estimation and testing stationarity for double-autoregressive models, *Journal of Royal Statistical Society Series B*, 66(1) 2004, 63–78.
7. Z. Psaradakis, Blockwise bootstrap testing for stationarity, *Statistics & Probability Letters*, 76(6) 2006, 562–570.
8. P. M. M. Rodrigues, and A. Rubia, A note on testing for nonstationarity in autoregressive processes with level dependent conditional heteroskedasticity, *Statistical Papers*, 49(3) 2008, 581–593.
9. A. I. Al-Shoshan, Time-varying modeling of a nonstationary signal, *Telecommunication Systems*, 12(4) 1999, 389–396.
10. O. Darné, and C. Diebolt, Non-stationarity tests in macroeconomic time series, In *New Trends in Macroeconomics*, Diebolt C., and Kyrtsou C. (eds.), Berlin, Heidelberg, 2005, pp. 173–194.
11. D. J. Sheskin, *Handbook of Parametric and Nonparametric Statistical Procedures*, 5th ed., CRC Press, Boca Raton, 2004.
12. P. Abry, and D. Veitch, Wavelet analysis of long-range dependent traffic, *IEEE Transactions on Information Theory*, 44(1) 1998, 2–15.
13. M. Li, Modeling autocorrelation functions of long-range dependent teletraffic series based on optimal approximation in Hilbert space: A further study, *Applied Mathematical Modelling*, 31(3) 2007, 625–631.
14. M. Li, Generalized fractional Gaussian noise and its application to traffic modeling, *Physica A*, 579, 2021, 1236137 (22 pp).
15. M. Li, Multi-fractional generalized Cauchy process and its application to teletraffic, *Physica A*, 550, 2020, 123982 (14 pp).
16. M. Li, Long-range dependence and self-similarity of teletraffic with different protocols at the large time scale of day in the duration of 12 years: Autocorrelation modeling, *Physica Scripta*, 95(4) 2020, 065222 (15 pp).
17. M. Li, Record length requirement of long-range dependent teletraffic, *Physica A*, 472, 2017, 164–187.
18. P. Loiseau, P. Gonçalves, G. Dewaele, P. Borgnat, P. Abry, and Vicat-Blanc Primet, Investigating self-similarity and heavy-tailed distributions on a large-scale experimental facility, *IEEE/ACM Transactions on Networking*, 18(4) 2010, 1261–1274.
19. R. Fontugne, P. Abry, K. Fukuda, D. Veitch, K. Cho, P. Borgnat, and H. Wendt, Scaling in internet traffic: A 14 year and 3 day longitudinal study, with multiscale analyses and random projections, *IEEE/ACM Transactions on Networking*, 25(4) 2017, 2152–2165.
20. H. Sheng, Y.-Q. Chen, and T. Qiu, On the robustness of Hurst estimators, *IET Signal Processing*, 5(2) 2011, 209–225.
21. H. Sheng, and Y.-Q. Chen, FARIMA with stable innovations model of Great Salt Lake elevation time series, *Signal Processing*, 91(3) 2011, 553–561.

22. M. H. Heydari, Z. Avazzadeh, and M. R. Mahmoudi, Chebyshev cardinal wavelets for nonlinear stochastic differential equations driven with variable-order fractional Brownian motion, *Chaos, Solitons & Fractals*, 124, 2019, 105–124.

23. M. H. Heydari, M. R. Mahmoudi, A. Shakiba, and Z. Avazzadeh, Chebyshev cardinal wavelets and their application in solving nonlinear stochastic differential equations with fractional Brownian motion, *Communications in Nonlinear Science and Numerical Simulation*, 64, 2018, 98–121.

24. M. Pinchas, Residual ISI obtained by blind adaptive equalizers and fractional noise, *Mathematical Problems in Engineering*, 2013, 2013, Article ID 972174 (11 pp).

25. M. Pinchas, Cooperative multi PTP slaves for timing improvement in an fGn environment, *IEEE Communications Letters*, 22(7) 2018, 1366–1369.

26. Y. Avraham, and M. Pinchas, Two novel one-way delay clock skew estimators and their performances for the fractional Gaussian noise/generalized fractional Gaussian noise environment applicable for the IEEE 1588v2 (PTP) case, *Frontiers in Physics*, 10, 2022, 867861.

27. Y. Avraham, and M. Pinchas, A novel clock skew estimator and its performance for the IEEE 1588v2 (PTP) case in fractional Gaussian noise/generalized fractional Gaussian noise environment, *Frontiers in Physics*, 9, 2021, 796811.

28. M. R. P. da Silva, and F. G. C. Rocha, Traffic modeling for communications networks: A multifractal approach based on few parameters, *Journal of the Franklin Institute*, 358(3) 2021, 2161–2177.

29. D. Nashat, and F. A. Hussain, Multifractal detrended fluctuation analysis based detection for SYN flooding attack, *Computers & Security*, 107, 2021, 10234.

30. M. Mitzenmacher, A brief history of generative models for power law and lognormal distributions, *Internet Mathematics*, 1(2) 2003, 226–251.

31. S. I. Tadaki, Long-term power-law fluctuation in internet traffic, *Journal of the Physics Society of Japan*, 76(4) 2007, 044001 (5 pp).

32. A. Ercan, and M. L. Kavvas, Time-space fractional governing equations of one-dimensional unsteady open channel flow process: Numerical solution and exploration, *Hydrological Processes*, 31(16) 2017, 2961–2971.

33. L. Jiang, N. Li, and X. Zhao, Scaling behaviors of precipitation over China, *Theoretical and Applied Climatology*, 128(1–2) 2017, 63–70.

34. M. Abolbashari, S. M. Kim, G. Babaie, B. Jonathan, and F. Faramarz, Fractional bispectrum transform: Definition and properties, *IET Signal Processing*, 11(8) 2017, 901–908.

35. C. Yang, Y.-K. Zhang, and X. Liang, Analysis of temporal variation and scaling of hydrological variables based on a numerical model of the Sagehen Creek watershed, *Stochastic Environmental Research and Risk Assessment*, 32(2) 2018, 357–368.

36. T. Gneiting, Power-law correlations, related models for long-range dependence and their simulation, *Journal of Applied Probability*, 37(4) 2000, 1104–1109.

37. J. Beran, *Statistics for Long-Memory Processes*, Chapman & Hall, New York, 1994.

38. B. B. Mandelbrot, *Gaussian Self-Affinity and Fractals*, Springer, New York, 2001.

39. H. E. Hurst, Long term storage capacity of reservoirs, *Transactions of the American Society of Civil Engineers*, 116, 1951, 770–799.

40. B. B. Mandelbrot, and J. W. van Ness, Fractional Brownian motions, fractional noises and applications, *SIAM Review*, 10(4) 1968, 422–437.

41. M. Li, Modified multifractional Gaussian noise and its application, *Physica Scripta*, 96(12) 2021, 125002 (12 pp).

42. M. Li, and S. C. Lim, A rigorous derivation of power spectrum of fractional Gaussian noise, *Fluctuation and Noise Letters*, 6(4) 2006, C33–36.

43. M. Li, Generation of teletraffic of generalized Cauchy type, *Physica Scripta*, 81(2) 2010, 025007 (10 pp).

44. M. Li, Fractal time series: A tutorial review, *Mathematical Problems in Engineering*, 2010, 2010, Article ID 157264 (26 pp).

45. M. Li, and J.-Y. Li, Generalized Cauchy model of sea level fluctuations with long-range dependence, *Physica A*, 484, 2017, 309–335.
46. M. Li, and S. C. Lim, Power spectrum of generalized Cauchy process, *Telecommunication Systems*, 43(3–4) 2010, 219–222.
47. M. Li, and S. C. Lim, Modeling network traffic using generalized Cauchy process, *Physica A*, 387(11) 2008, 2584–2594.
48. S. C. Lim, and M. Li, A generalized Cauchy process and its application to relaxation phenomena, *Journal of Physics A: Mathematical and General*, 39(12) 2006, 2935–2951.
49. T. Gneiting, and M. Schlather, Stochastic models that separate fractal dimension and Hurst effect, *SIAM Review*, 46(2) 2004, 269–282.
50. J. S. Bendat, and A. G. Piersol, *Random Data: Analysis and Measurement Procedure*, 3rd ed., John Wiley & Sons, New York, 2000.
51. G. E. P. Box, G. M. Jenkins, and G. C. Reinsel, *Time Series Analysis: Forecasting and Control*, 4th ed., John Wiley & Sons, New York, 2008.
52. M. Li, X. Sun, and X. Xiao, Revisiting fractional Gaussian noise, *Physica A*, 514, 2019, 56–62.
53. B. B. Mandelbrot, Note on the definition and the stationarity of fractional Gaussian noise, *Journal of Hydrology*, 30(4) 1976, 407–409.
54. M. Li, A method for requiring block size for spectrum measurement of ocean surface waves, *IEEE Transactions on Instrumentation and Measurement*, 55(6) 2006, 2207–224.
55. J. H. Laning, and R. H. Battin, *Random Processes in Automatic Control*, McGraw-Hill, New York, 1956.
56. A. Y. Khinchin, *Mathematical Foundations of Statistical Mechanics*, Dover, New York, 1949.
57. S. Resnick, G. Samorodnitsky, and F. Xue, How misleading can sample ACFs of stable mas be? (Very!), *The Annals of Applied Probability*, 9(3) 1999, 797–817.

CHAPTER 5

Predictability of Long-Range-Dependent Series

This chapter points out that the predictability analysis of conventional time series may in general be invalid for long-range-dependent (LRD) series since the conventional mean square error (MSE) may generally not exist for predicting LRD series. In order to make the MSE of the prediction of LRD series exist, we introduce a generalized MSE. With that, the proof of the predictability of LRD series is presented.

5.1 INTRODUCTION

Let $x(t)$ be a realization, which is a second-order random function for $t \in [0, \infty)$. Let $x_T(t)$ be a given sample of $x(t)$, for $0 \leq t \leq T$. Then, one of the important problems in time series is to predict or forecast $x(t)$ for $t > T$ based on the past data of $x_T(t)$ (see Clements and Hendry [1], Box et al. [2], and Fuller [3]).

A well-known case in the field of time series prediction refers to Yule's work for the analysis of Wolfer's sunspot numbers (Yule [4]). The early basic theory of predicting conventional two-order stationary random function refers to the work of Wiener [5] and Kolmogorov [6]. By conventional we mean that the stationary random functions Wiener and Kolmogorov considered are not long-range-dependent (LRD). In other words, the time series they studied are not heavy-tailed, as can be seen from Zadeh and Ragazzini [7], Bhansali [8, 9], Robinson [10], Papoulis [11], Vaidyanathan [12], Lyman et al. [13], Lyman and Edmonson [14], and Dokuchaev [15], just to mention a few.

The predictability of conventional time series has been well studied (see Papoulis [11], Vaidyanathan [12], Bhansali [9], Lyman et al. [13], Lyman and Edmonson [14], and Dokuchaev [15]). The basic idea in this regard is to use MSE as a constraint to obtain a prediction (see Harrison et al. [16], Bellegem and Sachs [17], Man [18], Clements and Hendry [19]). We shall note in next section that the conventional MSE may, in general, fail to be used for predicting LRD series.

LRD processes have gained increasing applications in various fields of sciences and technologies (see Beran [20], Mandelbrot [21], and Li [22–29]). Consequently, the prediction is

DOI: 10.1201/9781003354987-6

desired for LRD series. The literature regarding the prediction of LRD series appears to increasing (see Brodsky and Hurvich [30], Reisen and Lopes [31], Bisaglia and Bordignon [32], Bhansali and Kokoszka [33], Man [34], Bayraktar et al. [35], Man and Tiao [36], Bhansali [37], Godet [38], and Gooijer and Hyndman [39]). However, unfortunately, suitable MSE used for predicting LRD series may be overlooked, leaving a pitfall in this respect. We shall present a generalized MSE in the domain of generalized functions for the purpose of proving the existence of LRD series prediction.

The rest of this chapter is arranged as follows: Section 5.2 will point out the pitfall of prediction of time series based on traditional MSE. The proof of the predictability of LRD series will be proposed in Section 5.3, which is followed by summary.

5.2 PROBLEM STATEMENTS

Denote the autocorrelation function (ACF) of $x(t)$ by $r_{xx}(\tau)$, where $r_{xx}(\tau) = \mathrm{E}[x(t)x(t + \tau)]$. Then, $x(t)$ is called short-range-dependent (SRD) series if $r_{xx}(\tau)$ is integrable (Beran [20]), i.e.,

$$\int_0^\infty r_{xx}(\tau)d\tau < \infty. \tag{5.1}$$

On the other side, $x(t)$ is LRD if $r_{xx}(\tau)$ is non-integrable, that is,

$$\int_0^\infty r_{xx}(\tau)d\tau = \infty. \tag{5.2}$$

A typical form of such an ACF has the following asymptotic expression:

$$r_{xx}(\tau) \sim c|\tau|^{-\beta} \, (\tau \to \infty), \tag{5.3}$$

where $c > 0$ is a constant and $0 < \beta < 1$.

Denote the probability density function (PDF) of $x(t)$ by $p(x)$. Then, the ACF of $x(t)$ can be expressed by

$$r_{xx}(\tau) = \int_{-\infty}^\infty x(t)x(t+\tau)p(x)dx. \tag{5.4}$$

Considering that $r_{xx}(\tau)$ is non-integrable, we see that a heavy-tailed PDF is a consequence of LRD series (see Li [23], Resnick [40], Heath et al. [41], Paxson and Floyd [42], Li and Li [43], Abry et al. [44], and Adler et al. [45]).

Denote the mean of $x(t)$ by μ_x. Then,

$$\mu_x = \int_{-\infty}^\infty xp(x)dx. \tag{5.5}$$

The variance of $x(t)$ is given by

$$\text{Var}(x) = \int_{-\infty}^{\infty} (x - \mu_x)^2 p(x) dx. \tag{5.6}$$

One remarkable thing in LRD series is that the tail of $p(x)$ may be so heavy that the above integral, either (5.5) or (5.6), may not always exist (see Bassingthwaighte et al. [46], Doukhan et al. [47], and Li [48]). To explain this, we utilize the Pareto distribution. Denote the PDF of the Pareto distribution by $p_{\text{Pareto}}(x)$. Then (Korn and Korn [49]),

$$p_{\text{Pareto}}(x) = \frac{ab}{x^{a+1}}, \tag{5.7}$$

where $x \geq a$. The mean and variance of $x(t)$ that follows $p_{\text{Pareto}}(x)$ are respectively given by

$$\mu_{\text{Pareto}} = \frac{ab}{a-1}, \tag{5.8}$$

and

$$\text{Var}(x)_{\text{Pareto}} = \frac{ab^2}{(a-1)^2 (a-2)}. \tag{5.9}$$

It can be easily seen that μ_{Pareto} and $\text{Var}(x)_{\text{Pareto}}$ do not exist if $a = 1$.

Following the work of Kolmogorov, a linear prediction can be expressed as follows. Given $n > 0$ and $m \geq 0$, the selection of proper real coefficient a_s is such that the following linear combination of random variables $x(t-1)$, $x(t-2)$, ..., $x(t-n)$ given by

$$L = \sum_{i=1}^{n} a_i x(t-i) \tag{5.10}$$

can approximate $x(t+m)$ as accurately as possible (Kolmogorov [6]). The following MSE is usually chosen as the prediction criterion for (5.10)

$$\sigma^2 = \sigma^2(n,m) = E[x(t+m) - L]^2. \tag{5.11}$$

By minimizing equation (5.11), one has the desired a_i in (5.10). Wiener well studied that criterion for both prediction and filtering (see Levinson [50, 51]). A predictor following (5.10) and (5.11) can be regarded in the class of Wiener–Kolmogorov predictors.

Various forms of linear combination in terms of (5.10) have been developed, such as autoregressive moving average (ARMA) model, autoregressive (AR) model, moving average (MA) model, autoregressive integrated moving average (ARIMA) model (see Clements and Hendry [1], Box et al. [2], Lyman and Edmonson [13, 14], Wolff et al. [52], Bhansali [12, 53], Markhoul [54], Kohn and Ansley [55], Zimmerman and Cressie [56], Peiris and Perera [57], Kudritskii [58], Bisaglia and Bordignon [32], Kim [59], Cai [60], Harvill and Ray [61], Atal [62], Huang [63], Schick and Wefelmeyer [64], Jamalizadeh and Balakrishnan [65]). However, one thing in common for different forms of predictors is to minimize prediction error that in principle usually follows the form of (5.11).

Note that the necessary condition for the above-described Wiener–Kolmogorov predictor to be valid is that $E[x(t)]$ exists (Kolmogorov [6]). For LRD series, however, this condition may not always be satisfied. For instance, if an LRD series obeys the Pareto distribution, its mean does not exist for $a = 1$, see (5.8).

In addition to the fact that the mean of an LRD series may not exist, its variance may not exist either. The error in (5.11) can be generalized by

$$\sigma^2(n, m) = E[x(t+m) - L)]^2 = \int_{-\infty}^{\infty} [x(t+m) - L)]^2 p(x) dx. \tag{5.12}$$

Kolmogorov stated that the above $\sigma^2(n, m)$ does not increase as n increases [6]. However, the statement may be untrue if $x(t)$ is LRD.

It is worth noting that errors may be heavy-tailed (see Peng and Yao [66], Hall and Yao [67]). For instance, LRD teletraffic is heavy-tailed (Resnick [68], Michiel and Laevens [69]) and it is Gaussian at large time scales (Abry and Veitch [70], Scherrer et al. [71]). Therefore, it is quite reasonable to assume that $[x(t + m) - L]$ follows a heavy-tailed distribution, for the purpose of this presentation. If it obeys the Pareto distribution, then, the above expression approaches infinity for $a = 2$ (see (5.9)) no matter how large n is.

From the above discussions, we see that it may be unsuitable to use the conventional MSE as used in the class of conventional Wiener–Kolmogorov predictors to infer that LRD series is predictable. In the next section, we shall give the proof of the predictability of LRD series.

5.3 PREDICTABILITY OF LRD SERIES

Let $x(t + m) \in X$, where X is the set of LRD processes. Let $L \in \hat{X}$. Then, $\hat{X} \subseteq X$. We now consider the norms and inner products in $\hat{X}$ and X.

Definition 5.1. (Griffel [72]). A function of rapid decay is a smooth $\phi: \mathbf{R} \to \mathbf{C}$ such that $t^n \phi^{(r)}(t) \to 0$ as $t \to \pm \infty$ for all $n, r \geq 0$, where $\mathbf{C}$ is the space of complex numbers. The set of all functions of rapid decay is denoted by S. □

In the discrete case, the rapid decayed function is denoted by $\phi(n)$ and we still use the symbol $\mathcal{S}$ to specify the space it belongs to for simplicity without confusions.

Lemma 5.1 [72]. *Every function belonging to S is absolutely integrable in the continuous case or absolutely summable in the discrete case.* □

Now, define the norm of $x(t + m) \in X$ by

$$\|x(t+m)\|^2 = \langle x(t+m), x(t+m) \rangle = \int_{-\infty}^{\infty} x^2(t+m)p(x)g(x)dx, \qquad (5.13)$$

where $g \in S$. *Define the inner product of* $x(t + m) \in X$ *by*

$$\langle x(t+m), x(t+m) \rangle = \|x(t+m)\|^2. \qquad (5.14)$$

Then, combining any $x(t + m) \in X$ with its limit makes X a Hilbert space.
Note that

$$\|L^2\| = \langle L, L \rangle = \int_{-\infty}^{\infty} L^2 p(x)g(x)dx, \qquad (5.15)$$

then, $\hat{X}$ is the closed subset of X (Liu [73]).

Lemma 5.2 [72–74]. *(Existence of a unique minimizing element in Hilbert space). Let H be a Hilbert space and M be a closed convex subset of H. Let $x \in H$, $x \notin H$. Then, there exists a unique element $\hat{x} \in \mathcal{M}$ satisfying*

$$\|x - \hat{x}\| = \inf_{y \in \mathcal{M}} \|x - y\|. \qquad (5.16)$$

Theorem 5.1. **Let L be a linear combination of the past values of $x(n)$ according to (5.10). Then, there exists a unique $L \in \hat{X}$ such that $L - x(t+m) = \inf_{s \in \hat{X}} x(t+m) - s$.**

Proof. X is a Hilbert space. $\hat{X}$ is its closed subset. Thus, it is convex. According to Lemma 5.2, for any $x(t + m) \in X$, there exists a unique $L \in \hat{X} \subset X$ such that $\|L - x(t+m)\| = \inf_{s \in \hat{X}} \|x(t+m) - s\|$. □

Theorem 5.2. **Let L be a linear combination of the past values of LRD series $x(n)$ according to (5.10). Then, $x(n)$ is predicable.**

Proof. Let $g \in S$. *Following* Theorem 5.1, there exists a unique $L \in \hat{X}$ such that $\|L - x(t+m)\| = \inf_{s \in \hat{X}} \|x(t+m) - s\|$. When the square error expressed by (5.12) is generalized to the form

$$\sigma^2(n, m) = E[x(t+m) - L)]^2 = \int_{-\infty}^{\infty} [x(t+m) - L)]^2 p(x)g(x)dx, \qquad (5.17)$$

$\sigma^2(n, m)$ does not increase as n increases. Thus, LRD series $x(n)$ is predicable. □

5.4 SUMMARY

LRD series considerably differ from the conventional series (see Beran [20, 75], Adler et al. [45], Doukhan et al. [48], Künsch et al. [76]). Examples mentioned in this regard are regressions for fitting LRD models (Peng and Yao [66], Beran [77], Beran et al. [78]), variance analysis of autocorrelation estimation (Li [79]), stationarity test (Li et al. [80]), power spectra (Li and Lim [81, 82]), and prediction error study of LRD series (Li et al. [83]). In this chapter, we have addressed the particularity of the predictability of LRD series. We have given a proof of LRD series being predictable. As a side product obtained from the proof procedure, the MSE used by Kolmogorov as a criterion of LRD series prediction has been generalized to the form of expression (5.17).

REFERENCES

1. M. P. Clements, and D. F. Hendry, *Forecasting Economic Time Series*, Cambridge University Press, Cambridge, 1998.
2. G. E. P. Box, G. M. Jenkins, and G. C. Reinsel, *Time Series Analysis: Forecasting and Control*, Prentice Hall, New York, 1994.
3. W. A. Fuller, *Introduction to Statistical Time Series*, 2nd ed., John Wiley & Sons, New York, 1995.
4. G. Udny Yule, On a method of investigating periodicities in disturbed series, with special reference to Wolfer's sunspot numbers, *Philosophical Transactions of the Royal Society A*, 226 (636–646) 1927, 267–297.
5. N. Wiener, *Extrapolation, Interpolation and Smoothing of Stationary Time Series*, The MIT Press, New York, 1964.
6. A. N. Kolmogorov, Interpolation and extrapolation of stationary random sequences, *Izv. Akad. Nauk SSSR Ser. Mat.* 5, 1941, 3–14.
7. L. A. Zadeh, and J. R. Ragazzini, An extension of Wiener's theory of prediction, *Journal of Applied Physics*, 21(7) 1950, 645–655.
8. R. J. Bhansali, Asymptotic properties of the Wiener-Kolmogorov predictor. I, *Journal of the Royal Statistical Society. Series B*, 36(1) 1974, 61–73.
9. R. Bhansali, Linear prediction by autoregressive model fitting in the time domain, *Annals of Statistics*, 6(1) 1978, 224–231.
10. E. A. Robinson, A historical perspective of spectrum estimation, *Proceedings of the IEEE*, 70(9) 1982, 885–907.
11. A. Papoulis, A note on the predictability of band-limited processes, *Proceedings of the IEEE*, 73(8) 1985, 1332–1333.
12. P. P. Vaidyanathan, On predicting a band-limited signal based on past sample values, *Proceedings of the IEEE*, 75(8) 1987, 1125–1159.
13. R. J. Lyman, W. W. Edmonson, S. McCullough, and M. Rao, The predictability of continuous-time, bandlimited processes, *IEEE Transactions on Signal Processing*, 48(2) 2000, 311–316.
14. R. J. Lyman, and W. W. Edmonson, Linear prediction of bandlimited processes with flat spectral densities, *IEEE Transactions on Signal Processing*, 49(7) 2002, 311–316.
15. N. Dokuchaev, The predictability of band-limited, high-frequency and mixed processes in the presence of ideal low-pass filters, *Journal of Physics A: Mathematical and Theoretical*, 41(38) 2008, 382002 (7 pp).
16. R. Harrison, G. Kapetanios, and T. Yates, Forecasting with measurement errors in dynamic models, *International Journal of Forecasting*, 21(3) 2005, 595–607.
17. S. van Bellegem, and R. von Sachs, Forecasting economic time series with unconditional time-varying variance, *International Journal of Forecasting*, 20(4) 2004, 611–627.

18. K. S. Man, Linear prediction of temporal aggregates under model misspecification, *International Journal of Forecasting*, 20(4) 2004, 659–670.

19. M. P. Clements, and D. F. Hendry, Forecasting economic processes, *International Journal of Forecasting*, 14(1) 1998, 111–131.

20. J. Beran, *Statistics for Long-Memory Processes*, Chapman & Hall, New York, 1994.

21. B. B. Mandelbrot, *Gaussian Self-Affinity and Fractals*, Springer, New York, 2001.

22. M. Li, and A. Wang, Fractal teletraffic delay bounds in computer networks, *Physica A*, 557, 2020, 124903 (13 pp).

23. M. Li, Fractal time series: A tutorial review, *Mathematical Problems in Engineering*, 2010, 2010, Article ID 157264 (26 pp).

24. M. Li, Generation of teletraffic of generalized Cauchy type, *Physica Scripta*, 81(2) 2010, 025007 (10pp).

25. M. Li, Generalized fractional Gaussian noise and its application to traffic modeling, *Physica A*, 579, 2021, 1236137 (22 pp).

26. M. Li, An approach to reliably identifying signs of DDOS flood attacks based on LRD traffic pattern recognition, *Computers & Security*, 23(7) 2004, 549–558.

27. M. Li, Change trend of averaged Hurst parameter of traffic under DDOS flood attacks, *Computers & Security*, 25(3) 2006, 213–220.

28. M. Li, Long-range dependence and self-similarity of teletraffic with different protocols at the large time scale of day in the duration of 12 years: Autocorrelation modeling, *Physica Scripta*, 95(4) 2020, 065222 (15 pp).

29. M. Li, Multi-fractional generalized Cauchy process and its application to teletraffic, *Physica A*, 550, 2020, 123982 (14 pp).

30. J. Brodsky, and C. M. Hurvich, Multi-step forecasting for long-memory processes, *Journal of Forecasting*, 18(1) 1999, 59–75.

31. V. A. Reisen, and S. Lopes, Some simulations and applications of forecasting long-memory time-series models, *Journal of Statistical Planning and Inference*, 80(1–2) 1999, 269–2876.

32. L. Bisaglia, and S. Bordignon, Mean square prediction error for long-memory processes, *Statistical Papers*, 43(2) 2002, 161–175.

33. R. J. Bhansali, and P. S. Kokoszka, Prediction of long-memory time series: A tutorial review, *Springer Lecture Notes in Physics*, 621, 2003, 3–21.

34. K. S. Man, Long memory time series and short term forecasts, *International Journal of Forecasting*, 19(3) 2003, 467–475.

35. E. Bayraktar, H. V. Poor, and R. Rao, Prediction and tracking of long-range-dependent sequences, *Systems & Control Letters*, 54(11) 2005, 1083–1090.

36. K. S. Man, and G. C. Tiao, Aggregation effect and forecasting temporal aggregates of long memory processes, *International Journal of Forecasting*, 22(2) 2006, 267–281.

37. R. J. Bhansali, Forecasting long memory time series when occasional breaks occur, *Economics Letters*, 98(3) 2008, 253–258.

38. F. Godet, Prediction of long memory processes on same-realisation, *Journal of Statistical Planning and Inference*, 140(4) 2010, 907–926.

39. J. G. De Gooijer, and R. J. Hyndman, 25 years of time series forecasting, *International Journal of Forecasting*, 22(3) 2006, 267–281.

40. S. I. Resnick, *Heavy-Tail Phenomena: Probabilistic and Statistical Modeling*, Springer, New York, 2007.

41. D. Heath, S. Resnick, and G. Samorodnitsky, Heavy tails and long range dependence in on-off processes and associated fluid models, *Mathematics of Operations Research*, 23(1) 1998, 145–165.

42. V. Paxson, and S. Floyd, Wide area traffic: The failure of Poisson modeling, *IEEE/ACM Transactions on Networking*, 3(3) 1995, 226–244.

43. M. Li, and J.-Y. Li, Generalized Cauchy model of sea level fluctuations with long-range dependence, *Physica A*, 484, 2017, 309–335.
44. P. Abry, P. Borgnat, F. Ricciato, A. Scherrer, and D. Veitch, Revisiting an old friend: On the observability of the relation between long range dependence and heavy tail, preprint, *Telecommunication Systems*, 43(3-4) 2010, 147–165.
45. R. J. Adler, R. E. Feldman, and M. S. Taqqu (eds.), *A Practical Guide to Heavy Tails: Statistical Techniques and Applications*, Birkhäuser, Boston, 1998.
46. J. B. Bassingthwaighte, L. S. Liebovitch, and B. J. West, *Fractal Physiology*, Oxford University Press, Oxford, 1994.
47. P. Doukhan, G. Oppenheim, and M. S. Taqqu (eds.), *Theory and Applications of Long-Range Dependence*, Birkhäuser, Boston, 2003.
48. M. Li, *Fractal Teletraffic Modeling and Delay Bounds in Computer Communications*, CRC Press, Boca Raton, 2022.
49. G. A. Korn, and T. M. Korn, *Mathematical Handbook for Scientists and Engineers*, McGraw-Hill, New York, 1961.
50. N. Levinson, The Wiener RMS (root mean squares) error criterion in filter design and prediction, *Journal of Mathematical Physics*, 25, 1947, 261–278.
51. N. Levinson, A heuristic exposition of Wiener's mathematical theory of prediction and filtering, *Journal of Mathematical Physics*, 26, 1947, 110–119.
52. S. S. Wolff, J. L. Gastwirth, and J. B. Thomas, Linear optimum predictors, *IEEE Transactions on Information Theory*, 13(1) 1967, 30–32.
53. R. J. Bhansali, Asymptotic mean-square error of predicting more than one-step ahead using the regression method, *Journal of the Royal Statistical Society. Series C*, 23(1) 1974, 35–42.
54. J. Markhoul, Linear prediction: A tutorial review, *Proceedings of the IEEE*, 63(4) 1975, 561–580.
55. R. Kohn, and C. F. Ansley, Estimation, prediction, and interpolation for ARIMA models with missing data, *Journal of the American Statistical Association*, 81(395) 1986, 751–761.
56. D. L. Zimmerman, and N. Cressie, Mean squared prediction error in the spatial linear model with estimated covariance parameters, *Annals of the Institute of Statistical Mathematics*, 44(1) 1992, 203–217.
57. M. S. Peiris, and B. J. C. Perera, On prediction with fractionally differenced ARIMA models, *Journal of Time Series Analysis*, 9(3) 1988, 215–220.
58. V. D. Kudritskii, Optimal linear extrapolation algorithm for realization of a vector random sequence observed without errors, *Cybernetics and Systems Analysis*, 34(2) 1998, 406–412.
59. J. H. Kim, Forecasting autoregressive time series with bias-corrected parameter estimators, *International Journal of Forecasting*, 19(3) 2003, 493–502.
60. Y. Cai, A forecasting procedure for nonlinear autoregressive time series models, *Journal of Forecasting*, 24(5) 2005, 335–351.
61. J. L. Harvill, and B. K. Ray, A note on multi-step forecasting with functional coefficient autoregressive models, *International Journal of Forecasting*, 21(4) 2005, 717–727.
62. B. S. Atal, The history of linear prediction, *IEEE Signal Processing Magazine*, 23(2) 2006, 154–161.
63. D. Huang, Levinson-type recursive algorithms for least-squares autoregression, *Journal of Time Series Analysis*, 11(4) 2008, 295–315.
64. A. Schick, and W. Wefelmeyer, Prediction in moving average processes, *Journal of Statistical Planning and Inference*, 138(3) 2008, 694–707.
65. A. Jamalizadeh, and N. Balakrishnan, Prediction in a trivariate normal distribution via a linear combination of order statistics, *Statistics & Probability Letters*, 79(21) 2009, 2289–2296.
66. L. Peng, and Q.-W. Yao, Nonparametric regression under dependent errors with infinite variance, *Annals of the Institute of Statistical Mathematics*, 56(1) 2004, 73–86.
67. P. Hall, and Q.-W. Yao, Inference in ARCH and GARCH models with heavy-tailed errors, *Econometrica*, 71(1) 2003, 285–317.

68. S. I. Resnick, Heavy tail modeling and teletraffic data (with discussion), *The Annals of Statistics*, 25(5) 1997, 1805–1869.
69. H. Michiel, and K. Laevens, Teletraffic engineering in a broad-band era, *Proceedings of the IEEE*, 85(12) 1997, 2007–2033.
70. P. Abry, and D. Veitch, Wavelet analysis of long-range-dependent traffic, *IEEE Transactions on Information Theory*, 44 (1) 1998, 2–15.
71. A. Scherrer, N. Larrieu, P. Owezarski, P. Borgnat, and P. Abry, Non-Gaussian and long memory statistical characterisations for internet traffic with anomalies, *IEEE Transactions on Dependable and Secure Computing*, 4(1) 2007, 56–70.
72. D. H. Griffel, *Applied Functional Analysis*, John Wiley & Sons, New York, 1981.
73. C. K. Liu, *Applied Functional Analysis*, Defense Industry Press, Beijing, 1986. In Chinese.
74. M. Li, Modeling autocorrelation functions of long-range dependent teletraffic series based on optimal approximation in Hilbert space: A further study, *Applied Mathematical Modelling*, 31(3) 2007, 625–631.
75. J. Beran, Statistical methods for data with long-range dependence, *Statistical Science*, 7(4) 1992, 404–416.
76. H. Künsch, J. Beran, and F. Hampel, Contrasts under long-range correlations, *The Annals of Statistics*, 21(2) 1993, 943–964.
77. J. Beran, Fitting long-memory models by generalized linear regression, *Biometrika*, 80(4) 1993, 817–822.
78. J. Beran, S. Ghosh, and D. Schell, On least squares estimation for long-memory lattice processes, *Journal of Multivariate Analysis*, 100(10) 2009, 2178–2194.
79. M. Li, Record length requirement of long-range dependent teletraffic, *Physica A*, 472, 2017, 164–187.
80. M. Li, W.-S. Chen, and L. Han, Correlation matching method of the weak stationarity test of LRD traffic, *Telecommunication Systems*, 43(3–4) 2010, 181–195.
81. M. Li, and S. C. Lim, Power spectrum of generalized Cauchy process, *Telecommunication Systems*, 43(3–4) 2010, 219–222.
82. M. Li, and S. C. Lim, A rigorous derivation of power spectrum of fractional Gaussian noise, *Fluctuation and Noise Letters*, 6(4) 2006, C33–36.
83. M. Li, Y. C. Li, and J. X. Leng, Power-type functions of prediction error of sea level time series, *Entropy*, 17(7) 2015, 4809–4837.

II

Traffic Modeling and Traffic Data Processing

Long-Range Dependence and Self-Similarity of Daily Traffic with Different Protocols

Traffic modeling is crucial in the analysis and design of the infrastructure of cyber-physical network systems from a view of traffic engineering. However, reports regarding traffic modeling at the large time scale of day in the duration of years are rarely seen. This chapter expounds my finding in autocorrelation function modeling in the closed form of traffic at the time scale of day (daily traffic for short) in the duration of 12 years with different protocols. We shall show that the autocorrelation function of daily traffic takes the autocorrelation function form of the generalized Cauchy process based on studying the autocorrelation function modeling of daily traffic with real traffic data. Thus, the long-range dependence and local self-similarity of daily traffic are in general uncorrelated according to the theory of the generalized Cauchy process. In addition, we will exhibit that the concrete values of long-range dependence measure and local self-similarity measure of daily traffic may relate to protocol types.

6.1 BACKGROUND

Modeling of traffic with long-range dependence plays a role in traffic engineering [1]. Usually, real traffic data used for the research of modeling are at small time scales. By small time scales one means the time scales from microseconds to milliseconds in the durations from seconds to minutes or hours (see refs [2–30]). Traffic modeling at small time scales is needed in applications such as queuing, buffer design, traffic delay, anomaly detection, etc. (see refs [1–16, 30–53]). From the point of view of the study of network infrastructure for network management, design planning, and simulation, however, traffic modeling at a large time scale is particularly needed. By large time scale, in this research, we mean the time scale at the level of day in the duration of years. For simplicity, we call it daily traffic if it is at the large time scale. In this aspect, an autocorrelation function model of daily

DOI: 10.1201/9781003354987-8

traffic is greatly desired for the purpose of understanding how the current traffic will be correlated to the traffic someday in future. This chapter addresses my finding in the auto-correlation function model of daily traffic.

The related work on traffic modeling is described as follows. Let $x(t)$ and $r(\tau)$ be a traffic time series and its autocorrelation function, respectively. Usually, at the small time scales, the fractional Gaussian noise (fGn) is taken as a traffic model (see refs [1–21, 27–32, 55] to mention a few). Note that the autocorrelation function of the fGn is expressed by

$$r_{fGn}(\tau) = \frac{\sigma^2 \varepsilon^{2H-2}}{2}\left[\left(\frac{|\tau|}{\varepsilon}+1\right)^{2H}+\left|\frac{|\tau|}{\varepsilon}-1\right|^{2H}-2\left|\frac{\tau}{\varepsilon}\right|^{2H}\right], \qquad (6.1)$$

where $0 < H < 1$ is the Hurst parameter, $\sigma^2 = (H\pi)^{-1}\Gamma(1-2H)\cos(H\pi)$ is the intensity of fGn, where $\Gamma(\cdot)$ is the gamma function. The quantity $\varepsilon > 0$ is used for regularizing the fractional Brownian motion [22, 56]. Letting $\varepsilon \to 0$ yields a commonly used form $r_{fGn}(\tau)=0.5\sigma^2[(|\tau|+1)^{2H}+||\tau|-1|^{2H}-2|\tau|^{2H}]$ for the simplicity. Practically, a traffic trace is a discrete series. The autocorrelation function of the discrete fGn (dfGn) is given by

$$r_{dfGn}(k) = \frac{\sigma^2}{2}\left[(k+1)^{2H}+|k-1|^{2H}-2|k|^{2H}\right]. \qquad (6.2)$$

In the above, H is a measure of long-range dependence. In short, fGn is a single parameter model. Scientists noticed that the model of fGn does not well fit real traffic. In fact, Paxson and Floyd said that "it might be difficult to characterize the correlation of traffic with a single parameter H" [2, left column, p. 237]. They noticed that "further work is required to fully understand the correlation structure of traffic" [2, right column, p. 237]. Besides, Tsybakov and Georganas noted that "the model of fractional Gaussian noise is too narrow for modeling real traffic" [13, right column, p. 1713]. We explain the limitation of fGn in traffic modeling in the following text.

There are two important properties of traffic. One is its long-range dependence, which is a global property characterized by the Hurst parameter H. The other is its self-similarity that is measured by the fractal dimension denoted by D. According to the theory of fractal time series, H is generally independent of D and vice versa (Mandelbrot [57]). Briefly speaking, long-range dependence and self-similarity are two different concepts (see refs [23, 25, 34, 37, 57–60] and references therein). Specifically for fGn, nonetheless, H happens to have a relationship with D [57, 61]. The relationship is expressed by

$$D = 2 - H. \qquad (6.3)$$

Thus, in fGn, single parameter H is used for describing two different measures, namely, long-range dependence and self-similarity because equation (6.3) linearly relates D with H.

Therefore, the limitation of fGn in traffic modeling is that it is very difficult to use fGn to separately characterize the long-range dependence and the self-similarity of traffic. For this reason, a traffic model with H that is independent of D is desired so as to release the limitation of fGn in traffic modeling.

In June 2000, we proposed an autocorrelation function form with two parameters for modeling traffic. Its normalized form is given by

$$r(\tau) = \left(1 + |\tau|^a\right)^{2H-2}, \tag{6.4}$$

where $0 < a \leq 1$ and $0.5 < H < 1$ [62, Eq. (3–7)]. In December 2000, Gneiting presented an autocorrelation function in the form

$$r(\tau) = \left(1 + \left|\frac{\tau}{c}\right|^\alpha\right)^{-\frac{\beta}{\alpha}}, \tag{6.5}$$

where $0 < \alpha \leq 2$ and $\beta > 0$ [58, Eq. (5)]. The quantity $c > 0$ is a scale factor instead of a parameter for long-range dependence or self-similarity. Thus, it is a model with two parameters. He calls it the model of the Cauchy family. In 2004, Gneiting and Schlather rewrote (6.5) in [61] as

$$r(\tau) = \left(1 + |\tau|^\alpha\right)^{\frac{\beta}{\alpha}}, \tag{6.6}$$

which is termed as the autocorrelation function of the Cauchy class [61, Eq. (11)]. In 2006, we introduced the term "the generalized Cauchy process" to describe the processes that follow (6.6) [63]. Then, in 2008, we exhibited that the generalized Cauchy process well fits real traffic at the small time scales [23]. The key point of the generalized Cauchy process is its flexibility to separately characterize the long-range dependence and self-similarity of traffic. We shall further brief about the generalized Cauchy process as preliminaries in Section 6.3.

Note that the properties of self-similarity and long-range dependence for traffic are conventionally investigated at the small time scales in the duration of minutes or hours [2–30, 55, 64–66]. Thus, whether the traffic in the duration of years will keep those properties or not is an interesting research issue. In this regard, Fontugne et al. recently found that the traffic keeps the properties of self-similarity and long-range dependence in the duration of 14 years at the small time scales [67]. Nevertheless, their research did not mention the autocorrelation function model of the daily traffic in the closed form. Therefore, comes a research problem of what the autocorrelation function model of daily traffic in the closed form might be. This chapter aims at giving a solution to this problem. To be precise, I shall show my finding that the autocorrelation function of the generalized Cauchy process well fits with the autocorrelation function of the daily traffic traces studied. Thus, we also found that the autocorrelation function model of daily traffic remains same as that of traffic at small time scales.

The rest of chapter is organized as follows: We describe the real data of daily traffic used in this research in Section 6.2. In Section 6.3, we brief the preliminaries about the generalized Cauchy process. The modeling results are explained in Section 6.4. Discussions are given in Section 6.5, which is followed by summary.

6.2 DATA

Let $x(i)$ ($i = 0, 1, \ldots$) be a traffic trace measured from 2:00 PM to 2:15 PM on the ith day. Then, we call $x(i)$ the daily traffic for the simplicity in general. It may be the number of packets/bytes of total traffic or the average bytes per packet with a specific protocol, such as TCP (Transmission Control Protocol) or IP (Internet Protocol). The daily traffic data used in this research are provided by the MAWI (Measurement and Analysis on the WIDE Internet) Working Group Traffic Archive, Japan. The MAWI has collected daily traffic data of a fixed period of 15 minutes everyday from 14:00:00 PM to 14:15:00 PM since 1 July 2006 at the standard Japanese time. We use the data recorded from 1 January 2007 to 31 December 2018. They are accessible at http://mawi.wide.ad.jp/mawi/.

In this research, the length of $x(i)$ is 4340 data points. We use a specific symbol to denote a specific trace. For instance, $x\_ip1(i)$ stands for the number of IP packets measured from 2:00 PM to 2:15 PM on the ith day, $x\_ip2(i)$ implies the number of IP bytes recorded from 2:00 PM to 2:15 PM on the ith day, and $x\_ip3(i)$ denotes the average bytes per IP packet from 14:00 to 14:15 on the ith day. Table 6.1 gives the descriptions of the traffic traces used in the research, where HTTP means Hypertext Transfer Protocol, HTTPS implies Hypertext Transfer Protocol over Secure Socket Layer, and UDP is the abbreviation of User Datagram Protocol.

TABLE 6.1 Traffic Traces Used in the Research

Symbol	Duration	Meaning
$x\_ip1(i)$	01/07/2007–31/12/2018	Number of IP packets from 14:00 to 14:15 a day
$x\_ip2(i)$	01/07/2007–31/12/2018	Number of IP bytes from 14:00 to 14:15 a day
$x\_ip3(i)$	01/07/2007–31/12/2018	Average bytes per IP packet from 14:00 to 14:15 a day
$x\_tcp1(i)$	01/07/2007–31/12/2018	Number of TCP packets from 14:00 to 14:15 a day
$x\_tcp2(i)$	01/07/2007–31/12/2018	Number of TCP bytes from 14:00 to 14:15 a day
$x\_tcp3(i)$	01/07/2007–31/12/2018	Average TCP bytes per packet from 14:00 to 14:15 a day
$x\_http1(i)$	01/07/2007–31/12/2018	Number of HTTP packets from 14:00 to 14:15 a day
$x\_http2(i)$	01/07/2007–31/12/2018	Number of HTTP bytes from 14:00 to 14:15 a day
$x\_http3(i)$	01/07/2007–31/12/2018	Average bytes per HTTP packet from 14:00 to 14:15 a day
$x\_https1(i)$	01/07/2007–31/12/2018	Number of HTTPS packets from 14:00 to 14:15 a day
$x\_https2(i)$	01/07/2007–31/12/2018	Number of HTTPS bytes from 14:00 to 14:15 a day
$x\_https3(i)$	01/07/2007–31/12/2018	Average bytes per HTTPS packet from 14:00 to 14:15 a day
$x\_udp1(i)$	01/07/2007–31/12/2018	Number of UDP packets from 14:00 to 14:15 a day
$x\_udp2(i)$	01/07/2007–31/12/2018	Number of UDP bytes from 14:00 to 14:15 a day
$x\_udp3(i)$	01/07/2007–31/12/2018	Average bytes per UDP packet from 14:00 to 14:15 a day

6.3 PRELIMINARIES: BRIEF OF GENERALIZED CAUCHY PROCESS

6.3.1 Long-Range Dependence

Let $x(t)$ be a time series. Let $r(\tau)$ be its autocorrelation function. Then,

$$r(\tau) = E[x(t)x(t+\tau)], \tag{6.7}$$

where E stands for the operation of mathematical expectation and τ is lag.

The meaning of long-range dependence is given as follows. If $r(\tau)$ decays slowly such that

$$\int_{-\infty}^{\infty} r(\tau)d\tau = \infty, \tag{6.8}$$

one says that $x(t)$ is of long-range dependence or has long memory (Beran [68, 69]).

An asymptotic form of $r(\tau)$ for a series with long-range dependence is expressed by

$$r(\tau) \sim c\tau^{-\beta} \ (\tau \rightarrow \infty), \tag{6.9}$$

where $c > 0$ is a constant and $\beta \in (0, 1)$ is the index of long-range dependence. In order to memorize the work by H. E. Hurst [70], β is often written by the Hurst parameter H in the form:

$$\beta = 2 - 2H. \tag{6.10}$$

Therefore, (6.9) is rewritten as

$$r(\tau) \sim c\tau^{2H-2} \ (\tau \rightarrow \infty). \tag{6.11}$$

Owing to $0 < \beta < 1$, the condition of long-range dependence may be expressed by $0.5 < H < 1$.

There are two points that are worth noting with respect to long-range dependence. One is that it is a global property of $x(t)$. In fact, its measure H is attained under the condition of $\tau \rightarrow \infty$. The other is that the larger the H value, the stronger the long-range dependence of $x(t)$. Extremely, if $H \rightarrow 1$, $r(\tau) \sim c$ for $\tau \rightarrow \infty$.

6.3.2 Self-Similarity

If $x(t)$ satisfies

$$x(at) =_d a^v x(t), a > 0, \tag{6.12}$$

where $=_d$ denotes the equality in joint finite distribution, one says that $x(t)$ is a self-similar series with the self-similarity index v (Mandelbrot [56, 57, 71]).

The self-similarity described by (6.12) may be too restrictive in applications. For example, the self-similar scaling property of traffic may only hold locally [16–18, 23, 64]. From a view of applications, therefore, one is usually interested in the property of local self-similarity of traffic [34, 54, 55, 65, 66].

As far as the local self-similarity is concerned, we use the definition stated by refs [72–80]. If $r(\tau)$ is sufficiently smooth on $(0, \infty)$ and

$$r(0) - r(\tau) \sim c|\tau|^{\alpha}, \quad \tau \to 0, \tag{6.13}$$

the parameter $0 < \alpha \leq 2$ is called the fractal index.

The fractal index α is a measure of local self-similarity, which is a local property of $x(t)$ since it is obtained when $\tau \to 0$.

Following refs [72–80], the fractal dimension D is expressed by

$$D = 2 - \frac{\alpha}{2}, \tag{6.14}$$

where $1 \leq D < 2$.

In case of D, we note that the larger the value of D, the stronger the local self-similarity of $x(t)$. The term "local self-similarity" often refers to the meaning of local irregularity (Hall [73]) or local smoothness (Constantine and Hall [75], Wu and Lim [81]) or local roughness (Hall and Wood [72], Chan and Wood [74], Davies and Hall [76], Kent and Wood [77], Hall and Roy [78], Feuerverger et al. [79], Gneiting et al. [82], Taylor and Taylor [83]).

As can be seen from the above, long-range dependence and self-similarity are two different concepts. The former is for the global property of $x(t)$ while the latter is for the local one. The model called the generalized Cauchy process has two parameters for separately characterizing the local self-similarity and long-range dependence of $x(t)$ [23, 34, 61, 63, 84–89].

6.3.3 The Generalized Cauchy Process

Let $1 < \alpha \leq 2$ and $0 \leq \beta$. $X(t)$ is called the generalized Cauchy process if its autocorrelation function is in the form

$$C(\tau) = \psi^2 \left(1 + |\tau|^{\alpha}\right)^{-\frac{\beta}{\alpha}}, \tag{6.15}$$

where ψ^2 is the intensity of $X(t)$.

The generalized Cauchy process is of long-range dependence if $0 \leq \beta < 1$. It is of short-range dependence if $\beta > 1$. As a matter of fact, since $\lim_{\tau \to \infty} C(\tau) = \psi^2 |\tau|^{-\beta}$, we have $\int_{-\infty}^{\infty} |\tau|^{-\beta} d\tau = \infty$ for $0 \leq \beta < 1$. Thus, $0 \leq \beta < 1$ is the condition for the generalized Cauchy process to be of long-range dependence. On the other hand, for $\beta > 1$, we have $\int_{-\infty}^{\infty} (1 + |\tau|^{\alpha})^{-\frac{\beta}{\alpha}} d\tau = \frac{2}{\alpha} B\left(\frac{1}{\alpha}, \frac{\beta - \alpha}{\alpha}\right) < \infty$, where $B(\alpha, \beta)$ is the beta function. Hence, short-range dependence when $\beta > 1$.

One important thing is that the long-range dependence property of the generalized Cauchy process only relies on the index β irrelevant to α.

Asymptotically for $\tau \to 0$ (Vengadesh et al. [86], Lim and Teo [87]), one has

$$\lim_{\tau \to 0} C(\tau) = \psi^2 |\tau|^{\alpha}. \tag{6.16}$$

Another important thing is that the local self-similarity property of the generalized Cauchy process just depends on the fractal index α irrelevant to β.

Considering (6.10) and (6.14), $C(\tau)$ may be expressed by

$$C(\tau) = \psi^2 \left(1 + |\tau|^{4-2D}\right)^{-\frac{1-H}{2-D}}. \tag{6.17}$$

For the modeling purpose, we consider $\frac{X(t)}{\psi}$ or let $\Psi = 1$ without the generality losing in what follows. When $\Psi = 1$, we write (6.17) as

$$C(\tau) = \left(1 + |\tau|^{4-2D}\right)^{-\frac{1-H}{2-D}}. \tag{6.18}$$

The simulation method and the power spectrum density of the generalized Cauchy process refer to our previous work [88, 89].

6.4 MODELING RESULTS

When using the generalized Cauchy process to model the daily traffic $x(i)$, we need estimating H and D. There are many reports with respect to the estimators of H and/or D. Various estimators relate to different methods, such as R/S analysis, maximum likelihood method, variogram-based method, box counting, detrended fluctuation analysis, wavelet fractional Fourier transform, spectrum regression or periodogram regression, golden sectioning method, and others (see refs [28–30, 55–57, 66–81]). This research utilizes the approach of correlation-based regression for the estimations of D and H.

Denote the theoretic autocorrelation function of the generalized Cauchy process by $R(k)$. Let $r(k)$ be the measured autocorrelation function of a real daily traffic trace. The cost function is defined with the root mean square error (RMSE) as

$$\mathrm{RMSE}(D, H) = \sqrt{\mathrm{E}\left[R(k) - r(k)\right]^2}. \tag{6.19}$$

As there are two parameters D and H in $R(k)$, the above RMSE is with a pair of parameters (D, H). Let (D_0, H_0) be such that

$$\mathrm{RMSE}(D_0, H_0) \leq 10^{-3}, \tag{6.20}$$

we say that $x(i)$ is consistent with the generalized Cauchy model at degree of 10^{-3}.

In the research, we estimate the autocorrelation function of $x(i)$ on block-by-block basis. An autocorrelation function estimate is the average of the autocorrelation function of each block so as to reduce the error of an autocorrelation function estimate. The settings for the estimates are as follows. The block size is 256 and the average count is 16.

Figures 6.1–6.15 indicate the data modeling of the daily traffic traces described in Table 6.1. The estimates of (D_0, H_0) are summarized in Table 6.2.

Table 6.2 exhibits that the RMSE(D_0, H_0)s for the data fitting are in the order of magnitude of 10^{-3} or less. Using the conventional model of fGn, at small time scales, fitting error is in the magnitude of order of 10^{-2} in the sense of RMSE, see lines below Eq. (16) in [23]. Note that the fitting error in ref [23] is mean square error (MSE) while this chapter uses RMSE. If RMSE is in the order of 10^{-3}, its MSE may be in the order of around 10^{-6}. Thus, we infer that the daily traffic may be well modeled by the generalized Cauchy process. Thus, the autocorrelation function of daily traffic is expressed by

$$C(\tau) = \left(1 + |\tau|^{4-2D_0}\right)^{-\frac{1-H_0}{2-D_0}}. \tag{6.21}$$

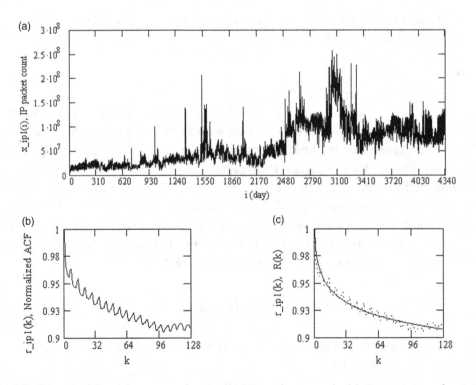

FIGURE 6.1 Modeling the trace of $x\_ip1(i)$ (IP packet count). (a) Data series of $x\_ip1(i)$. (b) The measured autocorrelation function of $x\_ip1(i)$ denoted by $r\_ip1(k)$. (c) Fitting the data of $r\_ip1(k)$.

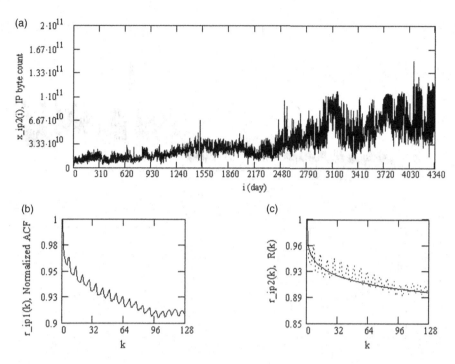

FIGURE 6.2 Modeling the trace of $x\_ip2(i)$ (IP byte count). (a) Data series of $x\_ip2(i)$. (b) The measured autocorrelation function of $x\_ip2(i)$ denoted by $r\_ip2(k)$. (c) Fitting the data of $r\_ip2(k)$.

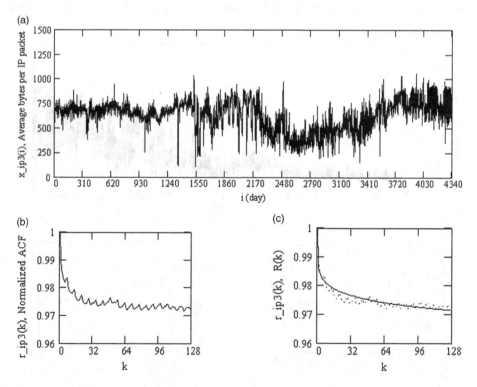

FIGURE 6.3 Modeling the trace of $x\_ip3(i)$ (average bytes per IP packet). (a) Data series of $x\_ip3(i)$. (b) The measured autocorrelation function of $x\_ip3(i)$ denoted by $r\_ip3(k)$. (c) Fitting the data of $r\_ip3(k)$.

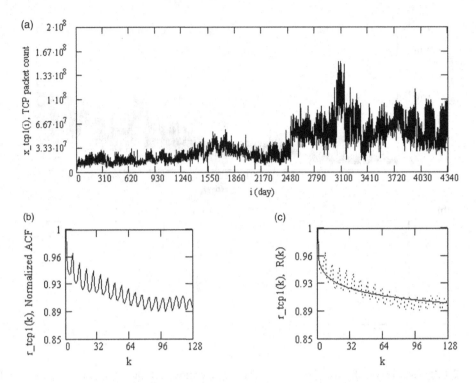

FIGURE 6.4 Modeling the trace of $x\_tcp1(i)$ (TCP packet count). (a) Data series of $x\_tcp1(i)$. (b) The measured autocorrelation function of $x\_tcp1(i)$ denoted by $r\_tcp1(k)$. (c) Fitting the data of $r\_tcp1(k)$.

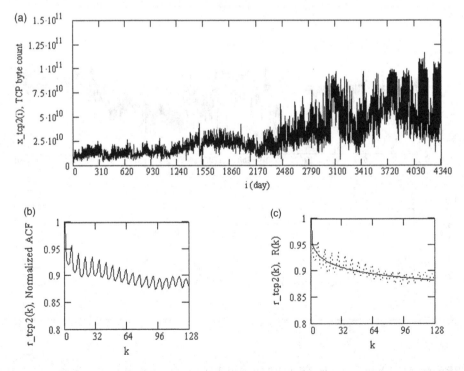

FIGURE 6.5 Modeling the trace of $x\_tcp2(i)$ (TCP byte count). (a) Data series of $x\_tcp2(i)$. (b) The measured autocorrelation function of $x\_tcp2(i)$ denoted by $r\_tcp2(k)$. (c) Fitting the data of $r\_tcp2(k)$.

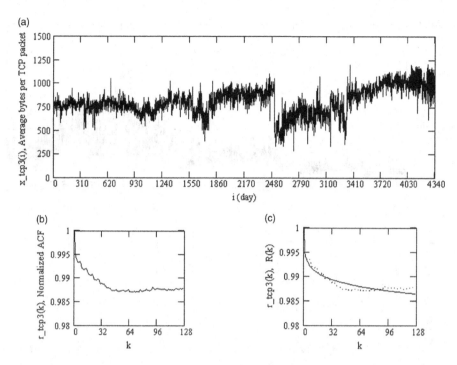

FIGURE 6.6 Modeling the trace of $x\_tcp3(i)$ (average bytes per TCP packet). (a) Data series of $x\_tcp3(i)$. (b) The measured autocorrelation function of $x\_tcp3(i)$ denoted by $r\_tcp3(k)$. (c) Fitting the data of $r\_tcp3(k)$.

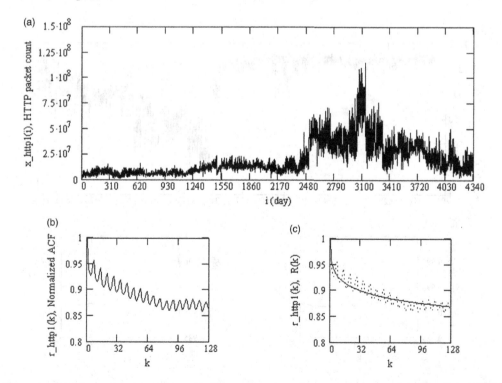

FIGURE 6.7 Modeling the trace of $x\_http1(i)$ (HTTP packet count). (a) Data series of $x\_http1(i)$. (b) The measured autocorrelation function of $x\_http1(i)$ denoted by $r\_http1(k)$. (c) Fitting the data of $r\_http1(k)$.

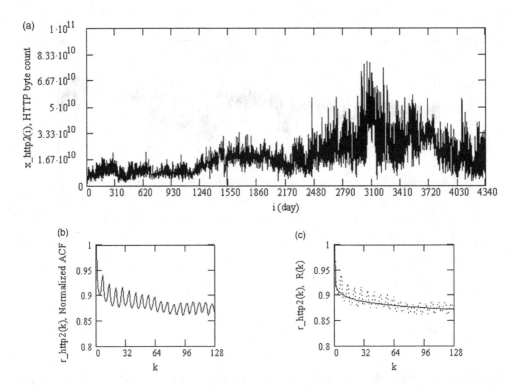

FIGURE 6.8 Modeling the trace of x_http2(i) (HTTP byte count). (a) Data series of x_http2(i). (b) The measured autocorrelation function of x_http2(i) denoted by r_http2(k). (c) Fitting the data of r_http2(k).

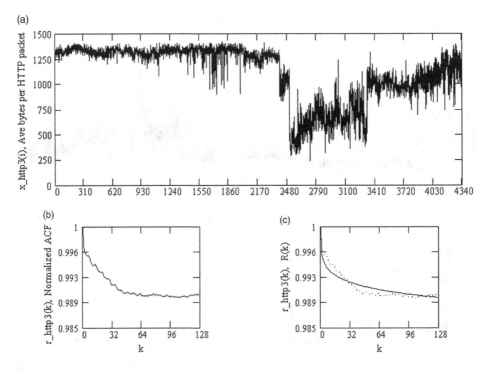

FIGURE 6.9 Modeling x_http3(i) (average bytes per HTTP packet). (a) Data series of x_http3(i). (b) The measured autocorrelation function of x_http3(i) denoted by r_http3(k). (c) Fitting the data.

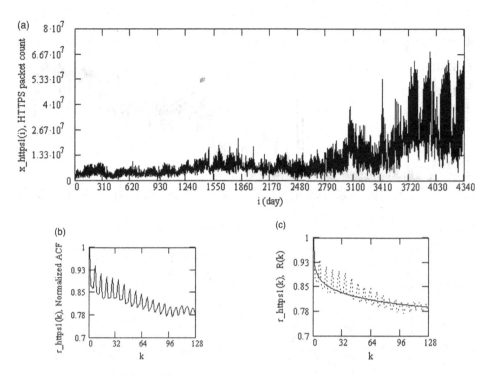

FIGURE 6.10 Modeling $x\_https1(i)$ (HTTPS packet count). (a) Data series of $x\_https1(i)$. (b) The measured autocorrelation function of $x\_https1(i)$ denoted by $r\_https1(k)$. (c) Fitting the data.

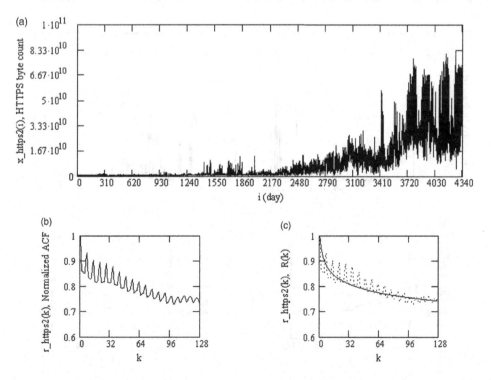

FIGURE 6.11 Modeling $x\_https2(i)$ (HTTPS byte count). (a) Data series of $x\_https2(i)$. (b) The measured autocorrelation function of $x\_https2(i)$ denoted by $r\_https2(k)$. (c) Fitting the data.

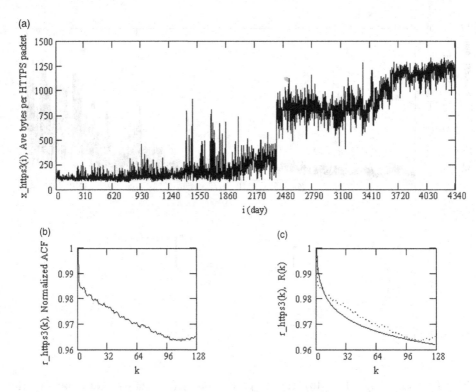

FIGURE 6.12 Modeling the trace of x_https3(i) (average bytes per HTTPS packet). (a) Data series of x_https3(i). (b) The measured autocorrelation function r_https3(k) (c). Fitting the data.

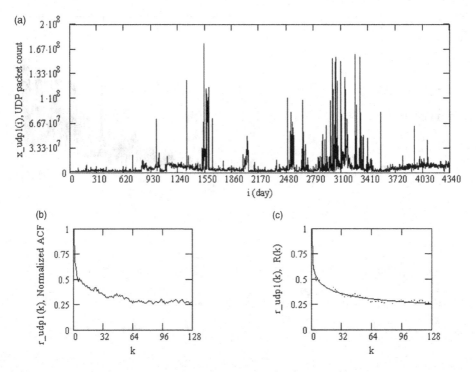

FIGURE 6.13 Modeling the trace of x_udp1(i) (UDP packet count). (a) Data series of x_udp1(i). (b) The measured autocorrelation function r_udp1(k). (c) Fitting the data of r_udp1(k).

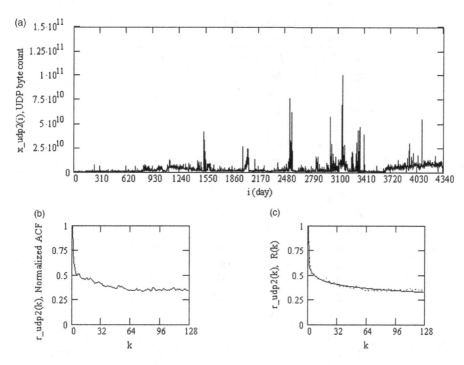

FIGURE 6.14 Modeling x_udp2(i) (UDP byte count). (a) Data series of x_udp2(i). (b) The measured autocorrelation function of x_udp2(i) denoted by r_udp2(k). (c) Fitting the data of r_udp2(k).

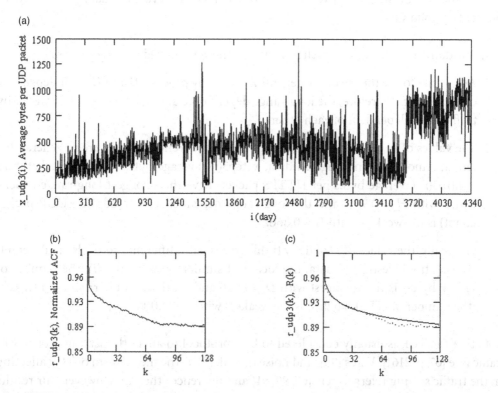

FIGURE 6.15 Modeling x_udp3(i) (average bytes per UDP packet). (a) Data series of x_udp3(i). (b) The measured autocorrelation function of x_udp3(i) denoted by r_udp3(k). (c) Fitting the data.

TABLE 6.2 Estimates of D and H of Traffic Traces Used in the Research

Trace	D_0	H_0	$\text{RMSE}(D_0, H_0)$
$x\_\text{ip1}(i)$	1.580	0.990	3.729×10^{-4}
$x\_\text{ip2}(i)$	1.730	0.989	9.325×10^{-4}
$x\_\text{ip3}(i)$	1.869	0.996	1.363×10^{-4}
$x\_\text{tcp1}(i)$	1.832	0.990	9.254×10^{-4}
$x\_\text{tcp2}(i)$	1.810	0.988	1.100×10^{-3}
$x\_\text{tcp3}(i)$	1.813	0.999	7.599×10^{-4}
$x\_\text{http1}(i)$	1.750	0.986	9.948×10^{-4}
$x\_\text{http2}(i)$	1.920	0.991	1.100×10^{-3}
$x\_\text{http3}(i)$	1.820	0.999	6.924×10^{-4}
$x\_\text{https1}(i)$	1.800	0.977	2.100×10^{-3}
$x\_\text{https2}(i)$	1.000	0.970	2.600×10^{-3}
$x\_\text{https3}(i)$	1.520	0.996	1.848×10^{-4}
$x\_\text{udp1}(i)$	1.773	0.866	1.718×10^{-3}
$x\_\text{udp2}(i)$	1.857	0.901	1.919×10^{-3}
$x\_\text{udp3}(i)$	1.800	0.989	5.040×10^{-4}

6.5 DISCUSSIONS

The significance of the present results in this chapter is to reveal the data driven facts in four folds as follows:

1. The daily traffic traces are well modeled by the generalized Cauchy process.

2. According to the theory of the generalized Cauchy process, D and H are independent of each other. Therefore, the long-range dependence and local self-similarity of daily traffic are independent of each other.

3. The values of H of the daily traffic traces with different protocols are in general different. Among the investigated data, $x\_\text{tcp3}(i)$ (average bytes per TCP packet) and $x\_\text{http3}(i)$ (average bytes per HTTPS packet) have the strongest long-range dependence with $H = 0.999$, and the long-range dependence of $x\_\text{udp1}(i)$ (UDP packet count) is the weakest with $H = 0.866$.

4. Generally, the values of D of daily traffic traces with different protocols are different. Among the investigated data, the local self-similarity of $x\_\text{http2}(i)$ (the number of HTTP bytes) is the strongest with $D = 1.920$ and local self-similarity of $x\_\text{https2}(i)$ (the number of HTTPS bytes) is the weakest with $D = 1.000$.

Note that the TCP is usually considered to be a protocol to affect the scaling property of traffic (see [67, p. 2161], Veres [90], and Loiseau et al. [91]). The mechanism of TCP affecting on the traffic scaling refers to refs [67, 90, 91] and references therein. However, our results in Table 6.2 propose the evidence that other types of protocols, such as UDP, IP, HTTP, and

HTTPS, may also affect the scaling behavior of traffic. Our future work will be on studying the multi-fractal behavior of daily traffic based on the multi-fractional generalized Cauchy model recently introduced by Li [34].

6.6 SUMMARY

We have shown that the autocorrelation function of daily traffic is well fitted by that of the generalized Cauchy process. Thus, we infer that the generalized Cauchy process may yet be a model of traffic at the large time scale of day in the duration of years. Protocols have effect on concrete values of H or D of daily traffic.

REFERENCES

1. H. Michiel, and K. Laevens, Teletraffic engineering in a broad-band era, *Proceedings of the IEEE*, 85(12) 1997, 2007–2033.
2. V. Paxson, and S. Floyd, Wide area traffic: The failure of Poisson modeling, *IEEE/ACM Transactions on Networking*, 3(3) 1995, 226–244.
3. J. Beran, R. Shernan, M. S. Taqqu, and W. Willinger, Long-range dependence in variable bit-rate video traffic, *IEEE Transactions on Communications*, 43(2–3–4) 1995, 1566–1579.
4. W. E. Leland, M. S. Taqqu, W. Willinger, and D. V. Wilson, On the self-similar nature of ethernet traffic (extended version), *IEEE/ACM Transactions on Networking*, 2(1) 1994, 1–15.
5. W.-B. Gong, Y. Liu, V. Misra, and D. Towsley, Self-similarity and long range dependence on the internet: A second look at the evidence, origins and implications, *Computer Networks*, 48(3) 2005, 377–399.
6. I. W. C. Lee, and A. O. Fapojuwo, Stochastic processes for computer network traffic modeling, *Computer Communications*, 29(1) 2005, 1–23.
7. I. Lokshina, Study on estimating probabilities of buffer overflow in high-speed communication networks, *Telecommunication Systems*, 62(2) 2016, 269–302.
8. J.-S. R. Lee, S.-K. Ye, and H.-D. J. Jeong, ATMSim: An anomaly teletraffic detection measurement analysis simulator, *Simulation Modelling Practice and Theory*, 49, 2014, 98–109.
9. M. Pinchas, Cooperative multi PTP slaves for timing improvement in an fGn environment, *IEEE Communications Letters*, 22(7) 2018, 1366–1369.
10. M. Pinchas, Residual ISI obtained by blind adaptive equalizers and fractional noise, *Mathematical Problems in Engineering*, 2013, 2013, Article ID 972174 (11 pp).
11. M. Pinchas, Symbol error rate for non-blind adaptive equalizers applicable for the SIMO and FGn case, *Mathematical Problems in Engineering*, 2014, 2014, Article ID 606843 (11 pp).
12. I. Tejado, S. H. Hosseinnia, B. M. Vinagre, X. Song, and Y. Q. Chen, Dealing with fractional dynamics of IP network delays, *International Journal of Bifurcation and Chaos*, 22(4) 2012, 1250089 (13 pp).
13. B. Tsybakov, and N. D. Georganas, Self-similar processes in communications networks, *IEEE Transactions on Information Theory*, 44(5) 1998, 1713–1725.
14. B. Tsybakov, and N. D. Georganas, Self-similar traffic and upper bounds to buffer-overflow probability in an ATM queue, *Performance Evaluation*, 32(1) 1998, 57–80.
15. S. Ma, and C. Ji, Modeling heterogeneous network traffic in wavelet domain, *IEEE/ACM Transactions on Networking*, 9(5) 2001, 634–649.
16. W. Willinger, M. S. Taqqu, R. Sherman, and D. V. Wilson, Self-similarity through high-variability: Statistical analysis of ethernet LAN traffic at the source level, *IEEE/ACM Transactions on Networking*, 5(1) 1997, 71–86.
17. A. Feldmann, A. C. Gilbert, W. Willinger, and T. G. Kurtz, The changing nature of network traffic: Scaling phenomena, *ACM SIGCOMM Computer Communication Review*, 28(2) 1998, 5–29.

18. A. Adas, Traffic models in broadband networks, *IEEE Communication Magazine*, 35(7) 1997, 82–89.
19. M. Roughan, D. Veitch, and P. Abry, Real time estimation of the parameters of long-range dependence, *IEEE/ACM Transactions on Networking*, 8(4) 2000, 467–478.
20. P. Abry, and D. Veitch, Wavelet analysis of long-range dependent traffic, *IEEE Transactions on Information Theory*, 44(1) 1998, 2–15.
21. C. S. Sastry, S. Rawat, A. K. Pujari, and V. P. Gulati, Network traffic analysis using singular value decomposition and multiscale transforms, *Information Sciences*, 177(2–3) 2007, 5275–5291.
22. M. Li, and S. C. Lim, A rigorous derivation of power spectrum of fractional Gaussian noise, *Fluctuation and Noise Letters*, 6(4) 2006, C33–36.
23. M. Li, and S. C. Lim, Modeling network traffic using generalized Cauchy process, *Physica A*, 387(11) 2008, 2584–2594.
24. M. Li, Modeling autocorrelation functions of long-range dependent teletraffic series based on optimal approximation in Hilbert space: -A further study, *Applied Mathematical Modelling*, 31(3) 2007, 625–631.
25. M. Li, Record length requirement of long-range dependent teletraffic, *Physica A*, 472, 2017, 164–187.
26. P. Borgnat, G. Dewaele, K. Fukuda, P. Abry, and K. Cho, Seven years and one day: Sketching the evolution of Internet traffic, *Proceedings of the 28th IEEE INFOCOM 2009*, Rio de Janeiro (Brazil), May 2009, 711–719.
27. O. Cappe, E. Moulines, J.-C. Pesquet, A. P. Petropulu, and X. S. Yang, Long-range dependence and heavy-tail modeling for teletraffic data, *IEEE Signal Processing Magazine*, 19(3) 2002, 14–27.
28. W. S. Cleveland, and D. X. Sun, Internet traffic data, *Journal of the American Statistical Association*, 95(451) 2000, 979–985.
29. W. Willinger, M. S. Taqqu, W. E. Leland, and D. V. Wilson, Self-similarity in high-speed packet traffic: Analysis and modeling of ethernet traffic measurements, *Statistical Science*, 10(1) 1995, 67–85.
30. C. Park, F. Hernández-Campos, L. Le, J. S. Marron, J. Park, V. Pipiras, F. D. Smith, R. L. Smith, M. Trovero, and Z. Zhu, Long-range dependence analysis of internet traffic, *Journal of Applied Statistics*, 38(7) 2011, 1407–1433.
31. Y. Avraham, and M. Pinchas, Two novel one-way delay clock skew estimators and their performances for the fractional Gaussian noise/generalized fractional Gaussian noise environment applicable for the IEEE 1588v2 (PTP) case, *Frontiers in Physics*, 10, 2022, 867861.
32. Y. Avraham, and M. Pinchas, A novel clock skew estimator and its performance for the IEEE 1588v2 (PTP) case in fractional Gaussian noise/generalized fractional Gaussian noise environment, *Frontiers in Physics*, 9, 2021, 796811.
33. S. Tadaki, Long-term power-law fluctuation in internet traffic, *Journal of the Physical Society of Japan*, 76(4) 2007, 044001 (5 pp).
34. M. Li, Multi-fractional generalized Cauchy process and its application to teletraffic, *Physica A*, 550, 2020, 123982 (14 pp).
35. M. Li, Generalized fractional Gaussian noise and its application to traffic modeling, *Physica A*, 579, 2021, 1236137 (22 pp).
36. J. Lévy-Véhel, Beyond multifractional Brownian motion: New stochastic models for geophysical modeling, *Nonlinear Processes in Geophysics*, 20(5) 2013, 643–655.
37. K. Kiyono, Establishing a direct connection between detrended fluctuation analysis and Fourier analysis, *Physical Review E*, 92(4) 2015, 042925.
38. Y. Gao, F. Villecco, M. Li, and W. Song, Multi-scale permutation entropy based on improved LMD and HMM for rolling bearing diagnosis, *Entropy*, 19(4) 2017, 176 (10 pp).

39. L. Arshadi, and A. H. Jahangir, An empirical study on TCP flow interarrival time distribution for normal and anomalous traffic, *International Journal of Communication Systems*, 30(1) 2017, e2881.

40. K. Cho, Recursive lattice search: Hierarchical heavy hitters revisited, *ACM IMC 2017*, London, UK, 1–3 Nov. 2017, 283–289.

41. M. Kato, K. Cho, M. Honda, and H. Tokuda, Monitoring the dynamics of network traffic by recursive multi-dimensional aggregation, *OSDI2012 MAD Workshop*, Hollywood, CA, 8–10 Oct. 2012 (7 pp).

42. H.-P. Schwefel, I. Antonios, and L. Lipsky, Understanding the relationship between network traffic correlation and queueing behavior: A review based on the N-Burst ON/OFF model, *Performance Evaluation*, 115, 2017, 68–91.

43. R. Delgado, A packet-switched network with On/Off sources and a fair bandwidth sharing policy: State space collapse and heavy-traffic, *Telecommunication Systems*, 62(2) 2016, 461–479.

44. A. Hajjar, J. E. Díaz-Verdejo, and J. Khalife, Network traffic application identification based on message size analysis, *Journal of Network and Computer Applications*, 58, 2015, 130–143.

45. C. Callegari, S. Giordano, and M. Pagano, An information-theoretic method for the detection of anomalies in network traffic, *Computers & Security*, 70, 2017, 351–365.

46. M. Li, Change trend of averaged Hurst parameter of traffic under DDOS flood attacks, *Computers & Security*, 25(3) 2006, 213–220.

47. M. Li, An approach to reliably identifying signs of DDOS flood attacks based on LRD traffic pattern recognition, *Computers & Security*, 23(7) 2004, 549–558.

48. M. Li, and A. Wang, Fractal teletraffic delay bounds in computer networks, *Physica A*, 557, 2020, 124903 (13 pp).

49. M. Marchetti, F. Pierazzi, M. Colajanni, and A. Guido, Analysis of high volumes of network traffic for advanced persistent threat detection, *Computer Networks*, 109, 2016, 127–141.

50. E. Kohler, J. Li, V. Paxson, and S. Shenker, Observed structure of addresses in IP traffic, *IEEE/ACM Transactions on Networking*, 14(6) 2006, 1207–1218.

51. Y. Zhang, L. Breslau, V. Paxson, and S. Shenker, On the characteristics and origins of internet flow rates, *ACM SIGCOMM Computer Communication Review*, 32(4) 2002, 309–322.

52. B. Hajek, and L. He, On variations of queue response for inputs with the same mean and autocorrelation function, *IEEE/ACM Transactions on Networking*, 6(5) 1998, 588–598.

53. N. L. S. Fonseca, G. S. Mayor, and C. A. V. Neto, On the equivalent bandwidth of self-similar source, *ACM Transactions on Modeling and Computer Simulation*, 10(2) 2000, 104–124.

54. W. Willinger, R. Govindan, S. Jamin, V. Paxson, and S. Shenker, Scaling phenomena in the internet critically, *Proceedings of the National Academy of Sciences of the United States of America*, 99(Suppl 1) 2002, 2573–2580.

55. P. Abry, R. Baraniuk, P. Flandrin, R. Riedi, and D. Veitch, Multiscale nature of network traffic, *IEEE Signal Processing Magazine*, 19(3) 2002, 28–46.

56. B. B. Mandelbrot, *Gaussian Self-Affinity and Fractals*, Springer, New York, 2001.

57. B. B. Mandelbrot, *The Fractal Geometry of Nature*, W. H. Freeman, New York, 1982.

58. T. Gneiting, Power-law correlations, related models for long-range dependence and their simulation, *Journal of Applied Probability*, 37(4) 2000, 1104–1109.

59. M. Li, Fractal time series: A tutorial review, *Mathematical Problems in Engineering*, 2010, 2010, Article ID 157264 (26 pp).

60. M. Li, Self-similarity and long-range dependence in teletraffic, *Proceedings of the 9th WSEAS International Conference on Multimedia Systems and Signal Processing*, Hangzhou, China, May 2009, 19–24.

61. T. Gneiting, and M. Schlather, Stochastic models that separate fractal dimension and Hurst effect, *SIAM Review*, 46(2) 2004, 269–282.

62. M. Li, W. Jia, and W. Zhao, A whole correlation structure of asymptotically self-similar traffic in communication networks, *Conf. Proc., IEEE WISE'2000*, 19-20 June 2000, Hong Kong, 461–466.

63. S. C. Lim, and M. Li, Generalized Cauchy process and its application to relaxation phenomena, *Journal of Physics A: Mathematical and General*, 39(12) 2006, 2935–2951.

64. H. Jiang, and C. Dovrolis, Why is the internet traffic bursty in short time scales? *ACM SIGMETRICS Performance Evaluation Review*, 33(1) 2005, 241–252.

65. L. O. Ostrowsky, N. L. S. da Fonseca, and C. A. V. Melo, A multiscaling traffic model for UDP streams, *Simulation Modelling Practice and Theory*, 26, 2012, 32–48.

66. F. H. T. Vieira, F. G. C. Rocha, and J. A. dos Santos, Loss probability estimation and control for OFDM/TDMA wireless systems considering multifractal traffic characteristics, *Computer Communications*, 35(2) 2012, 263–271.

67. R. Fontugne, P. Abry, K. Fukuda, D. Veitch, K. Cho, P. Borgnat, and H. Wendt, Scaling in internet traffic: A 14 year and 3 day longitudinal study, with multiscale analyses and random projections, *IEEE/ACM Transactions on Networking*, 25(4) 2017, 2152–2165.

68. J. Beran, *Statistics for Long-Memory Processes*, Chapman & Hall, New York, 994.

69. J. Beran, Statistical methods for data with long-range dependence, *Statistical Science*, 7(4) 1992, 404–416.

70. H. E. Hurst, Long term storage capacity of reservoirs, *Transactions of the American Society of Civil Engineers*, 116, 1951, 770–799.

71. B. B. Mandelbrot, *Multifractals and 1/f Noise*, Springer, New York, 1998.

72. P. Hall, and A. Wood, On the performance of box-counting estimators of fractal dimension, *Biometrika*, 80(1) 1993, 246–251.

73. P. Hall, On the effect of measuring a self-similar process, *SIAM Journal on Applied Mathematics*, 55(3) 1995, 800–808.

74. G. Chan, and A. T. A. Wood, Estimation of fractal dimension for a class of non-Gaussian stationary processes and fields, *The Annals of Statistics*, 32(3) 2004, 1222–1260.

75. A. G. Constantine, and P. Hall, Characterizing surface smoothness via estimation of effective fractal dimension, *Journal of the Royal Statistical Society B*, 56(1) 1994, 97–113.

76. S. Davies, and P. Hall, Fractal analysis of surface roughness by using spatial data, *Journal of the Royal Statistical Society B*, 61(1) 1999, 3–37.

77. J. T. Kent, and A. T. Wood, Estimating the fractal dimension of a locally self-similar Gaussian process by using increments, *Journal of the Royal Statistical Society B*, 59(3) 1997, 679–699.

78. P. Hall, and R. Roy, On the relationship between fractal dimension and fractal index for stationary stochastic processes, *The Annals of Applied Probability*, 4(1) 1994, 241–253.

79. A. Feuerverger, P. Hall, and A. T. A. Wood, Estimation of fractal index and fractal dimension of a Gaussian process by counting the number of level crossings, *Journal of Time Series Analysis*, 15(6) 1994, 587–606.

80. R. J. Adler, *The Geometry of Random Fields*, John Wiley & Sons, New York, 1981.

81. W.-Y. Wu, and C. Y. Lim, Estimation of smoothness of a stationary Gaussian random field, *Statistica Sinica*, 26(4) 2016, 1729–1745.

82. T. Gneiting, H. Ševčíková, and D. B. Percival, Estimators of fractal dimension: Assessing the roughness of time series and spatial data, *Statistical Science*, 27(2) 2012, 247–277.

83. C. C. Taylor, and S. J. Taylor, Estimating the dimension of a fractal, *Journal of the Royal Statistical Society B*, 53(2) 1991, 353–364.

84. M. Li, Evidence of a two-parameter correlation of internet traffic, In *Internet Policies and Issues*, Vol. 8, B. G. Kutais, (ed.), Nova Science Publishers, Inc., USA, 2011, pp. 103–140.

85. M. Li, and J.-Y. Li, Generalized Cauchy model of sea level fluctuations with long-range dependence, *Physica A*, 484, 2017, 309–335.

86. P. Vengadesh, S. V. Muniandy, and W. H. Abd. Majid, Fractal morphological analysis of bacteriorhodopsin (bR) layers deposited onto indium tin oxide (ITO) electrodes, *Materials Science and Engineering: C*, 29(5) 2009, 1621–1626.

87. S. C. Lim, and L. P. Teo, Gaussian fields and Gaussian sheets with generalized Cauchy covariance structure, *Stochastic Processes and Their Applications*, 119(4) 2009, 1325–1356.

88. M. Li, Generation of teletraffic of generalized Cauchy type, *Physica Scripta*, 81(2) 2010, 025007 (10 pp).

89. M. Li, and S. C. Lim, Power spectrum of generalized Cauchy process, *Telecommunication Systems*, 43(3–4) 2010, 219–222.

90. A. Veres, Z. Kenesi, S. Molnár, and G. Vattay, TCP's role in the propagation of self-similarity in the Internet, *Computer Communications*, 26(8) 2003, 899–913.

91. P. Loiseau, P. Gonçalves, G. Dewaele, P. Borgnat, P. V.-B. Abry, and Primet, Investigating self-similarity and heavy-tailed distributions on a large-scale experimental facility, *IEEE/ACM Transactions on Networking*, 18(4) 2010, 1261–1274.

CHAPTER 7

Stationarity Test of Traffic

Testing the stationarity of long-range dependent (LRD) mono-fractal traffic used to be an open problem. In Chapter 4, we have proved that LRD traffic in the mono-fractal case is ergodic. Accordingly, the stationarity of LRD mono-fractal traffic is a consequence of its ergodicity. Because the ergodicity discussed in Chapter 4 is in the mono-fractal sense but traffic has the multi-fractal property in nature and practical traffic series is of finite length, the research regarding the stationarity test of LRD multi-fractal traffic with finite length is still greatly desired. This chapter discourses a method of the weak stationarity test of a single-history LRD multi-fractal traffic series of finite length. How to apply this method to testing the stationarity of real traffic is demonstrated here. The results in this chapter suggest that there may be no general conclusion that traffic is either stationary or non-stationary since the stationarity of traffic is observation-scale-dependent. Some of the investigated real traffic traces that are stationary in an observation scale may be non-stationary in a larger observation scale.

7.1 BACKGROUND

Let $\{x_l(t)\}$ ($-\infty < t < \infty$) be a Gaussian process, where $x_l(t)$ is a sample function ($l = 1, 2, \ldots$). Let E be the mean operator. Then,

$$\mu_x(t) = \mathrm{E}\big[x_l(t)\big] \tag{7.1}$$

is the ensemble mean at arbitrary fixed values of t and

$$r(t_1, t_2) = \mathrm{E}\big[x_l(t_1)x_l(t_2)\big] \tag{7.2}$$

is called autocorrelation function (ACF).

For a weak stationary (stationary for short) process, $\mu_x(t)$ and $r(t_1, t_2)$ are independent of time such that

$$\mu_x(t) = \mathrm{const}, \tag{7.3}$$

DOI: 10.1201/9781003354987-9

and

$$r(t_1, t_2) = \mathrm{E}[x_l(t+\tau)x_l(t)] = r(\tau), \tag{7.4}$$

where $\tau = t_1 - t_2$ is time lag. Therefore, for a stationary Gaussian process $\{x_l(t)\}$, its ACF or its power spectrum density function (PSD)

$$S_{xx}(\omega) = \int_{-\infty}^{\infty} r_{xx}(\tau)e^{-j\omega\tau}d\tau, \tag{7.5}$$

is enough to characterize its statistics [1].

Nevertheless, if $\{x_l(t)\}$ is non-stationary, $\mu_x(t)$ and $r(t_1, t_2)$ are dependent on time (Priestley [2], Al-Shoshan [3]). Its PSD in this case can be analyzed in a time-frequency plane (Bendat and Piersol [4]). That is,

$$S_{xx}(t, \omega) = \int_{-\infty}^{\infty} r_{xx}(t, \tau)e^{-j\omega\tau}d\tau. \tag{7.6}$$

Therefore, the stationarity test of $\{x_l(t)\}$ plays a fundamental role in processing $\{x_l(t)\}$. As a matter of fact, the stationarity test of $\{x_l(t)\}$ is the first issue one has to consider before processing $\{x_l(t)\}$.

A difficulty regarding the stationarity test is that a traffic series to be processed in practice is a single history with finite length, making the mathematical definition of the stationarity fail to test the stationarity of traffic.

There are various methods in the field of stationarity test. Von Sachs and Neumann [5] used wavelet as a tool for the stationarity test, Ling [6] studied the stationarity test of a process with the double-autoregressive model, Psaradakis [7] adopted the blockwise bootstrap method for the stationarity test, Rodrigues and Rubia [8] utilized the level-dependent conditional heteroskedasticity to test the stationarity of autoregressive processes, and Borgnat and Flandrin [9] tested the stationarity of a time series by introducing the concept of the relative stationarity in the sense of different observation scales with the tool of stationarized surrogate data. These methods of the stationarity test are suitable for ordinary processes that have the property such that both $\mathrm{E}[x_l(t)]$ and $\mathrm{Var}[x_l(t)]$ exist. However, these tests may be inappropriate for the LRD processes because either $\mathrm{E}[x_l(t)]$ or $\mathrm{Var}[x_l(t)]$ may not exist.

Note that a commonly used model of LRD traffic is fractional Gaussian noise (fGn), see Tsybakov and Georganas [10], Nogueira et al. [11], Li [12], Leland et al. [13], Paxson and Floyd [14], Beran [15], Jin and Min [16]. Since the mean for LRD fGn does not exist (Mandelbrot [17]), there also does not exist mean for LRD traffic. Thus, the stationarity test of LRD traffic becomes a tough issue. In this aspect, Vaton [18] generalized the test of sphericity as a test of stationarity for time series and concluded that LRD traffic is stationary; Abry and Veitch [19] stated a particular test method specifically for LRD traffic by investigating the time invariability of the Hurst parameter; Zhang et al. [20] proposed the concept of operational stationarity to test the stationarity of traffic; and Li et al. [21] introduced a correlation matrix to test the stationarity of LRD traffic by investigating the variation of ACF. In general, however, classical statistical approaches for the stationarity test no longer hold for LRD traffic (see Abry and Veitch [19, Sentence 1, Section 3, pp. 7], Heyman and Lakshman [22, Sentence 1, Paragraph 3, Section 1, pp. 629]).

Due to lack of sound methods to test the stationarity of real traffic and paucity of evidences about whether LRD traffic is stationary or not, computer scientists have not reached the consensus if the real LRD traffic is stationary or not. For instance, some imply that LRD traffic is stationary (see refs [10–14]). Others consider that LRD traffic is non-stationary (see Liu et al. [23], Rincón and Sallent [24]). This chapter aims at extending our work described in ref [21] in two aspects. One is to complete the explanation of the method by investigating more real LRD traffic traces. The other is to correct the inappropriate conclusion that LRD traffic is generally stationary as stated in ref [21].

The remaining chapter is organized as follows: A method for the stationarity test of a single-history LRD traffic series for finite length is addressed in Section 7.2. Case study with real traffic traces is given in Section 7.3. Discussions are given in Section 7.4 and summary is presented in Section 7.5.

7.2 CORRELATION METHOD FOR STATIONARITY TEST OF LRD TRAFFIC

Since the PSD of an LRD series is a generalized function while its ACF is an ordinary function, I use the ACF defined by

$$r(t, t+\tau) = \lim_{N\to\infty} \frac{1}{N} \sum_{l=1}^{N} x_l(t) x_l(t+\tau). \tag{7.7}$$

A key point worth noting is that any real traffic series to be processed is of single history and of finite length. Consequently, the limitation $N \to \infty$ in (7.7) can never be achieved in practice. It is only used in the definition of ACF.

Let $x(i)$ be a traffic series of length N. Divide $x(i)$ into M non-overlapped sections. Each section is of length L. Denote

$$x_l(i) = x(i) \text{ for } i \in \left[(l-1)L, lL-1\right], l=1, 2, \ldots, M. \tag{7.8}$$

Then, $x_l(i)$ represents the lth sample series and $\{x_l\}$ a traffic process. Now, divide $x_l(i)$ into S non-overlapped blocks. The block size is B. Denote $x_{l,s}(i)$ as the series in the sth block ($s = 1, 2, \ldots, S$) in the lth section. Denote the ACF of $x_{l,s}(i)$ by $r(k; l)_s$. Then, $r(k;l)_s = \frac{1}{B}\sum_{i=(s-1)B}^{sB-1} x_{l,s}(i) x_{l,s}(i+k)$. The ACF of $x_l(i)$ is estimated and denoted by

$$r(k;l) =_{\mathrm{d}} rl = \frac{1}{S}\sum_{s=1}^{S} r(k;l)_s, \tag{7.9}$$

where $r(k; l)$ is a series of length M over the index l. The purpose of average is to reduce the estimation variance (Mitra and Kaiser [25]).

In the case that the mean of LRD traffic does not exist, we propose to test the stationarity of LRD traffic by checking the fluctuation of $r(k; l)$. Usually, $r(k; l) \neq r(k; n)$ for $l \neq n$. However, the above never implies that $r(k; l)$ is time-varying because x_i is of finite length, the number of samples is finite and there are errors in measurement or computation. As a matter of fact, one may never obtain an exact time-invarying ACF from a real time series. Owing to this, we say that a single-history traffic series of finite length is referred to as being stationary if $r(k; l)$ does not vary significantly as l changes. Here, that $r(k; l)$ does not vary *significantly* implies that $r(k; l)$ and $r(k; n)$ are similar in the sense of pattern matching according to a certain rule for all l and n. The crucial thing herein is to define a measure for characterizing the fluctuation of $r(k; l)$.

Now, we introduce a correlation matrix consisting of sample ACFs. Without losing the generality, ACFs considered here and below are normalized. As known, correlation matching is a commonly used technique in pattern matching (see Fu [26] and Li [27]). Denote the correlation coefficient between two sample ACFs r_l and r_n by

$$\text{corr}\left[r_l, r_n\right] = c_{ln}. \tag{7.10}$$

Then,

$$C = \left[c_{ln}\right] = \begin{bmatrix} c_{11} & c_{12} & ... & c_{1M} \\ c_{21} & c_{22} & ... & c_{2M} \\ ... & ... & ... & ... \\ c_{M1} & c_{M2} & ... & c_{MM} \end{bmatrix} \tag{7.11}$$

is an $M \times M$ matrix. For the matrix C, $|c_{ln}| = 1$ if $l = n$ and $|c_{ln}| \leq 1$ for $l \neq n$.

Let γ be the threshold regarding pattern similarity. Then, saying that $x(i)$ is considered to be stationary means that r_l does not vary significantly as l varies in the sense of $|c_{ln}| \geq \gamma$ for all l and n. That $x(i)$ is non-stationary implies that r_l is time-varying in the sense that $|c_{ln}| \geq \gamma$ does not hold for all l and n. In practical terms, the patterns of r_l and r_n are reasonably similar in engineering for $\gamma = 0.7$ and quite satisfactorily similar for $\gamma = 0.8$. This chapter adopts $\gamma = 0.7$. Therefore, we say that $x(i)$ is stationary if $|c_{ln}| \geq 0.7$ for all l and n. Otherwise, it is non-stationary.

7.3 CASE STUDY

Real traffic data used in this chapter consist of ten series. They are four traces of Ethernet traffic in Table 7.1 and six traces of TCP traffic in Table 7.2. The series with the prefix BC were collected on Ethernet at the Bellcore Morristown Research and Engineering Facility, those with the prefix DEC were measured at Digital Equipment Corporation, and those with Lbl were recorded at the Lawrence Berkeley Laboratory. They are available from Danzig et al. [28]. In Table 7.1 and Table 7.2, the first column stands for series name, the second for record date, and the third for series length. These real traffic traces have been widely used in the research of traffic analysis and modeling (Leland et al. [13], Paxson and Floyd [14], Beran [15], Abry and Veitch [19], Li [29]).

TABLE 7.1   Four Series of Ethernet Traffic

Series Name	Record Date	Series Length
BC-pAug89.TL	11:25AM, 29Aug89	10^6
BC-pOct89.TL	11:00AM, 05Oct89	10^6
BC-OctExt89.TL	11:46PM, 03Oct89	10^6
BC-Oct89Ext4.TL	2:37PM, 10Oct89	10^6

TABLE 7.2   Six Series of TCP Traffic

Series Name	Record Date	Series Length
DEC-pkt-1.TCP	08Mar95	3.3×10^6
DEC-pkt-2.TCP	08Mar95	3.9×10^6
DEC-pkt-3.TCP	08Mar95	4.3×10^6
DEC-pkt-4.TCP	08Mar95	5.7×10^6
Lbl-pkt-4.TCP	21Jan94	862946
Lbl-pkt-5.TCP	28Jan94	710614

7.3.1 Demonstration I: Stationary Case

In what follows, traffic $x(i)$ implies the number of bytes in the ith packet ($i = 0, 1, \ldots$). We compute the sample ACFs of $x(i)$ section by section. Starting from the first point data of $x(i)$, in each section, we set the block size $B = 1024$ and average count $S = 20$. Computing the sample ACFs from the 1st section to the 30th section yields 30 sample ACFs, namely, r1, r2, …, r30. Thus, we shall investigate the following matrix:

$$C = \left[r_{ln} \right]_{30 \times 30},\qquad (7.12)$$

where $r_{ln} = \text{corr}[rl, rn]$ represents the correlation coefficient between the lth ACF and the nth ACF, where $l, n = 1, 2, \ldots, 30$. Figure 7.1 indicates the first 1024 points of the series BC-pAug89.TL. Figures 7.2 (r1), (r2), …, (r30) give the plots of 30 sample ACFs from r1 to r30, respectively. Although $rl \neq rn$ for $l \neq n$ as described in Figure 7.2, one cannot say that the ACF is time-varying. Whether the ACF is time-varying or not relies on the fact that if min[C] is less than 0.7 or not. Figures 7.3 (a) and (b) show the surface plot and the contour

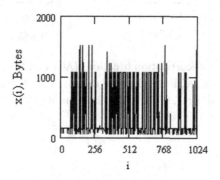

FIGURE 7.1   Real traffic series BC-pAug89.TL.

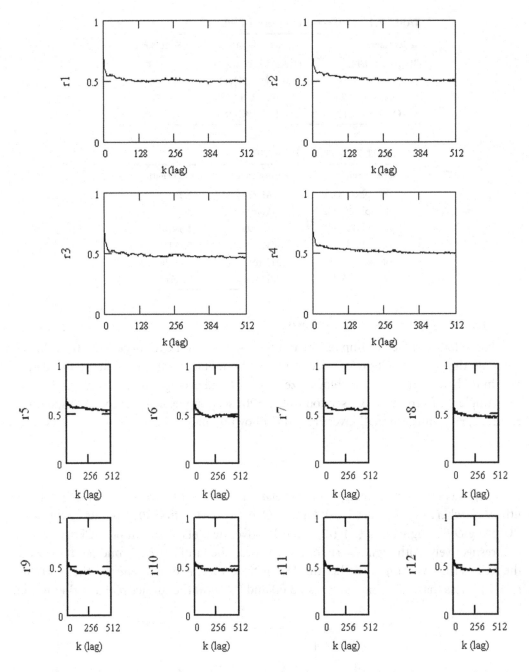

FIGURE 7.2 Thirty sample autocorrelation functions (ACFs) of the traffic BC-pAug89. (r1). ACF r1. (r2). ACF r2.·(r3). ACF r3.·(r4). ACF r4.·(r5). ACF r5. (r6). ACF r6. (r7). ACF r7. (r8). ACF r8. (r9). ACF r9. (r10). ACF r10. (r11). ACF r11. (r12). ACF r12. *(Continued)*

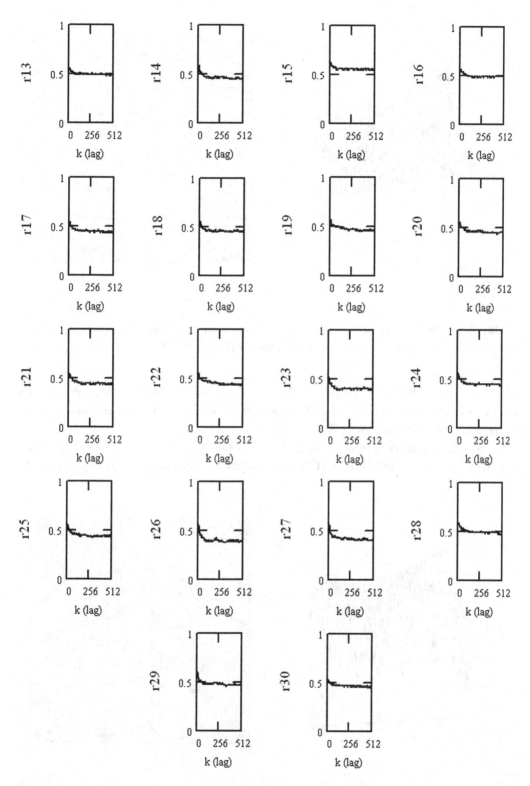

FIGURE 7.2 *(Continued)* (r13). ACF r13. (r14). ACF r14. (r15). ACF r15. (r16). ACF r16. (r17). ACF r17. (r18). ACF r18. (r19). ACF r19. (r20). ACF r20. (r21). ACF r21. (r22). ACF r22. (r23). ACF r23. (r24). ACF r24. (r25). ACF r25. (r26). ACF r26. (r27). ACF r27. (r28). ACF r28. (r29). ACF r29. (r30). ACF r30.

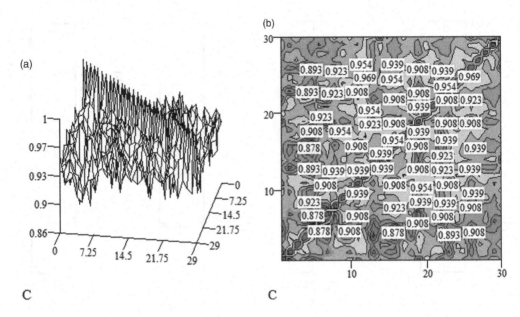

FIGURE 7.3 Correlation matrix consisting of 30 sample autocorrelation functions of the traffic BC-pAug89.TL with min[C] = 0.862. (a) Surface plot. (b) Contour plot.

one of C, respectively. Since min[C] = 0.862, see Figure 7.3 (a), the trace BC-pAug89.TL is stationary in the sense of min[C] > 0.7 in the first 30 sections. The locations of corr[rl, rn] are shown in the contour plot in Figure 7.3 (b).

The correlation matrices of other six traces (BC-pOct89.TL, DEC-pkt-1.TCP, DEC-pkt-2. TCP, DEC-pkt-3.TCP, DEC-pkt-4.TCP, and Lbl-pkt-4.TCP) are given in Figures 7.4–7.9.

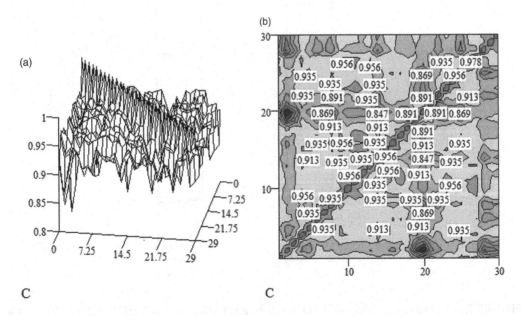

FIGURE 7.4 Correlation matrix consisting of 30 sample autocorrelation functions of the traffic BC-pOct89.TL with min[C] = 0.804. (a) Surface plot. (b) Contour plot.

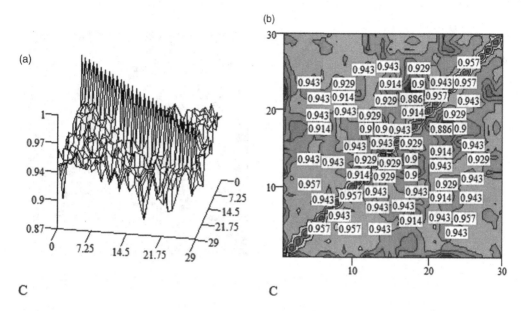

FIGURE 7.5 Correlation matrix consisting of 30 sample autocorrelation functions of the traffic DEC-pkt-1.TCP with min[C] = 0.871. (a) Surface plot. (b) Contour plot.

Since min[C] > 0.7 holds for all those traces, they are stationary in the sense of min[C] > 0.7 in the first 30 sections.

7.3.2 Demonstration II: Non-Stationary Case

The first 4096 points of the trace BC-OctExt89.TL are plotted in Figure 7.10. Starting from the first point data of $x(i)$, in each section, we set the block size $B = 4096$ and average count

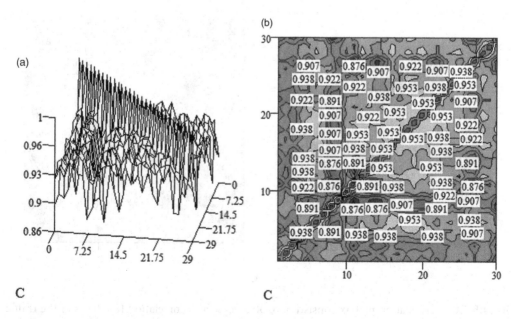

FIGURE 7.6 Correlation matrix consisting of 30 sample autocorrelation functions of the traffic DEC-pkt-2.TCP with min[C] = 0.860. (a) Surface plot. (b) Contour plot.

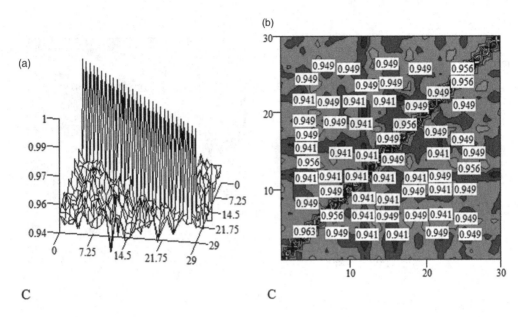

FIGURE 7.7 Correlation matrix consisting of 30 sample autocorrelation functions of the traffic DEC-pkt-3.TCP with min[C] = 0.943. (a) Surface plot. (b) Contour plot.

$S = 5$. Thirty sample ACFs from r1 to r30 of that trace are indicated in Figures 7.11 (r1), (r2), …, (r30), respectively. The surface plot and the contour plot of the matrix C for that series are shown in Figures 7.12 (a) and (b), respectively. For this series, min[C] = 0.438. For the trace BC-Oct89Ext4.TL, min[C] = 0.227 (Figure 7.13), and for Lbl-pkt-5.TCP, min[C] = 0.581 (Figure 7.14). Consequently, all three traces are non-stationary in the first 30 sections.

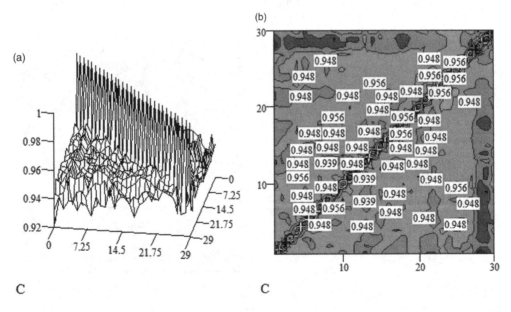

FIGURE 7.8 Correlation matrix consisting of 30 sample autocorrelation functions of the traffic DEC-pkt-4.TCP with min[C] = 0.921. (a) Surface plot. (b) Contour plot.

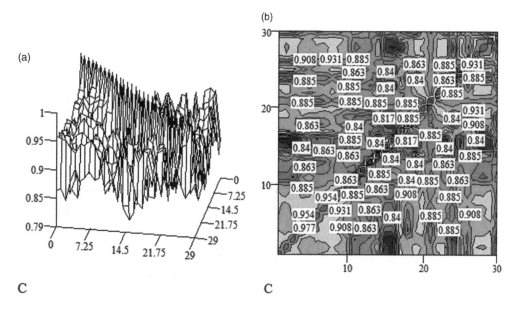

FIGURE 7.9 Correlation matrix consisting of 30 sample autocorrelation functions of the traffic Lbl-pkt-4.TCP with min[C] = 0.794. (a) Surface plot. (b) Contour plot.

7.4 DISCUSSIONS

It is worth noting that both the stationary and the non-stationary cases can be observed from the same traffic series, depending on the observation scale. To explain this point of view, we denote

$$C11 = [r_{ln}]_{10 \times 10,} \tag{7.13}$$

where $l, n = 1, 2, ..., 10$. That is, C11 is the correlation matrix obtained by computing ten sample ACFs from the 1st section to the 10th section, namely, r1, r2, ..., r10. We further denote

$$C12 = [r_{ln}]_{10 \times 10,} \tag{7.14}$$

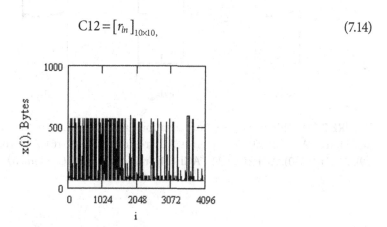

FIGURE 7.10 Real traffic trace BC-OctExt89.TL.

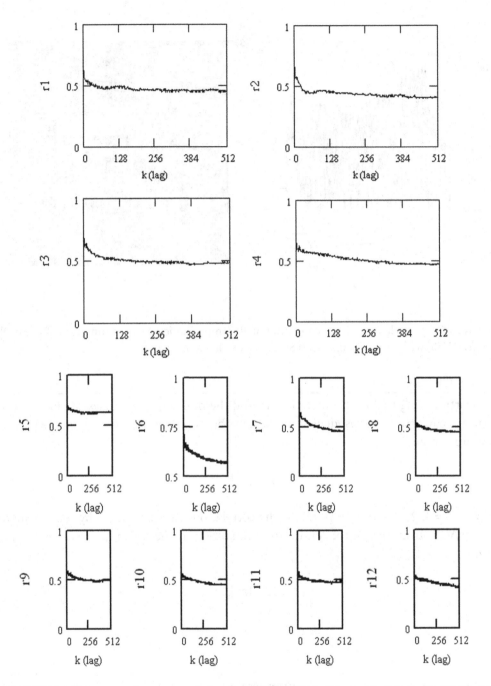

FIGURE 7.11 Thirty sample autocorrelation functions (ACFs) of the traffic BC-OctExt89.TL. (r1). ACF r1. (r2). ACF r2.·(r3). ACF r3.·(r4). ACF r4.·(r5). ACF r5. (r6). ACF r6. (r7). ACF r7. (r8). ACF r8. (r9). ACF r9. (r10). ACF r10. (r11). ACF r11. (r12). ACF r12. *(Continued)*

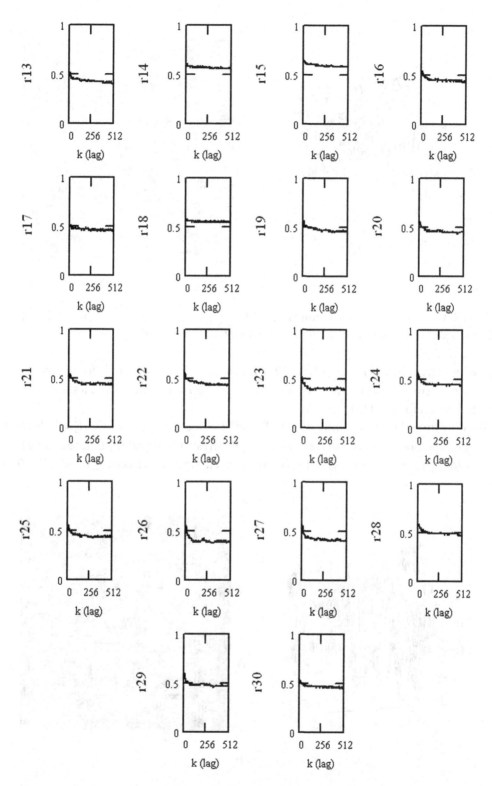

FIGURE 7.11 *(Continued)* (r13). ACF r13. (r14). ACF r14. (r15). ACF r15. (r16). ACF r16. (r17). ACF r17. (r18). ACF r18. (r19). ACF r19. (r20). ACF r20. (r21). ACF r21. (r22). ACF r22. (r23). ACF r23. (r24). ACF r24. (r25). ACF r25. (r26). ACF r26. (r27). ACF r27. (r28). ACF r28. (r29). ACF r29. (r30). ACF r30.

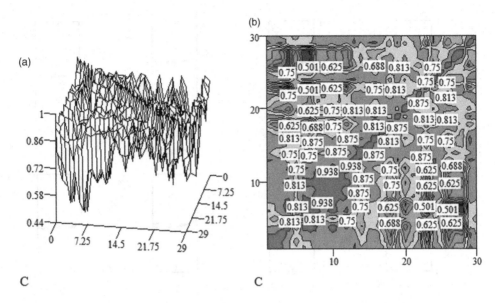

FIGURE 7.12 Correlation matrix consisting of 30 sample autocorrelation functions of the traffic BC-OctExt89.TL with min[C] = 0.438. (a) Surface plot. (b) Contour plot.

where $l = 1, 2, ..., 10$ and $n = 11, 12, ..., 20$. The element r_{ln} in (7.14) stands for the correlation coefficient between a sample ACF in the lth section for $l = 1, 2, ..., 10$ and a sample ACF in the nth section for $n = 11, 12, ..., 20$.

We now investigate the traffic BC-OctExt89.TL. This trace has min[C11] = 0.754 > 0.7. Hence, it is stationary in the first ten sections. Figures 7.15 (a) and (b) indicate the surface and contour plots of C11 for this series, respectively. However, this trace has min[C12] = 0.567,

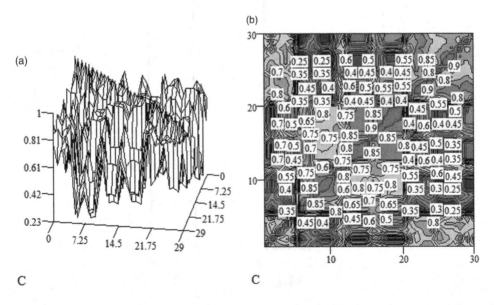

FIGURE 7.13 Correlation matrix consisting of 30 sample autocorrelation functions of the traffic BC-Oct89Ext4.TL with min[C] = 0.227. (a) Surface plot. (b) Contour plot.

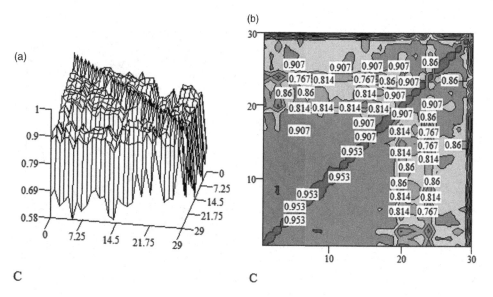

FIGURE 7.14 Correlation matrix consisting of 30 sample autocorrelation functions of the traffic Lbl-pkt-5.TCP with min[C] = 0.581. (a) Surface plot. (b) Contour plot.

and as a matter of fact, corr[r7, r18] = 0.567. Thus, it is non-stationary in the first 20 sections, see Figure 7.16. Therefore, the trace BC-OctExt89.TL is stationary in the first 10 sections but non-stationary in the first 20 sections. Hence, the stationarity of LRD traffic is observation-scale-dependent.

That the stationarity of LRD traffic is observation-scale-dependent is not a special case obtained from the trace BC-OctExt89.TL. To be precise, it is a common phenomenon in

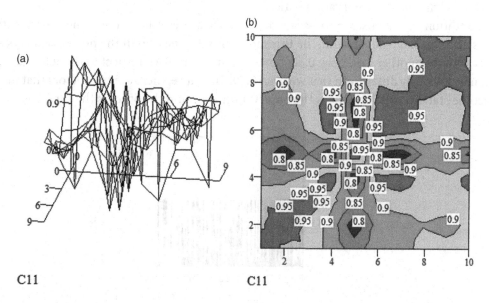

FIGURE 7.15 Correlation matrix consisting of ten sample autocorrelation functions of the traffic BC-OctExt89.TL with min[C11] = 0.754. (a) Surface plot. (b) Contour plot.

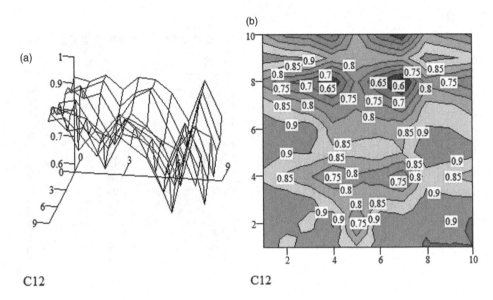

FIGURE 7.16 Correlation matrix consisting of 20 sample autocorrelation functions of the traffic BC-OctExt89.TL with min[C12] = 0.567. (a) Surface plot. (b) Contour plot.

this regard. For instance, for the trace Lbl-pkt-5.TCP (Figure 7.17), it has min[C11] = 0.921 (Figure 7.18) and min[C12] = 0.758 (Figure 7.19). Thus, it is stationary in the first 20 sections but non-stationary in the first 30 sections, see Figure 7.14. In fact, min[C13] = 0.581 due to corr[r6, r30] = 0.581 as indicated in Figure 7.20.

The previous discussions exhibit that one cannot say that LRD traffic is either stationary or non-stationary in the general sense because the stationarity of LRD traffic is a relative concept, relying on the observation scale.

In addition to the observation-scale dependence as described above, the stationarity test is also threshold dependent. The threshold γ is set to be 0.70 in the previous analysis. If γ is set to be another value, say 0.80, however, the result with respect to the stationarity test may obviously differ from that with $\gamma = 0.70$. Therefore, there are two factors that may affect the stationarity test. One is the observation scale and the other is the threshold.

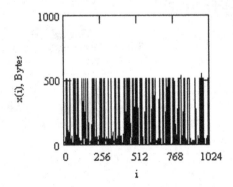

FIGURE 7.17 Real traffic Lbl-pkt-5.TCP in packet size.

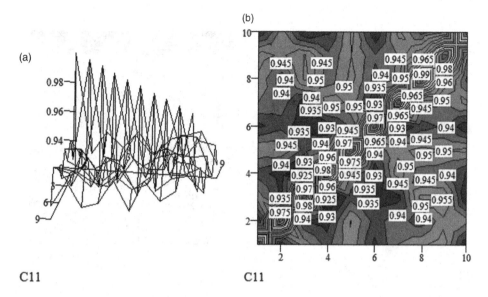

FIGURE 7.18 Correlation matrix consisting of ten sample autocorrelation functions of the traffic Lbl-pkt-5.TCP with min[C11] = 0.921. (a) Surface plot. (b) Contour plot.

Let $\text{II}(s, \gamma)$ be a two-value function, where s is the scale index. Suppose $\text{II}(s, \gamma) = 1$ implies the stationarity of a tested series while $\text{II}(s, \gamma) = 0$ represents the non-stationarity. Then, my future work is to infer an analytical expression of $\text{II}(s, \gamma)$ so that one can research and refine the result of the stationarity test, which is dependent on two variables, namely, the scale and the threshold.

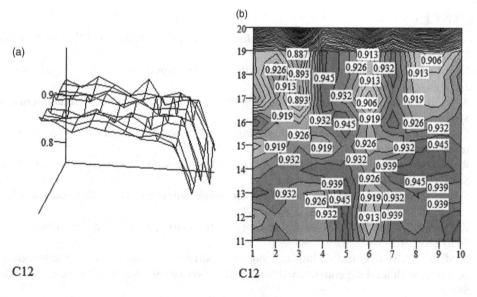

FIGURE 7.19 Correlation matrix consisting of 20 sample autocorrelation functions of the traffic Lbl-pkt-5.TCP with min[C12] = 0.758. (a) Surface plot. (b) Contour plot.

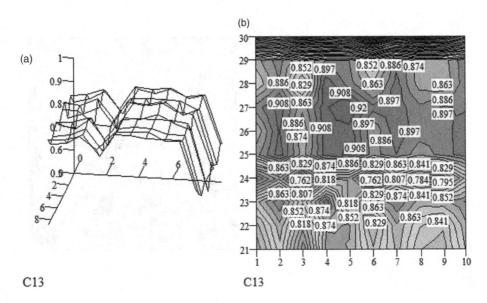

FIGURE 7.20 Correlation matrix consisting of 20 sample autocorrelation functions of the traffic Lbl-pkt-5.TCP with min[C13] = 0.581. (a) Surface plot. (b) Contour plot.

7.5 SUMMARY

In this chapter, we have discussed a method to test the weak stationarity of a single-history LTD traffic of finite length. A particular point presented is that the stationarity of LRD traffic is observation-scale-dependent. An LRD traffic trace may be stationary in one observation scale but non-stationary in a larger observation scale. In the qualitative sense, traffic may be non-stationary in large observation scale but stationary in small observation scale.

REFERENCES

1. A. Papoulis, *Probability, Random Variables, and Stochastic Processes*, McGraw-Hill, New York, 1984.
2. M. B. Priestley, Evolutionary spectra and non-stationary processes, *Journal of Royal Statistical Society Series B*, 27(2) 1965, 204–237.
3. A. I. Al-Shoshan, Time-varying modeling of a nonstationary signal, *Telecommunication Systems*, 12(4) 1999, 389–396.
4. J. S. Bendat, and A. G. Piersol, *Random Data: Analysis and Measurement Procedure*, 3rd ed., John Wiley & Sons, New York, 2000.
5. R. Von Sachs, and M. H. Neumann, A wavelet-based test for stationarity, *Journal of Time Series Analysis*, 21(5) 2000, 597–613.
6. S. Ling, Estimation and testing stationarity for double-autoregressive models, *Journal of Royal Statistical Society Series B*, 66(1) 2004, 63–78.
7. Z. Psaradakis, Blockwise bootstrap testing for stationarity, *Statistics & Probability Letters*, 76(6) 2006, 562–570.
8. P. M. M. Rodrigues, and A. Rubia, A note on testing for nonstationarity in autoregressive processes with level dependent conditional heteroskedasticity, *Statistical Papers*, 49(3) 2008, 581–593.
9. P. Borgnat, and P. Flandrin, Stationarization via surrogates, *Journal of Statistical Mechanics: Theory and Experiment*, 29, 2009, 1–14.

10. B. Tsybakov, and N. D. Georganas, Self-similar processes in communications networks, *IEEE Transactions on Information Theory*, 44(5) 1998, 1713–1725.

11. A. Nogueira, P. Salvador, R. Valadas, and A. Pacheco, Modeling network traffic with multi-fractal behavior, *Telecommunication Systems*, 24(2–3) 2003, 339–362.

12. M. Li, Modeling autocorrelation functions of long-range dependent teletraffic series based on optimal approximation in Hilbert space-a further study, *Applied Mathematical Modelling*, 31(3) 2007, 625–631.

13. W. E. Leland, M. S. Taqqu, W. Willinger, and D. V. Wilson, On the self-similar nature of ethernet traffic (extended version), *IEEE/ACM Transactions on Networking*, 2(1) 1994, 1–15.

14. V. Paxson, and S. Floyd, Wide area traffic: The failure of Poisson modeling, *IEEE/ACM Transactions on Networking*, 3(3) 1995, 226–244.

15. J. Beran, *Statistics for Long-Memory Processes*, Chapman & Hall, New York, 1994.

16. X. Jin, and G. Min, Performance modelling of generalized processor sharing systems with multiple self-similar traffic flows, *Telecommunication Systems*, 38(3–4) 2008, 111–120.

17. B. B. Mandelbrot, Note on the definition and the stationarity of fractional Gaussian noise, *Journal of Hydrology*, 30(4) 1976, 407–409.

18. S. Vaton, A new test of stationarity and its application to teletraffic data, *Proceedings of the 1998 IEEE International Conference on Acoustics, Speech and Signal Processing*, ICASSP'98, vol. 6, Seattle, 1998, 3449–3452.

19. P. Abry, and D. Veitch, Wavelet analysis of long-range dependent traffic, *IEEE Transactions on Information Theory*, 44(1) 1998, 2–15.

20. Y. Zhang, V. Paxson, and S. Shenker, The stationarity of Internet path properties: Routing, loss, and throughput, ACIRI Technical Report, USA, May 2000.

21. M. Li, Y.-Y. Zhang, and W. Zhao, A practical method for weak stationarity test of network traffic with long-range dependence, *International Journal of Mathematics and Computers in Simulation*, 1(4) 2007, 307–311.

22. D. P. Heyman, and T. V. Lakshman, On the relevance of long-range dependence in network traffic, *IEEE/ACM Transactions on Networking*, 7(5) 1999, 629–640.

23. C. Liu, S. V. Wiel, and J. Yang, A nonstationary traffic train model for fine scale inference from coarse scale counts, *IEEE Journal of Selected Areas in Communications*, 21(6) 2003, 895–907.

24. D. Rincón, and S. Sallent, On-line segmentation of non-stationary fractal network traffic with wavelet transforms and log-likelihood-based statistics, *Springer LNCS*, 3375, 2005, 110–123.

25. S. K. Mitra, and J. F. Kaiser (eds.), *Handbook for Digital Signal Processing*, John Wiley & Sons, New York, 1993.

26. K. S. Fu, Ed., *Digital Pattern Recognition*, Springer, New York, 1976.

27. M. Li, An iteration method to adjusting random loading for a laboratory fatigue test, *International Journal of Fatigue*, 27(7) 2005, 73–78.

28. P. Danzig, J. Mogul, V. Paxson, and M. Schwartz, The internet traffic archive, 2000. ftp://ita.ee.lbl.gov/traces/. [dataset].

29. M. Li, *Fractal Teletraffic Modeling and Delay Bounds in Computer Communications*, CRC Press, Boca Raton, 2022.

CHAPTER 8

Record Length Requirement of LRD Traffic

The highlights we contribute in this chapter are mainly in two folds. First, we present a formula to compute the upper bound of the variance of the correlation periodogram measurement of traffic with long-range dependence (LRD) for a given record length T in Theorem 8.1 and for a given value of the Hurst parameter H in Theorem 8.2. Second, we propose two formulas for the computation of the variance upper bound of the correlation periodogram measurement of traffic of fractional Gaussian noise (fGn) type in Corollary 8.1 and the generalized Cauchy (GC) type in Corollary 8.2, respectively. They may constitute a reference guideline of record length requirement of traffic with LRD. In addition, record length requirement for the correlation periodogram measurement of traffic with either the Schuster type or the Bartlett one is studied and the present results about it show that both types of periodograms may be used for the correlation measurement of traffic with a pre-desired variance bound of correlation estimation. Moreover, real traffic in the Internet Archive by the Special Interest Group on Data Communication under the Association for Computing Machinery of US (ACM SIGCOMM) is analyzed in the case study.

8.1 BACKGROUND AND PROBLEM STATEMENTS

Time series with long-range dependence (LRD) or long memory remains interesting to scholars in various fields, ranging from physiology to computer science (see refs [1–32], just to mention a few).

Traffic with LRD plays a role in the Internet, which is an essential infrastructure of modern society, in many aspects, such as network management, performance analysis, computer security, and so forth (see refs [21–46]). In both practice and theory, the statistical model of traffic in the form of either autocorrelation function (ACF) or power spectrum density (PSD) is crucial (Roberts [44], Adas [45], Michiel and Laevens [46]). That is particularly the case when ACF or PSD is considered in the input model of a server or a queuing system in computer networks, see Li and Hwang [47, 48], Livny et al. [49], Sun et al. [50],

DOI: 10.1201/9781003354987-10

Hajek and He [51], letting along those that desire traffic pattern recognition in computer security, see Li [52].

Denote the size of the *i*th packet of traffic at time t_i ($i = 1, 2, \ldots$) by $x(t_i)$. Without the generality losing, we write $x(t_i)$ by $x(i)$ to represent the size of the *i*th packet of traffic from a view of discrete time series. Traffic has two common properties. One is

$$x(i) \geq 0, \tag{8.1}$$

because the traffic is directional for either arrival traffic or departure one (Gibson [53], Pitts and Schormans [54], McDysan [55], Tanenbaum [56], Stalling [57], Yue et al. [58]). The other is

$$0 \leq x(i) \leq P, \tag{8.2}$$

where P is the maximum size defined by communication protocols (Stalling [59]). For instance, the Ethernet protocol forces all packets to have at least a minimum size of 64 bytes and at most the maximum size of 1518 bytes, which is specified by the IEEE standard without technical reason and without considering the Ethernet preamble and header [59]. We assume that $x(i)$ satisfies the Nyquist rule in the research.

The particular property of traffic is LRD. Denote the ACF of traffic by $r(k) = E[x(i)x(i + k)]$. Then, the property of LRD implies

$$r(k) \sim ck^{-\beta}, \quad (0 < \beta < 1, \quad k \to \infty), \tag{8.3}$$

where c is a constant. The above means that $r(k)$ is non-summable over the interval $(0, \infty)$ [60–83]. That is,

$$\sum_{k=0}^{\infty} r(k) = \infty. \tag{8.4}$$

By using the Hurst parameter H, one may express β by

$$\beta = 2 - 2H.$$

Thus, with H, the LRD condition is expressed as

$$0.5 < H < 1. \tag{8.5}$$

The larger the H value, the stronger the LRD or long-range persistence.

Theoretically, $r(k)$ is given by (Priestley [84, 85], Bendat and Piersol [86–88])

$$r(k) = \int_{-\pi}^{\pi} S(\omega) \cos k\omega \, d\omega, \tag{8.6}$$

where $S(\omega)$ is the PSD of $x(i)$ in the form

$$S(\omega) = \sum_{k=-\infty}^{\infty} r(k)\cos k\omega, \quad (-\pi \leq \omega \leq \pi). \tag{8.7}$$

However, the exact form of either $r(k)$ in (8.6) or $S(\omega)$ in (8.7) may never be achieved practically due to the fact that measured record of $x(i)$ in engineering is always of finite length. Assume that the record length of $x(i)$ is T. In terms of practice, given a sample $x(i)$ for $i = 1$, 2, ..., T, therefore, the estimate of $S(\omega)$, denoted by $S_T(\omega)$, is

$$S_T(\omega) = 2\sum_{k=0}^{T-1} r_T(k)\cos k\omega, \quad -\pi \leq \omega \leq \pi, \tag{8.8}$$

where $r_T(k)$ is the sample ACF given by

$$r_T(k) = \frac{1}{T-k}\sum_{i=1}^{T-k} x(i)x(i+k), \quad 0 \leq k \leq T. \tag{8.9}$$

As a matter of fact, the sample ACF $r_T(k)$ is an estimate of true ACF $r(k)$. Hence, comes the issue of spectrum estimation or correlation estimation (Priestley [84, 85, 89, 90], Parzen [91, 92], Jenkins [93], Robinson [94], Thomson [95], Mitra and Kaiser [96], Adnani et al. [97]). In engineering, spectrum/correlation measurement is taken as the synonym of spectrum/correlation estimation in statistics (Blackman and Tukey [98, 99], Tukey [100], Garden [101]). Considering the Wiener–Khinchin theorem, $r_T(k)$ and $S_T(\omega)$ are a pair of Fourier transforms (Jenkins [93], Robinson [94], Mitra and Kaiser [96], Papoulis and Pillai [102], Mortensen [103], Li [104]), $r_T(k)$ keeps its all information in the time domain, including estimation error, as that by $S_T(\omega)$ represented in the frequency domain. Thus, the estimation quality of $r_T(k)$ is same as that of $S_T(\omega)$ and vice versa. For this reason, the meaning of spectrum estimation may usually be the same as that of correlation estimation without the necessity to distinguish between two.

Note that $r_T(k)$ is a random variable (Bendat and Piersol [86–88], Li [104], Zhu [105], Wylton et al. [106]), which was considered to be fluctuation in spectrum measurement by Slepian [107]. Therefore, small value of $\mathrm{Var}[r_T(k)]$ is desired in measurement or estimation of correlation. Tukey used to emphasize the importance of variance analysis in this regard (Blackman and Tukey [98, 99], Tukey [100, 108–110], Brillinger [111, p. 1603]). In fact, $S_T(\omega)$ in (8.8) is the periodogram of Schuster's (Robinson [94], Schuster [112]). Nowadays, the term periodogram is usually in the sense of Bartlett almost as a default [113–116]. In order to brief the concept of periodogram of Bartlett type, let me consider a series of measurement records $(mT, (m+1)T)$ for $m = 0, 1, ..., (M-1)$. In this way, the estimate of the PSD of the mth sample record is in the form

$$S_T(\omega; m) = 2\sum_{k=mT}^{(m+1)T-1} r_T(k; m)\cos k\omega, \quad -\pi \leq \omega \leq \pi, \tag{8.10}$$

where $r_T(k; m)$ is the ACF of the mth sample record given by

$$r_T(k;m) = \frac{1}{T-k} \sum_{i=mT+1}^{(m+1)T-k} x(i)x(i+k), \quad 0 \le k \le T. \tag{8.11}$$

The above $r_T(k; m)$ or $S_T(\omega, m)$ evidently exhibits that the estimate $r_T(k; m)$ or $S_T(\omega, m)$ is a random variable in terms of m.

The periodogram of Bartlett type is in principle about averaged periodogram given by

$$I_T(\omega; M) = \frac{1}{M} \sum_{m=0}^{M-1} S_T(\omega; m), \tag{8.12}$$

or

$$g_T(k; M) = \frac{1}{M} \sum_{m=0}^{M-1} r_T(k; m). \tag{8.13}$$

Again, either $g_T(k; M)$ or $I_T(\omega, M)$ is a random variable. Denote the variances of $g_T(k; M)$ and $I_T(\omega, M)$ by Var[$g_T(k; M)$] and Var[$I_T(\omega, M)$], respectively. The significance of the periodogram of Bartlett type is Var[$g_T(k; M)$] or Var[$I_T(\omega, M)$] may be reduced by average count M (Lomnicki and Zaremba [117], Daniels [118], Dahlhaus [119], Welch [120], Hall and Li [121]). That is,

$$\mathrm{Var}\left[I_T(\omega; M) \right] \sim \frac{1}{M} \mathrm{Var}\left[S_T(\omega; m) \right], \tag{8.14}$$

and

$$\mathrm{Var}\left[g_T(k; M) \right] \sim \frac{1}{M} \mathrm{Var}\left[r_T(k; m) \right]. \tag{8.15}$$

Therefore, one may qualitatively assure of small value of Var[$g_T(k; M)$] or Var[$I_T(\omega, M)$] provided M is large enough even if Var[$r_T(k; m)$] or Var[$S_T(\omega, m)$] is unknown. This is an advantage of the periodogram of Bartlett type in engineering. The sample size T, in the sense of the periodogram of Bartlett type, is usually called the block size (Mitra and Kaiser [96], Adnani et al. [97], Li [104], Harris [122], Schlumberger Electronics Ltd [123]). The above expression (8.14) or (8.15) is actually an application of the central limit theorem to the issue of correlation or spectrum measurement/estimation, referring Lindgren and McElrath [124], Meyer [125], Rice [126] for the central limit theorem. Accordingly, the total record length L is required by

$$L = M \times T + \Delta, \tag{8.16}$$

where

$$\Delta \approx (M \times T)10\% \qquad (8.17)$$

is suggested in measurement techniques by taking the measurement transition into account (see Schlumberger Electronics Ltd. [123], Li [127]). In the subsequent text, we ignore Δ unless otherwise stated and that may not affect the generality of the research discussed in this chapter.

The shortage about (8.16) for the requirement of measurement record length L is lacking the guideline of how to quantitatively demand T for a given degree of $\mathrm{Var}[r_T(k;\ m)]$ or $\mathrm{Var}[S_T(\omega;\ m)]$. In this aspect, as a matter of fact, T, in most cases, can usually be set by roughly demanding that T should be large as can be seen from refs [84, 85, 89–123]. That is particularly true when $x(i)$ is of LRD though there are reports regarding the issue of traffic sampling (Raspall [128], Carela-Español et al. [129], Elbiaze et al. [130], Chabchoub et al. [131], Fernandes et al. [132], Choi and Zhang [133], Duffield [134], He and Hou [135]).

Denote the standard deviation of either $r_T(k;\ m)$ or $S_T(\omega;\ m)$ by $\mathrm{std}(T,\ H)$. Then, $\mathrm{std}(T,H)=\sqrt{\mathrm{Var}\left[r_T(k;m)\right]}$ or $\mathrm{std}(T,H)=\sqrt{\mathrm{Var}\left[S_T(\omega;m)\right]}$. The main contribution of this chapter is in two folds. One is to present an upper bound (bound for short) of $\mathrm{std}(T,\ H)$ for traffic in general for a given sample size T and a given value of H. The other is about two specific forms of $\mathrm{std}(T,\ H)$ when two types of traffic, namely, fGn and the GC process, are taken into account. With $\mathrm{std}(T,\ H)$, we may quantitatively know what the maximum error is for a given sample size T and a given value of H. As $L = T$ is for the periodogram of Schuster type and $L = M \times T$ is for the periodogram of Bartlett type, the total record length L may be quantitatively known if T is determined for a given degree of $\mathrm{Var}(r_T)$ of traffic.

In the rest of chapter, the theoretic results regarding $\mathrm{std}(T,\ H)$ and its specific forms for traffic of both fGn type and GC one are explained in Section 8.2. Section 8.3 expounds practical considerations about $\mathrm{std}(T,\ H)$. In Section 8.4, case study with the real data of traffic is dissertated. Discussions are arranged in Section 8.5, which is followed by summary.

8.2 THEORETICAL RESULTS

Since $r_T(k;\ m)$ is a random series in terms of m, one has

$$r_T(k;m) \neq r_T(k;n)\ \text{for}\ m \neq n. \qquad (8.18)$$

Nevertheless, either $r_T(k;\ m)$ or $r_T(k;\ n)$ is an unbiased estimate of $r(k)$ because

$$E[r_T(k;m)] = E[r_T(k;n)] = E[r(k)]. \qquad (8.19)$$

In fact,

$$E[r_T(k;m)] = \frac{1}{T-k} \sum_{i=mT+1}^{(m+1)T-k} E[x(i)x(i+k)] = \frac{1}{T-k} \sum_{i=mT+1}^{(m+1)T-k} r(k) = E[r(k)]. \qquad (8.20)$$

Therefore, when considering the mean square error of $r_T(k; m)$, which is denoted by $\mathrm{M}^2[r_T(k;m)]$, one has

$$\mathrm{M}^2[r_T(k;m)] = \mathrm{Var}[r_T(k;m)]. \tag{8.21}$$

The above may be used to interpret the importance of variance analysis in correlation measurement, which was stressed by Tukey in many of his reports (see refs [100, 108–110], and also other scholars like Thomson [95].

There are two things worth noting. One is that traffic is Gaussian at large time scales (see [1–3, 28–46, 52, 54–57, 61–83], Scherrer et al. [136], Feldmann et al. [137], Dahl and Willemain [138], Smith [139], Ledesma and Liu [140], López-Ardao et al. [141], Robert and Boudec [142], Paxon [143], Li and Lim [144], Li [145]). The other is that main statistical properties revealed in the early stage of the Internet, such as LRD, are retained even today (Karagiannis et al. [146], Borgnat et al. [147]). Without the generality losing, we consider $r_T(k;m)\big|_{m=0} = r_T(k)$ in what follows for the purpose of simplicity. Based on those, we mention a proposition below.

Proposition 8.1. Let $x(i)$ be a traffic series with LRD. Denote the true ACF of $x(i)$ by $r(k)$. Let T be the number of data points of a sample record. Let $r_T(k)$ be an estimate of $r(k)$. Suppose that $x(i)$ is Gaussian. Then, the variance bound of $r_T(k)$ is given by

$$\mathrm{Var}\big[r_T(k)\big] \le \frac{4}{T} \sum_{i=0}^{\infty} r^2(i). \tag{8.22}$$

Proof. As $\mathrm{Var}(r_T) = \mathrm{E}\{[r_T - \mathrm{E}(r_T)]^2\} = \mathrm{E}\big[(r_T)^2\big] - \mathrm{E}^2(r)$, one has

$$\mathrm{Var}(r_T) = \mathrm{E}\{[r_T - \mathrm{E}(r_T)]^2\} = \mathrm{E}\big[(r_T)^2\big] - r^2. \tag{8.23}$$

Expanding $\mathrm{E}[(r_T)^2]$ yields

$$\mathrm{E}\big[(r_T)^2\big] = \mathrm{E}\left\{\left[\frac{1}{T-k}\sum_{i=1}^{T-k}x(i)x(i+k)\right]^2\right\}$$

$$= \mathrm{E}\left[\frac{1}{(T-k)^2}\sum_{i_1=1}^{T-k}x(i_1)x(i_1+k)\sum_{i_2=1}^{T-k}x(i_2)x(i_2+k)\right]$$

$$= \mathrm{E}\left[\frac{1}{(T-k)^2}\sum_{i_2=1}^{T-k}\sum_{i_1=1}^{T-k}x(i_1)x(i_2)x(i_1+k)x(i_2+k)\right]$$

$$= \frac{1}{(T-k)^2}\sum_{i_2=1}^{T-k}\sum_{i_1=1}^{T-k}\mathrm{E}[x(i_1)x(i_2)x(i_1+k)x(i_2+k)]. \tag{8.24}$$

Therefore,

$$\text{Var}(r_T) = \frac{1}{(T-k)^2} \sum_{i_2=1}^{T-k} \sum_{i_1=1}^{T-k} \text{E}[x(i_1)x(i_2)x(i_1+k)x(i_2+k)] - r^2(k). \tag{8.25}$$

Let

$$\begin{cases} X_1 = x(n_1), \\ X_2 = x(n_2), \\ X_3 = x(n_1+k), \\ X_4 = x(n_2+k). \end{cases} \tag{8.26}$$

Then, we have

$$\text{E}[x(n_1)x(n_2)x(n_1+k)x(n_2+k)] = \text{E}(X_1X_2X_3X_4). \tag{8.27}$$

Since x is Gaussian, following Isserlis [148], there is a joint-normal distribution for four random variables X_1, X_2, X_3 and X_4, in the form

$$\text{E}(X_1X_2X_3X_4) = m_{12}m_{34} + m_{13}m_{24} + m_{14}m_{23}, \tag{8.28}$$

where

$$\begin{cases} m_{12} = \text{E}[x(n_1)x(n_2)] = r(n_2 - n_1), \\ m_{13} = \text{E}[x(n_1)x(n_1+k)] = r(k), \\ m_{14} = \text{E}[x(n_1)x(n_2+k)] = r(n_2 - n_1 + k), \\ m_{23} = \text{E}[x(n_2)x(n_1+k)] = r(n_1 - n_2 + k), \\ m_{24} = \text{E}[x(n_2)x(n_2+k)] = r(k), \\ m_{34} = \text{E}[x(n_1)x(n_2+k)] = r(n_2 - n_1). \end{cases} \tag{8.29}$$

Therefore,

$$\frac{1}{(T-k)^2} \sum_{1}^{T-k} \sum_{1}^{T-k} \text{E}[x(i_1)x(i_2)x(i_1+k)x(i_2+k)] = \frac{1}{(T-k)^2} \sum_{1}^{T-k} \sum_{1}^{T-k} \text{E}(X_1X_2X_3X_4)$$

$$= \frac{1}{(T-k)^2} \sum_{1}^{T-k} \sum_{1}^{T-k} (m_{12}m_{34} + m_{13}m_{24} + m_{14}m_{23})$$

$$= \frac{1}{(T-k)^2} \sum_{1}^{T-k} \sum_{1}^{T-k} r^2(i_2 - i_1) + r^2(k) + r(i_2 - i_1 + k)r(i_1 - i_2 + k)$$

$$= \frac{1}{(T-k)^2} \sum_{1}^{T-k} \sum_{1}^{T-k} r^2(i_2 - i_1) + r(i_2 - i_1 + k)r(i_1 - i_2 + k) + r^2(k). \tag{8.30}$$

The variance is expressed, according to (8.25), by

$$\text{Var}(r_T) = \frac{1}{(T-k)^2} \sum_{1}^{T-k} \sum_{1}^{T-k} r^2(i_2 - i_1) + r(i_2 - i_1 + k)r(i_1 - i_2 + k). \tag{8.31}$$

Replacing $(i_2 - i_1)$ with i in the above expression yields

$$\text{Var}(r_T) = \frac{1}{(T-k)^2} \sum_{i_1=1}^{T-k} \sum_{i=1}^{T-k-i_1} r^2(i) + r(i+k)r(-i+k) = \frac{1}{(T-k)^2} \sum_{i_1=1}^{T-k} \sum_{i=1}^{T-k-i_1} f(i), \tag{8.32}$$

where

$$f(i) = r^2(i) + r(i+k)r(-i+k). \tag{8.33}$$

With (8.32), (8.31) becomes

$$\text{Var}(r_T) = \frac{1}{(T-k)^2} \sum_{i=1}^{T-k} (T-i)f(i) + \frac{1}{(T-k)^2} \sum_{i=-(T-k)}^{-1} (T+i)f(i). \tag{8.34}$$

Since ACF is an even function, we rewrite the above as

$$\text{Var}(r_T) = \frac{2}{(T-k)^2} \sum_{i=1}^{T-k} (T-i)f(i) = \frac{2}{(T-k)^2} \sum_{i=1}^{T-k} (T-i)[r^2(i) + r(i+k)r(-i+k)]$$

$$\leq \frac{2}{T-k} \sum_{i=1}^{T-k} |1 - i/(T-k)| |r^2(i) + r(i+k)r(-i+k)|$$

$$\leq \frac{2}{T-k} \sum_{i=1}^{T-k} |r^2(i) + r(i+k)r(-i+k)| \approx \frac{2}{T} \sum_{i=1}^{T-k} |r^2(i) + r(i+k)r(-i+k)|. \tag{8.35}$$

In the above expression, $\text{Var}(r_T)$ is a function in terms of the lag k. Variance fluctuation is a topic in statistics, see Herbst [149], but the possible guideline of record length requirement of traffic desires that a bound of $\text{Var}(r_T)$ may be irrelevant of k from a view of the convenience use in engineering. In doing so, we replace $\sum_{i=1}^{T-k}$ with $\sum_{i=1}^{T}$. Thus, we have an inequality given by

$$\text{Var}(r_T) \leq \frac{2}{T} \sum_{i=1}^{T} |r^2(i) + r(i+k)r(-i+k)|. \tag{8.36}$$

Applying the Schwarz inequality to the right side of the above produces

$$\text{Var}(r_T) \leq \frac{2}{T} \sum_{i=1}^{T} |r(i+k)r(-i+k)| \leq \frac{2}{T} \sqrt{\sum_{i=1}^{T} r^2(i+k) \sum_{i=1}^{T} r^2(-i+k)}. \qquad (8.37)$$

In the above, there is still an argument k in the bound of $\text{Var}(r_T)$. Fortunately, ACF of traffic at large scales is monotonously decreasing, see (8.3). Thus, we have

$$\sum_{i=1}^{T} r^2(i+k) \leq \sum_{i=1}^{\infty} r^2(i),$$

$$\sum_{i=1}^{T} r^2(-i+k) \leq \sum_{i=1}^{\infty} r^2(i). \qquad (8.38)$$

Therefore, we have a bound of $\text{Var}(r_T)$, which is irrelevant of k in the form

$$\text{Var}(r_T) \leq \frac{4}{T} \sum_{i=0}^{\infty} r^2(i). \qquad (8.39)$$

Thus, Proposition 8.1 results. □

The analog of the above proposition to the continuous function $x(t)$ refers to my previous work [104, Appendix A] or Laning and Battin [150]. It is valid and useful for a time series that satisfies the condition $\sum_{i=0}^{\infty} r^2(i) < \infty$. For traffic, however, due to LRD, Proposition 8.1 is not enough unfortunately. In fact, for LRD traffic, $r(k) \sim ck^{-\beta}$ for $0 < \beta < 1$, or equivalently, $r(k) \sim ck^{2H-2}$ for $0.5 < H < 1$. Consequently, for $0 < \beta < 0.5$ or $0.75 < H < 1$, we have

$$\frac{4}{T} \sum_{i=1}^{\infty} r^2(i) = \infty. \qquad (8.40)$$

It is obvious that ∞ is a trivial bound of error.

The expression (8.40) exhibits the difficulty resulted from LRD with respect to the derivation of a bound of $\text{Var}(r_T)$. Thus, in order to find a non-trivial bound of $\text{Var}(r_T)$, I have to refine the proposition as follows.

Theorem 8.1. Let $x(i)$ be a traffic series with LRD. Denote the true ACF of $x(i)$ by $r(k)$. Let T be the number of data points of a sample record. Let $r_T(k)$ be an estimate of $r(k)$. Suppose that $x(i)$ is Gaussian. Then, the variance bound of $r_T(k)$ is given by

$$\text{Var}[r_T(k)] \leq \frac{4}{T} \sum_{i=1}^{T} r^2(i). \qquad (8.41)$$

Proof. Note that an ACF is an even function. Besides, it is monotonously decreasing in terms of k owing to $r(k) \sim ck^{-\beta}$. Thus, I rewrite (8.38) as

$$\sum_{i=1}^{T} r^2(i+k) \le \sum_{i=1}^{T} r^2(i),$$
$$\sum_{i=1}^{T} r^2(-i+k) \le \sum_{i=1}^{T} r^2(i). \qquad (8.42)$$

In that way, (8.36) may be refined by

$$\mathrm{Var}(r_T) \le \frac{2}{T} \sum_{i=1}^{T} \left| r^2(i) + r(i+k)r(-i+k) \right| \le \frac{2}{T} \sum_{i=1}^{T} \left| r^2(i) + r^2(i) \right| = \frac{4}{T} \sum_{i=1}^{T} r^2(i). \qquad (8.43)$$

This completes the proof. □

Note 8.1. The bound in Theorem 8.1 is tighter than that in Proposition 8.1. □

Note 8.2. Theorem 8.1 exhibits that the bound of $\mathrm{Var}(r_T)$ expressed by (8.41) is a function with two arguments. One is the record length T and the other is the Hurst parameter H, irrelevant of the lag k. □

Theorem 8.2. Let $x(i)$ be a traffic series with LRD. Denote the true ACF of $x(i)$ by $r(k)$. Let T be the number of data points of a sample record. Let $r_T(k)$ be an estimate of $r(k)$. Suppose that $x(i)$ is Gaussian. Then, the bound of variance of $r_T(k)$ is given by

$$\mathrm{Var}(r_T) \le \frac{4}{T} \sum_{i=1}^{T} r^2(i) \sim \frac{4}{T} \sum_{i=1}^{T} c^2 i^{2(2H-2)} = \frac{4c^2}{T} \sum_{i=1}^{T} i^{4H-4}, \quad 0.5 < H < 1. \qquad (8.44)$$

where the value of the constant c relies on a specific form of ACF.

Proof. For LRD traffic, we have $r(k) \sim ck^{2H-2}$ for $0.5 < H < 1$. Replacing that into the right side on (8.41) results in Theorem 8.2. □

FGn is a commonly used fractal random function in various fields [1–4], including traffic (see Willinger and Paxson [29], Sousa-Vieira et al. [31], Adas [45], Michiel and Laevens [46], Pitts and Schormans [54], McDysan [55], Stalling [57], Beran [60], Leland et al. [61], Paxson and Floyd [62], Beran et al. [63], Tsybakov and Georganas [64], Abry and Veitch [65], Lee and Fapojuwo [71], Li [145], Karagiannis et al. [146]). From Theorem 8.2, we have the corollary below.

Corollary 8.1. Let $x(i)$ be a traffic series with LRD. Denote the ACF of fGn by $r(k)$. Let T be the number of data points of a sample record. Let $r_T(k)$ be an estimate of $r(k)$. Then, for $0.5 < H < 1$, the variance bound of $r_T(k)$ is given by

$$\mathrm{Var}(r_T) \le \frac{4}{T} \sum_{i=1}^{T} r^2(i) \approx \frac{4\sigma^4 H^2 (2H-1)^2}{T} \sum_{i=1}^{T} i^{4H-4}, \qquad (8.45)$$

where $\sigma^2 = (H\pi)^{-1}\Gamma(1-2H)\cos(H\pi)$.

Proof. Following Mandelbrot [151], Li and Chi [152], Li [153], for the ACF of fGn, one has

$$c = \sigma^2 H(2H-1). \tag{8.46}$$

Substituting it to the right side of (8.44) yields (8.45). The proof finishes. $\square$

A recent model of traffic is the GC process (Li [153], Li and Lim [144, 154], Lim and Li [155], Lim and Teo [156]). When the traffic of GC type is considered, the variance bound of $r_T(k)$ may be obtained by the corollary below.

Corollary 8.2. Let $x(i)$ be a traffic series with LRD. Denote the ACF of GC process by $r(k)$. Let T be the number of data points of a sample record. Let $r_T(k)$ be an estimate of $r(k)$. Then, for $0.5 < H < 1$, the variance bound of $r_T(k)$ is given by

$$\text{Var}(r_T) \leq \frac{4}{T}\sum_{i=1}^{T} r^2(i) \approx \frac{4}{T}\sum_{i=1}^{T} i^{4H-4}. \tag{8.47}$$

Proof. Note that the GC process is Gaussian (Lim and Li [155]). For the ACF of GC process, one has

$$c = 1. \tag{8.48}$$

Replacing 1 on the right side of (8.44) by that in (8.48) produces (8.47). Hence, the proof completes. $\square$

8.3 PRACTICAL CONSIDERATIONS

Denote by std(T, H) the standard deviation in general for the ACF estimation of traffic of either GC type or fGn one. Denote the standard deviation of the ACF estimation of traffic of fGn type and GC type by std_fGn(T, H) and std_Cauchy(T, H), respectively. Then, for fGn type traffic, we have

$$\text{std\_fGn}(T, H) \leq \sqrt{\frac{4\sigma^4 H^2 (2H-1)^2}{T}\sum_{i=1}^{T} i^{4H-4}}$$

$$= \frac{2\Gamma(1-2H)\cos(H\pi)|2H-1|}{\pi}\sqrt{\frac{1}{T}\sum_{i=1}^{T}(i)^{4H-4}}. \tag{8.49}$$

On the other hand, for GC type, we have

$$\text{std\_Cauchy}(T, H) \leq \sqrt{\frac{4}{T}\sum_{i=1}^{T} i^{4H-4}}. \tag{8.50}$$

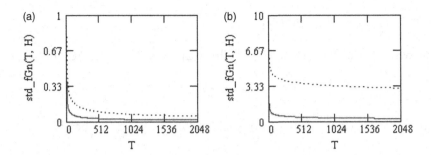

FIGURE 8.1 Illustrations of std_fGn(T, H) for $T = 1, 2, \ldots, 2048$. (a) Solid line: $H = 0.65$. Dot line: $H = 0.75$. (b) Solid line: $H = 0.85$. Dot line: $H = 0.95$.

Figures 8.1 and 8.2 illustrate std_fGn(T, H) and std_Cauchy(T, H) with fixed Hs, respectively.

Note 8.3 (Length effect). From Figures 8.1 and 8.2, we see that std(T, H) is a decreasing function in terms of T. This is natural but significance is that they describe std(T, H) quantitatively. $\square$

Figure 8.3 shows std_fGn(T, H) and std_Cauchy(T, H) with fixed Ts.

Note 8.4 (Hurst effect). Figure 8.3 exhibits that std(T, H) is an increasing function with respect to $H \in (0.5, 1)$. This is a novel phenomenon of time series with LRD, which is revealed in this research, in addition to LRD traffic, in the aspect of correlation/spectrum measurement. To be precise, the larger the H, or the stronger the LRD, the larger the error of correlation measurement if the record length T is fixed. $\square$

An interesting phenomenon in this research is that std_Cauchy(T, H) ≤ 2 is irrelevant of the record length T in the extreme case of $H = 1$. In fact,

$$\text{std\_Cauchy}(T, 1) \leq \sqrt{\frac{4}{T} \sum_{i=1}^{T} i^{4-4}} = 2. \tag{8.51}$$

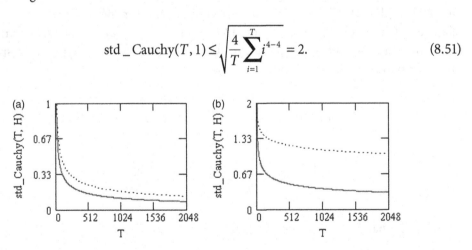

FIGURE 8.2 Illustrations of std_Cauchy(T, H) for $T = 1, 2, \ldots, 2048$. (a) Solid line: $H = 0.65$. Dot line: $H = 0.75$. (b) Solid line: $H = 0.85$. Dot line: $H = 0.95$.

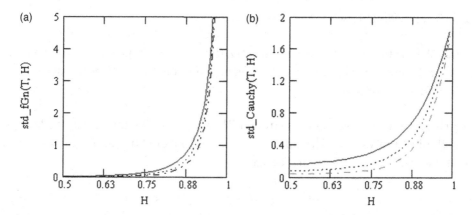

FIGURE 8.3 Plots of std_fGn(T, H) and std_Cauchy(T, H) with fixed Ts. (a) std_fGn(T, H) for $H = 0.5, ..., 1$. Solid line: $T = 256$. Dot line: $T = 1024$. Dadot line: $T = 4096$. (b). Std_Cauchy(T, H) for $H = 0.5, ..., 1$. Solid line: $T = 256$. Dot line: $T = 1024$. Dadot line: $T = 4096$.

For std_fGn(T, H) in the extreme case of $H = 1$, on the other side, we consider its limit for $H \rightarrow 1$ owing to the divergence of gamma function for $H = 1$. Thus, we have

$$\lim_{H \rightarrow 1} std\_fGn(T, H) \leq \lim_{H \rightarrow 1} -\frac{4\Gamma(1-2H)}{\pi}. \tag{8.52}$$

In both cases of (8.51) and (8.52), the bounds of std(T, H) do not decrease as T increases. Actually, they are irrelevant of T when $H \rightarrow 1$. This is consistent with what is described in Note 8.4.

Note that the Hurst effect is quite serious when H is large in addition to the extreme case for $H \rightarrow 1$. For instance, when $H = 0.51$ (weak LRD) and $T = 128, 259, 512$, we have

$$std\_fGn(128, 0.51) \leq 2.314 \times 10^{-3},$$

$$std\_fGn(256, 0.51) \leq 1.639 \times 10^{-3},$$

$$std\_fGn(512, 0.51) \leq 1.116 \times 10^{-3}, \tag{8.53}$$

which implies that the correlation measurement based on the periodogram of Schuster type may be satisfactory at the degree of the bound of std_fGn(T, 0.51) being in the magnitude of 10^{-3} when LRD is weak as 3 error bounds on the right side of (8.53) are in the same order of magnitude if T changes from 128 to 512 and when traffic of fGn type is considered. For traffic of GC type, things may be slightly different, see below.

$$std\_Cauchy(128, 0.51) \leq 0.229,$$

$$std\_Cacuhy(256, 0.51) \leq 0.162,$$

$$std\_Cauchy(512, 0.51) \leq 0.115, \tag{8.54}$$

Thus, we have the following note.

Note 8.5 (Weak Hurst effect). For weak LRD, the correlation measurement based on the periodogram of Schuster type may be satisfactory provided $T > 512$ at the degree of std_fGn(512, 0.51) $\leq 1.116 \times 10^{-3}$. □

Unfortunately, traffic is usually of strong LRD with $H > 0.95$ in most cases (Paxson and Floyd [62], Abry and Veitch [65], Feldmann et al. [137], Li and Lim [144]). Suppose $H = 0.95$. Then, in the case of fGn type traffic, we have

$$\text{std\_Cauchy}\left(2^{26},\ 0.95\right) \leq 1.103,$$

$$\text{std\_fGn}(2^{27},\ 0.95) \leq 1.029. \tag{8.55}$$

On the other side, for the GC type traffic, we have

$$\text{std\_Cauchy}(2^{26},\ 0.95) \leq 0.369,$$

$$\text{std\_Cauchy}(2^{27},\ 0.95) \leq 0.344. \tag{8.56}$$

Note that real traffic data provided by Bellcore Morristown Research and Engineering (BC) (Danzig et al. [157]) are all with $L = 10^6$. Thus, in the case of $H = 0.95$ and $T = 10^6$, and if the correlation periodogram of Schuster type is considered, we have

$$\text{std\_fGn}\left(10^6,\ 0.95\right) \leq 1.680,$$

$$\text{std\_Cauchy}(10^6,\ 0.95) \leq 0.562. \tag{8.57}$$

In addition, real wide-area traffic data provided by Digital Equipment Corporation (DEC) and the Lawrence Berkeley Laboratory (LBL) are with the maximum L being 5.7×10^6 (Danzig et al. [157]). In the case of $H = 0.95$ and $T = 5.7 \times 10^6$, and when we consider the correlation periodogram of Schuster type, however, one has

$$\text{std\_fGn}(5.7 \times 10^6,\ 0.95) \leq 1.412,$$

$$\text{std\_Cauchy}(5.7 \times 10^6,\ 0.95) \leq 0.472, \tag{8.58}$$

which are in the same order of magnitude with those in (8.57).

In passing, we note that the real traffic data provided by BC, DEC, and LBL in work of Danzig et al. [157] play a role in teletraffic engineering from the pioneering work of traffic analysis to the current (Leland et al. [61], Paxson and Floyd [62], Beran et al. [63], Abry and Veitch [65], Scherrer et al. [136], Feldmann et al. [137], Paxson [143], Li and Lim [144], Li [145], Borgnat et al. [147], Li [158–160]).

The previous exhibits that $H = 1$ represents the largest error bound of spectrum measurement without relation to the record length T. Thus, we obtain the following estimation representation with respect to spectrum measurement when traffic is considered to be the GC type without the generality losing.

$$\text{Var}\left[I_T(\omega; M)\right] \le \frac{4}{M},$$

(8.59)

and

$$\text{Var}\left[g_T(k; M)\right] \le \frac{4}{M}.$$

(8.60)

In the general case of $0.5 < H < 1$, we have

$$\text{Var}\left[g_T(k; M)\right] \le \frac{1}{M}\left[\text{std\_fGn}(T, H)\right]^2 \le \frac{4\sigma^4 H^2 (2H-1)^2}{T}\sum_{i=1}^{T} i^{4H-4}$$

$$= \frac{\left[2\Gamma(1-2H)\cos(H\pi)|2H-1|\right]^2}{M\pi^2}\frac{1}{T}\sum_{i=1}^{T}(i)^{4H-4},$$

(8.61)

and

$$\text{Var}\left[g_T(k; M)\right] \le \frac{1}{M}\left[\text{std\_Cauchy}(T, H)\right]^2 \le \frac{4}{MT}\sum_{i=1}^{T} i^{4H-4}.$$

(8.62)

The expression (8.61) or (8.62) implies a quantitative guideline of measurement record of traffic when using the periodogram of Bartlett type. By considering (8.16), that is, $L \approx M \times T$, error bound of correlation measurement can be predetermined when one is aware of T and M in advance. Large T is, of course, helpful for reducing error of spectrum measurement. If T is large enough, a correlation measurement with small error is achievable with $M = 1$, that is, with the periodogram of Schuster type.

In the end of this section, we note that periodogram is used in time series with LRD, including traffic, for either modeling or parameter estimation (see Li and Lim [144], Li [145], Li et al. [160], Robinson [161], Chan et al. [162], Lobato and Robinson [163], Sela and Hurvich [164], Reisen et al. [165], Sergides and Paparoditis [166], Velasco [167], Raymond et al. [168], Henry [169], Robinson and Marinucci [170], Moulines and Soulier [171], Kokoszka and Mikosch [172], Lobato [173], Bhansali [174], Reisen [175], Hurvich and Beltrao [176], Li and Zhao [177, 178]).

8.4 CASE STUDY

The real traffic data mentioned in the previous section are listed in Table 8.1 and Table 8.2. We select pAug.TL in our case study to demonstrate the effect of its correlation periodogram of Schuster type for different record length T and its correlation periodogram of Bartlett type for different average count M.

TABLE 8.1   Four Real Traffic Traces on the Ethernet [157]

Trace Name	Starting Time of Measurement	Duration	Number of Packets
pAug.TL	11:25AM, 29Aug89	52 minutes	1 million
pOct.TL	11:00AM, 05Oct89	29 minutes	1 million
OctExt.TL	11:46PM, 03Oct89	34.111 h	1 million
OctExt4.TL	2:37PM, 10Oct89	21.095 h	1 million

TABLE 8.2   Six Wide-Area TCP Traces [158, 159]

Trace Name	Date of Measurement	Duration	Number of Packets
dec-pkt-1.TCP	08Mar95	10 PM–11 PM	3.3 million
dec-pkt-2.TCP	09Mar95	2 AM–3 AM	3.9 million
dec-pkt-3.TCP	09Mar95	10 AM–11 AM	4.3 million
dec-pkt-4.TCP	09Mar95	2 PM–3 PM	5.7 million
Lbl-pkt-4.TCP	21Jan94	2 AM–3 AM	1.3 million
Lbl-pkt-5.TCP	28Jan94	2 AM–3 AM	1.3 million

8.4.1 Correlation Periodogram of Schuster Type of pAug.TL

The packet size of the first 4000 data points of pAug.TL is shown in Figure 8.4.

We now investigate the correlation periodogram of Schuster type. By Schuster type we mean $M = 1$. Since the GC process of traffic is more flexible and reasonable (Li and Lim [144], Lim and Li [155], Lim and Teo [156], Gneiting and Schlather [179]), we use std_Cauchy(T, H) in the case study.

The Hurst parameter of pAug.TL is 0.957 (Li and Lim [144, Table 8.1], Li [145, Table 8.2]). Figure 8.5 gives the plot of sample ACF with $T = 2048$. Figures 8.6 (a)–10 (a) show the sample ACFs with $T = 4096, 8192, 16384, 32768, 65536$, respectively. In order to compare the effect of reducing error, we plot the first 1024 points of the sample ACFs with $T = 4096$, 8192, 16384, 32768, 65536, respectively, in Figures 8.6 (b)–8.10 (b), where it is clear that estimation error is reduced when T increases. It is considerably reduced when T is up to 32,768 as can be seen from Figure 8.9.

The effect of reducing the fluctuation of sample ACFs in the sense of correlation periodogram of Schuster type can also be observed in Figure 8.11 by the log-log plots of sample ACFs. Again, we see that the estimation error is reduced significantly if $T = 32768$, see Figure 8.11(e).

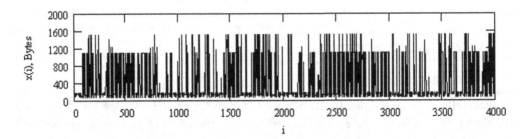

FIGURE 8.4   Illustration of real traffic trace of pAug.TL.

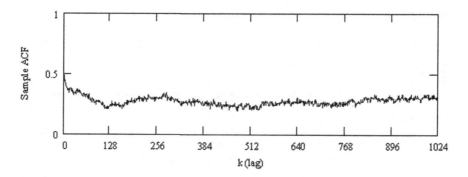

FIGURE 8.5 Sample correlation periodogram of Schuster type of pAug.TL with $T = 2048$. std_ Cauchy(2048, 0.957) ≤ 1.14.

Note 8.6. The spectrum/correlation measurement based on the periodogram of Schuster type may be useful so long as T is large enough. For LRD traffic, $T = 32768$ may be an economic selection as a reference value at the degree of std_Cauchy(32768, 0.957) ≤ 0.899. □

8.4.2 Correlation Periodogram of Bartlett Type of Traffic pAug.TL

Consider $T = 2048$. Then, observe the correlation periodogram of Bartlett type of traffic pAug.TL for $M = 1, 8, 16, 32, 40, 50, 60, 70, 80, 90, 100, 110$, respectively in Figure 8.12. By eye, we see that it is quite satisfactory if $M = 16$, see Figure 8.12(c). In log-log plots,

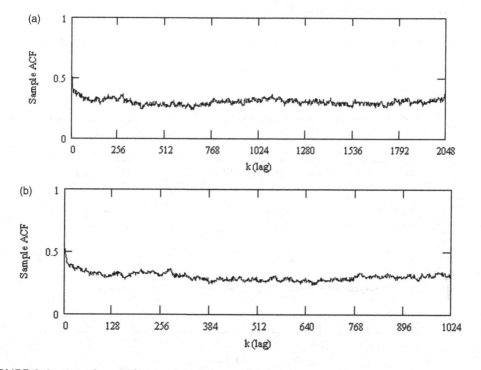

FIGURE 8.6 Sample correlation periodogram of Schuster type of pAug.TL with $T = 4096$. std_ Cauchy(4096, 0.957) ≤ 1.075. (a) Viewing the ACF estimate with $T = 4096$. (b). Viewing the ACF estimate with the corresponding first 1024 points.

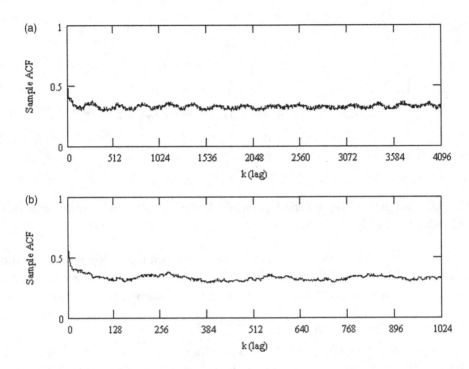

FIGURE 8.7 Sample correlation periodogram of Schuster type of pAug.TL with $T = 8192$. std_ Cauchy(8192, 0.957) ≤ 1.013. (a) Viewing the ACF estimate with $T = 8192$. (b) Viewing the ACF estimate with the corresponding first 1024 points.

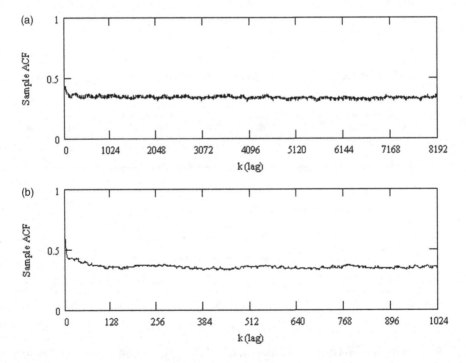

FIGURE 8.8 Sample correlation periodogram of Schuster type of pAug.TL with $T = 16384$. std_ Cauchy(16384, 0.957) ≤ 0.954. (a) Viewing the ACF estimate with $T = 16384$. (b) Viewing the ACF estimate with the corresponding first 1024 points.

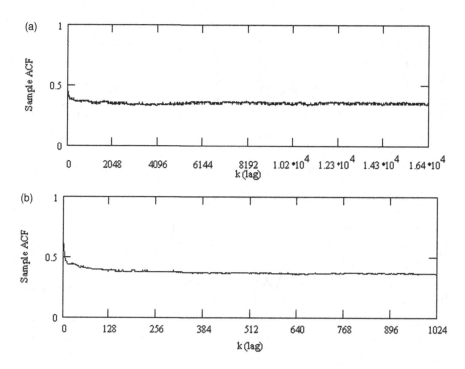

FIGURE 8.9   Sample correlation periodogram of Schuster type of pAug.TL with $T = 32768$. std_ Cauchy(32768, 0.957) ≤ 0.899. (a) Viewing the ACF estimate with $T = 32768$. (b) Viewing the ACF estimate with the corresponding first 1024 points.

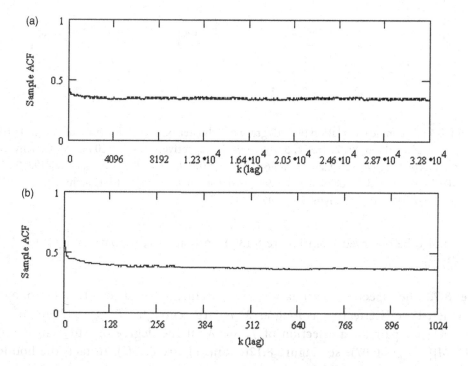

FIGURE 8.10   Sample correlation periodogram of Schuster type of pAug.TL with $T = 65536$. std_ Cauchy(65536, 0.957) ≤ 0.847. (a) Viewing the ACF estimate with $T = 65536$. (b) Viewing the ACF estimate with the corresponding first 1024 points.

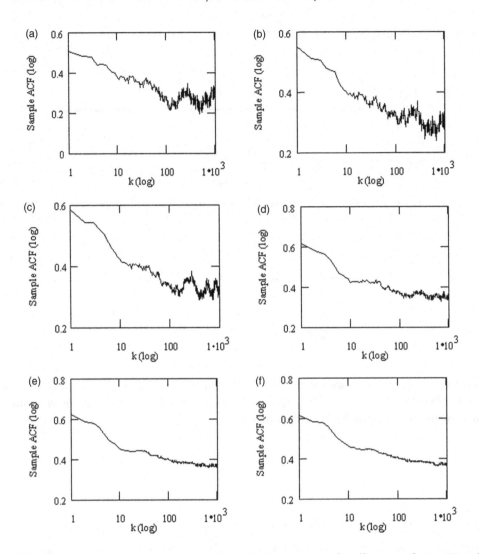

FIGURE 8.11 Sample correlation periodogram of Schuster type of traffic trace of pAug.TL in log-log with T = 2048, 4096, 8192, 16384, 32768, 65536, respectively. (a) T = 2048. std_Cauchy(2048, 0.957) ≤ 1.14. (b) T = 4096. std_Cauchy(4096, 0.957) ≤ 1.075. (c) T = 8192. std_Cauchy(8192, 0.957) ≤ 1.013. (d) T = 16384. std_Cauchy(16384, 0.957) ≤ 0.954. (e) T = 32768. std_Cauchy(32768, 0.957) ≤ 0.899. (f) T = 65536. std_Cauchy(65536, 0.957) ≤ 0.847.

that can also be observed from Figure 8.13(c). In that case, we have {std_Cauchy(2048, 0.957)/M}$|_{M=16}$ ≤ 0.071.

Note 8.7. The spectrum/correlation measurement based on the periodogram of Bartlett type is effective as long as M is large enough. (T, M) = (2048, 16) may be a reference pair for a selection of T and M at the degree of {std_Cauchy(2048, 0.957)/M}$|_{M=16}$ ≤ 0.071, see Figure 8.12(c) and Figure 8.13(c). In fact, the bound of std_Cauchy(2048, 0.957)/M for M = 16 is in the same order of magnitude when M changes from 16 to 110. □

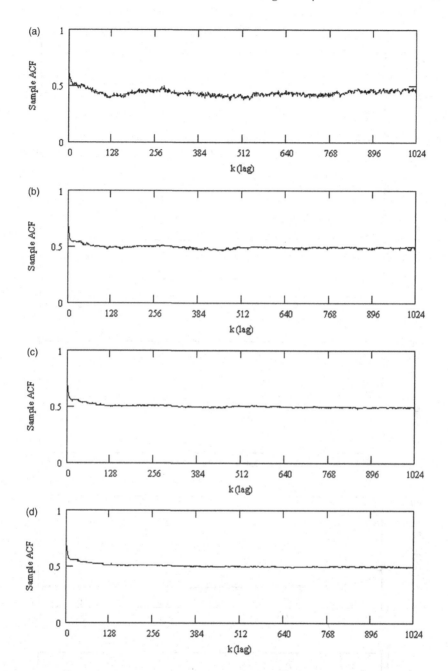

FIGURE 8.12 Sample correlation periodogram of Bartlett type of traffic pAug.TL. (a) $M = 1$. Error bound: std_Cauchy(2048, 0.957)/$M \leq 1.14$. (b) $M = 8$ ($L = 16{,}384$). Error bound: std_Cauchy(2048, 0.957)/$M \leq 0.143$. (c) $M = 16$ ($L = 32{,}768$). std_Cauchy(2048, 0.957)/$M \leq 0.071$. (d) $M = 32$ ($L = 65{,}536$). std_Cauchy(2048, 0.957)/$M \leq 0.036$. (e) $M = 40$ ($L = 81{,}920$). std_Cauchy(2048, 0.957)/$M \leq 0.023$. (f) $M = 50$ ($L = 102{,}400$). std_Cauchy(2048, 0.957)/$M \leq 0.023$. (g) $M = 60$ ($L = 122{,}880$). std_Cauchy(2048, 0.957)/$M \leq 0.019$. (h) $M = 70$ ($L = 143{,}360$). std_Cauchy(2048, 0.957)/$M \leq 0.016$. (i) $M = 80$ ($L = 163{,}840$). std_Cauchy(2048, 0.957)/$M \leq 0.014$. (j) $M = 90$ ($L = 184{,}320$). std_Cauchy(2048, 0.957)/$M \leq 0.013$. (k) $M = 100$ ($L = 204{,}800$). std_Cauchy(2048, 0.957)/$M \leq 0.011$. (l) $M = 110$ ($L = 225{,}280$). std_Cauchy(2048, 0.957)/$M \leq 0.010$. *(Continued)*

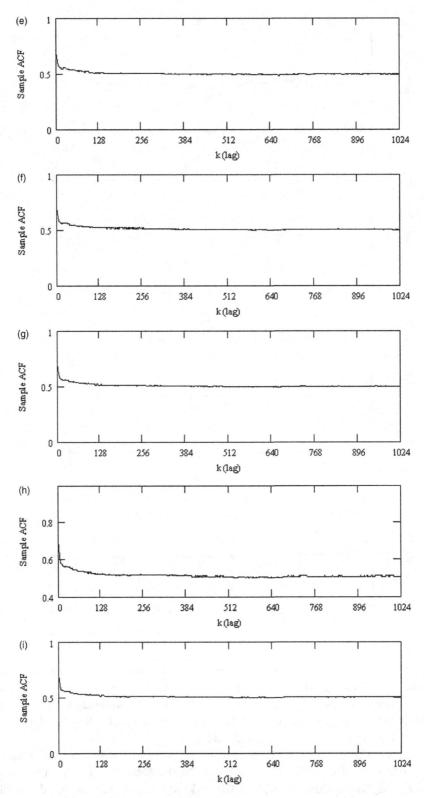

FIGURE 8.12 *(Continued)*

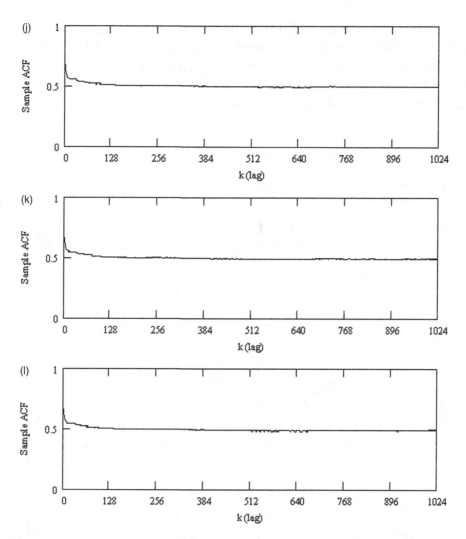

FIGURE 8.12 *(Continued)*

8.5 DISCUSSIONS

The previous case study exhibits that either the periodogram of Bartlett type or Schuster one is appropriate for the correlation/spectrum measurement of traffic with LRD. In addition, the real traffic data in the Internet Traffic Archive provided by ACM SIGCOMM are long enough for good correlation/spectrum periodogram estimation of either Schuster type or Bartlett one.

By comparing Figure 8.11(e) to Figure 8.13(j), by eye, we see they are similar in error regarding correlation measurement. However, the total record length for Figure 8.11(e) is $T = L = 32,768$ but for Figure 8.13(j) is $L = M \times T = 184,320$. The former is much shorter than the latter. By computation, however, the error bound for Figure 8.11(e) is std_Cauchy(32768,

0.957) $\leq$ 0.899 but for Figure 8.13(j) is {std_Cauchy(2048, 0.957)/M}$|_{M = 90}$ $\leq$ 0.013. The latter is much smaller than the former. Therefore, the question of which type of periodogram, Bartlett type or Schuster one, may be more preferred in the aspect of record length requirement for the purpose of correlation measurement of traffic appears challenging. The answer to it will be considered in my future work.

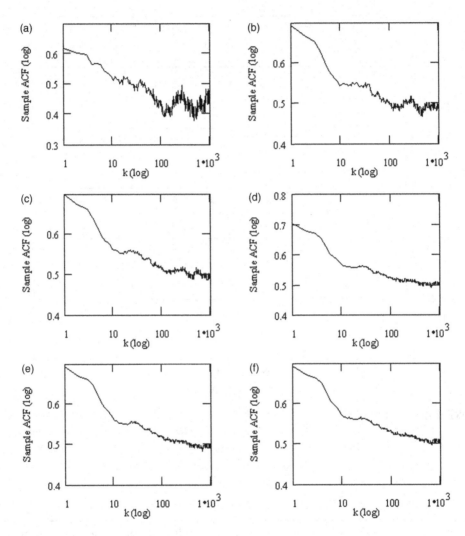

FIGURE 8.13 Sample correlation periodogram of Bartlett type of traffic pAug.TL in log-log. (a) $M = 1$. Error bound: std_Cauchy(2048, 0.957)/M $\leq$ 1.14. (b) $M = 8$ ($L = 16,384$). Error bound: std_Cauchy(2048, 0.957)/M $\leq$ 0.143. (c) $M = 16$ ($L = 32,768$). std_Cauchy(2048, 0.957)/M $\leq$ 0.071. (d) $M = 32$ ($L = 65536$). std_Cauchy(2048, 0.957)/$M \leq 0.036$. (e) $M = 40$ ($L = 81,920$). std_Cauchy(2048, 0.957)/M $\leq$ 0.023. (f) $M = 50$ ($L = 102,400$). std_Cauchy(2048, 0.957)/M $\leq$ 0.023. (g) $M = 60$ ($L = 122,880$). std_Cauchy(2048, 0.957)/M $\leq$ 0.019. (h) $M = 70$ ($L = 143,360$). std_Cauchy(2048, 0.957)/M $\leq$ 0.016. (i) $M = 80$ ($L = 163,840$). std_Cauchy(2048, 0.957)/M $\leq$ 0.014. (j) $M = 90$ ($L = 184,320$). std_Cauchy(2048, 0.957)/M $\leq$ 0.013. (k) $M = 100$ ($L = 204,800$). std_Cauchy(2048, 0.957)/$M \leq 0.011$. (l) $M = 110$ ($L = 225,280$). std_Cauchy(2048, 0.957)/$M \leq 0.010$. *(Continued)*

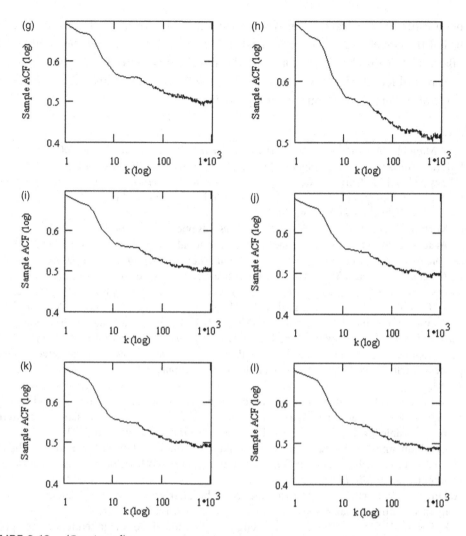

FIGURE 8.13 *(Continued)*

8.6 SUMMARY

In this chapter, we have proposed a representation of the error bound of correlation measurement of LRD traffic for a given record length and given value of the Hurst parameter. In addition, we have presented two expressions for the calculation of variance bounds of correlation measurement of LRD traffic. One is for the traffic of fGn type and the other is for the GC type traffic. They may serve as a guideline tool for requiring the record length of LRD traffic for its correlation measurement in the sense of periodogram estimation for either the Schuster type or the Bartlett one. The results may also be suitable for time series with LRD in general in addition to traffic. Since a record length required by using correlation measurement may also satisfy with other statistical models or parameters, such as probability density function (Bendat and Piersol [87, p. 313]), for LRD traffic, therefore, the record length requirement discussed in this chapter also satisfies with the Hurst parameter estimation or fractal dimension estimation. Moreover, as a byproduct, we have shown that the correlation periodogram estimation of traffic for either the Schuster type or the Bartlett type may be satisfactory with desired degree of error bound. Note that measurement record length is an issue of traffic at large time scales. Thus, the present results relate to H of GC process without relation to its D.

REFERENCES

1. B. B. Mandelbrot, *Gaussian Self-Affinity and Fractals*, Springer, New York, 2001.
2. J. Beran, *Statistics for Long-Memory Processes*, Chapman & Hall, New York, 1994.
3. J. Levy-Vehel, E. Lutton, and C. Tricot (eds.), *Fractals in Engineering*, Springer, Berlin, 1997.
4. P. Doukhan, G. Oppenheim, and M. S. Taqqu (eds.), *Theory and Applications of Long-Range Dependence*, Birkhauser, Boston, 2002.
5. G. Korvin, *Fractal Models in the Earth Sciences*, Elsevier, The Netherlands, 1992.
6. A. Ayache, F. Roueff, and Y. M. Xiao, Linear fractional stable sheets: Wavelet expansion and sample path properties, *Stochastic Processes and Their Applications*, 119(4) 2009, 1168–1197.
7. A. Ayache, D. Wu, and Y. M. Xiao, Joint continuity of the local times of fractional Brownian sheets, *Annales de Institut Henri Poincare – Probabilities et Statistiques*, 44(3) 2008, 727–748.
8. A. L. Goldberger, L. A. N. Amaral, J. M. Hausdorff, P. Ivanov Ch., C.-K. Peng, and H. E. Stanley, Fractal dynamics in physiology: Alterations with disease and aging, *Proceedings of the National Academy of Sciences of the United States of America*, 99(Suppl 1) 2002, 2466–2472.
9. B. Podobnik, D. F. Fu, H. E. Stanley, and P. Ivanov Ch., Power-law autocorrelated stochastic processes with long-range cross-correlations, *The European Physical Journal B*, 56(1) 2007, 47–52.
10. Y. Avraham, and M. Pinchas, Two novel one-way delay clock skew estimators and their performances for the fractional Gaussian noise/generalized fractional Gaussian noise environment applicable for the IEEE 1588v2 (PTP) case, *Frontiers in Physics*, 10, 2022, 867861.
11. Y. Avraham, and M. Pinchas, A novel clock skew estimator and its performance for the IEEE 1588v2 (PTP) case in fractional Gaussian noise/generalized fractional Gaussian noise environment, *Frontiers in Physics*, 9, 2021, 796811.
12. A. Dmitriev, V. Kornilov, V. Dmitriev, and N. Abbas, Early warning signals for critical transitions in sandpile cellular automata, *Frontiers in Physics*, 10, 2022, 839383.
13. S. X. Hu, Z. W. Liao, H. Rong, and C. Peng, Giant panda video image sequence and application in 3D reconstruction, *Frontiers in Physics*, 10, 2022, 839582.
14. S. X. Hu, Z. W. Liao, H. Rong, and C. Peng, Characteristic sequence analysis of giant panda voiceprint, *Frontiers in Physics*, 10, 2022, 839699.
15. J. Lévy Véhel, Beyond multifractional Brownian motion: New stochastic models for geophysical modeling, *Nonlinear Processes in Geophysics*, 20(5) 2013, 643–655.
16. K. Kiyono, Establishing a direct connection between detrended fluctuation analysis and Fourier analysis, *Physical Review E*, 92(4) 2015, 042925.
17. M. Pinchas, Residual ISI obtained by blind adaptive equalizers and fractional noise, *Mathematical Problems in Engineering*, 2013, 2013, Article ID 972174 (11 pp).
18. M. Pinchas, Symbol error rate for non-blind adaptive equalizers applicable for the SIMO and fGn case, *Mathematical Problems in Engineering*, 2014, 2014, Article ID 606843 (11 pp).
19. S. V. Muniandy, W. X. Chew, and C. S. Wong, Fractional dynamics in the light scattering intensity fluctuation in dusty plasma, *Physics of Plasmas*, 18(1) 2011, 013701.
20. X. Gabaix, P. Gopikrishnan, V. Plerou, and H. E. Stanley, A theory of power-law distributions in financial market fluctuations, *Nature*, 423(6937) 2003, 267–270.
21. I. Tejado, S. H. Hosseinnia, B. M. Vinagre, X. Song, and Y. Q. Chen, Dealing with fractional dynamics of IP network delays, *International Journal of Bifurcation and Chaos*, 22(4) 2012, 1250089 (13 pp).

22. S. I. Resnick, Heavy tail modeling and teletraffic data, *The Annals of Statistics*, 25(5) 1997, 1805–1849.

23. J. Adler, Discussion: Heavy tail modeling and teletraffic data, *The Annals of Statistics*, 25(5) 1997, 1849–1852.

24. J. Beran, Discussion: Heavy tail modeling and teletraffic data, *The Annals of Statistics*, 25(5) 1997, 1852–1856.

25. K. J. Brennan (ed.), *Handbook on Classification and Application of Fractals*, Nova Science Publishers, Inc, New York, 2012.

26. M. Takayasu, H. Takayasu, and T. Sato, Critical behaviors and $1/f$ noise in information traffic, *Physica A*, 233(3–4) 1996, 824–834.

27. R. G. V. Baker, Towards a physics of Internet traffic in a geographic network, *Physica A*, 391(4) 2012, 954–965.

28. W. S. Cleveland, and D. X. Sun, Internet traffic data, *Journal of the American Statistical Association*, 95(451) 2000, 979–985.

29. W. Willinger, and V. Paxson, Where mathematics meets the Internet, *Notices of the AMS*, 45(8) 1998, 961–970.

30. W. Willinger, R. Govindan, S. Jamin, V. Paxson, and S. Shenker, Scaling phenomena in the Internet critically, *Proceedings of the National Academy of Sciences of the United States of America*, 99(Suppl 1) 2002, 2573–2580.

31. M. E. Sousa-Vieira, A. Suárez-González, M. Fernández-Veiga, J. C. López-Ardao, and C. López-García, Model selection for long-memory processes in the spectral domain, *Computer Communications*, 36(13) 2013, 1436–1449.

32. R. Delgado, A packet-switched network with On/Off sources and a fair bandwidth sharing policy: State space collapse and heavy-traffic, *Telecommunication Systems*, 62(2) 2016, 461–479.

33. I. Lokshina, Study on estimating probabilities of buffer overflow in high-speed communication networks, *Telecommunication Systems*, 62(2) 2016, 269–302.

34. M. Owczarczuk, Long memory in patterns of mobile phone usage, *Physica A*, 391(4) 2012, 1428–1433.

35. R. V. Sole, and S. Valverde, Information transfer and phase transitions in a model of Internet traffic, *Physica A* 289(3–4) 2001, 595–605.

36. T. Huisinga, R. Barlovic, W. Knospe, A. Schadschneider, and M. Schreckenberg, A microscopic model for packet transport in the Internet, *Physica A*, 294(1–2) 2001, 249–256.

37. J. Yuan, Y. Ren, and X. Shan, Self-organized criticality in a computer network model, *Physical Review E*, 61(2) 2000, 1067–1071.

38. I. Csabai, $1/f$ noise in computer network traffic, *Journal of Physics A: Mathematical and General*, 27(2) 1994, L417–L421.

39. M. Masugi, and T. Takuma, Multi-fractal analysis of IP-network traffic for assessing time variations in scaling properties, *Physica D*, 225(2) 2007, 119–126.

40. A. H. Taherinia, and M. Jamzad, A two-step watermarking attack using long-range correlation image restoration, *Security and Communication Networks*, 5(6) 2012, 625–635.

41. L. Kaklauskas, and L. Sakalauskas, Study of on-line measurement of traffic self-similarity, *Central European Journal of Operations Research*, 21(1) 2013, 63–84.

42. A. Abdelkefi, Y. Jiang, B. E. Helvik, G. Biczók, and A. Calu, Assessing the service quality of an Internet path through end-to-end measurement, *Computer Networks*, 70, 2014, 30–44.

43. J.-S. R. Lee, S.-K. Ye, and H.-D. J. Jeong, ATMSim: An anomaly teletraffic detection measurement analysis simulator, *Simulation Modelling Practice and Theory*, 49, 2014, 98–109.

44. J. W. Roberts, Traffic theory and the Internet, *IEEE Communications Magazine*, 39(1) 2001, 94–99.

45. A. Adas, Traffic models in broadband networks, *IEEE Communications Magazine*, 35(7) 1997, 82–89.

46. H. Michiel, and K. Laevens, Teletraffic engineering in a broad-band era, *Proceedings of the IEEE*, 85(12) 1997, 2007–2033.
47. S. Q. Li, and C. L. Hwang, Queue response to input correlation functions discrete spectral analysis, *IEEE/ACM Transactions on Networking*, 1(5) 1993, 522–533.
48. S. Q. Li, and C. L. Hwang, Queue response to input correlation functions continuous spectral analysis, *IEEE/ACM Transactions on Networking*, 1(6) 1993, 678–692.
49. M. Livny, B. Melamed, and A. K. Tsiolis, The impact of autocorrelation on queuing systems, *Management Science*, 39(3) 1993, 322–339.
50. J.-T. Sun, S.-J. Wang, Z.-G. Huang, and Y.-H. Wang, Effect of degree correlations on networked traffic dynamics, *Physica A*, 388(15–16) 2009, 3244–3248.
51. B. Hajek, and L. He, On variations of queue response for inputs with the same mean and autocorrelation function, *IEEE/ACM Trans. Networking*, 6(5) 1998, 588–598.
52. M. Li, An approach to reliably identifying signs of DDOS flood attacks based on LRD traffic pattern recognition, *Computers & Security*, 23(7) 2004, 549–558.
53. J. D. Gibson (ed.), *The Communications Handbook*, IEEE Press, New York, 1997.
54. J. M. Pitts, and J. A. Schormans, *Introduction to IP and ATM Design and Performance: With Applications and Analysis Software*, 2nd ed., John Wiley & Sons, New York, 2000.
55. D. McDysan, *QoS & Traffic Management in IP & ATM Networks*, McGraw-Hill, New York, 2000.
56. A. S. Tanenbaum, *Computer Networks*, 4th ed., Pearson Education, New York, 2003.
57. W. Stalling, *Data and Computer Communications*, 7th ed., Pearson Education, New York, 2006.
58. W. Yue, H. Takagi, and Y. Takahashi, *Advances in Queueing Theory and Network Applications*, Springer, Berlin, 2009.
59. W. Stalling, *Data and Computer Communications*, 4th ed., Macmillan, New York, 1994.
60. J. Beran, Statistical method for data with long-range dependence, *Statistical Science*, 7(4) 1992, 404–416.
61. W. E. Leland, M. S. Taqqu, W. Willinger, and D. V. Wilson, On the self-similar nature of ethernet traffic (extended version), *IEEE/ACM Trans. Networking*, 2(1) 1994, 1–15.
62. V. Paxson, and S. Floyd, Wide area traffic: The failure of Poisson modeling, *IEEE/ACM Transactions on Networking*, 3(3) 1995, 226–244.
63. J. Beran, R. Shernan, M. S. Taqqu, and W. Willinger, Long-range dependence in variable bit-rate video traffic, *IEEE Transactions on Communications*, 43(2/3/4) 1995, 1566–1579.
64. B. Tsybakov, and N. D. Georganas, Self-similar processes in communications networks, *IEEE Transactions on Information Theory*, 44(5) 1998, 1713–1725.
65. P. Abry, and D. Veitch, Wavelet analysis of long-range dependent traffic, *IEEE Transactions on Information Theory*, 44(1) 1998, 2–15.
66. R. Roughan, D. Veitch, and P. Abry, Real-time estimation of the parameters of long-range dependence, *IEEE/ACM Transactions on Networking*, 8(4) 2000, 467–478.
67. X. Yang, and P. Petropulu, The extended alternating fractal renewal process for modeling traffic in high-speed communication networks, *IEEE Transactions on Signal Processing*, 49(7) 2001, 1349–1361.
68. O. Cappe, E. Moulines, J.-C. Pesquet, A. P. Petropulu, and X. Yang, Long-range dependence and heavy-tail modeling for teletraffic data, *IEEE Signal Processing Magazine*, 19(3) 2002, 14–27.
69. P. Abry, R. Baraniuk, P. Flandrin, R. Riedi, and D. Veitch, Multiscale nature of network traffic, *IEEE Signal Processing Magazine*, 19(3) 2002, 28–46.
70. A. R. Erramilli, M. Roughan, D. Veitch, and W. Willinger, Self-similar traffic and network dynamics, *Proceedings of the IEEE*, 90(5) 2002, 800–819.
71. I. W. C. Lee, and A. O. Fapojuwo, Stochastic processes for computer network traffic modeling, *Computer Communications*, 29(1) 2005, 1–23.

72. W.-B. Gong, Y. Liu, V. Misra, and D. Towsley, Self-similarity and long range dependence on the Internet: A second look at the evidence, origins and implications, *Computer Networks*, 48(3) 2005, 377–399.

73. S. Bregni, and L. Jmoda, Accurate estimation of the Hurst parameter of long-range dependent traffic using modified Allan and Hadamard variances, *IEEE Transactions on Communications*, 56(11) 2008, 1900–1906.

74. P. Abry, P. Borgnat, F. Ricciato, A. Scherrer, and D. Veitch, Revisiting an old friend: On the observability of the relation between long range dependence and heavy tail, *Telecommunication Systems*, 43(3–4) 2010, 147–165.

75. C. Park, F. Hernández-Campos, L. Le, J. S. Marron, J. Park, V. Pipiras, F. D. Smith, R. L. Smith, M. Trovero, and Z. Zhu, Long-range dependence analysis of Internet traffic, *Journal of Applied Statistics*, 38(7) 2011, 1407–1433.

76. P. Loiseau, P. V.-B. Primet, and P. Gonçalves, A long-range dependent model for network traffic with flow-scale correlations, *Stochastic Models*, 27(2) 2011, 333–361.

77. S. Zheng, and J. S. Baras, Sequential anomaly detection in wireless sensor networks and effects of long-range dependent data, *Sequential Analysis*, 31(4) 2012, 458–480.

78. S. Bregni, P. Giacomazzi, and G. Saddemi, Long-range dependence of traffic across schedulers with multiple service classes, *Computer Communications*, 35(7) 2012, 842–848.

79. A. Rizk, and M. Fidler, Non-asymptotic end-to-end performance bounds for networks with long range dependent fBm cross traffic, *Computer Networks*, 56(1) 2012, 127–141.

80. D. Al-Shammary, I. Khalil, and Z. Tari, A distributed aggregation and fast fractal clustering approach for SOAP traffic, *Journal of Network and Computer Applications*, 41, 2014, 1–14.

81. J. W. G. Stênico, and L. L. Ling, General solution to the losses estimation for multifractal traffic, *Journal of the Franklin Institute*, 351(10) 2014, 4904–4922.

82. A. Carbone, and K. Kiyono, Detrending moving average algorithm: Frequency response and scaling performances, *Physical Review E*, 93(6) 2016, 063309.

83. M. Li, Change trend of averaged Hurst parameter of traffic under DDOS flood attacks, *Computers & Security*, 25(3) 2006, 213–220.

84. M. B. Priestley, *Spectral Analysis and Time Series*, vol. 1, Academic Press, New York, 1981.

85. M. B. Priestley, *Spectral Analysis and Time Series*, vol. 2, Academic Press, New York, 1981.

86. J. S. Bendat, and A. G. Piersol, *Random Data: Analysis and Measurement Procedure*, John Wiley & Sons, New York, 1971.

87. J. S. Bendat, and A. G. Piersol, *Random Data: Analysis and Measurement Procedure*, 3rd ed., John Wiley & Sons, New York, 2000.

88. J. S. Bendat, and A. G. Piersol, *Random Data: Analysis and Measurement Procedure*, 4th ed., §8.6, John Wiley & Sons, New York, 2010.

89. M. B. Priestley, Basic considerations in the estimation of spectra, *Technometrics*, 4(4) 1962, 551–564.

90. M. B. Priestley, The role of bandwidth in spectral analysis, *Journal of the Royal Statistical Society C*, 14(1) 1965, 33–47

91. E. Parzen, Mathematical considerations in the estimation of spectra, *Technometrics*, 3(2) 1961, 167–190.

92. E. Parzen, On choosing an estimate of the spectral density function of a stationary time series, *The Annals of Mathematical Statistics*, 28(4) 1957, 921–932.

93. G. M. Jenkins, A survey of spectral analysis, *Journal of the Royal Statistical Society C*, 14(1) 1965, 2–32.

94. E. A. Robinson, A historical perspective of spectrum estimation, *Proc. the IEEE*, 70(9) 1982, 885–907.

95. D. J. Thomson, Spectrum estimation and harmonic analysis, *Proc. the IEEE*, 70(9) 1982, 1055–1096.

96. S. K. Mitra, and J. F. Kaiser, *Handbook for Digital Signal Processing*, John Wiley & Sons, New York, 1993.

97. A. Adnani, J. Duplicy, and L. Philips, Spectrum analyzers today and tomorrow part 1: Towards filterbanks-enabled real-time spectrum analysis, *IEEE Instrumentation & Measurement Magazine*, 16(5) 2013, 6–11.

98. R. B. Blackman, and J. W. Tukey, The measurement of power spectra from the point of view of communications engineering: part I, *Bell System Technical Journal*, 37(1) 1958, 185–282.

99. R. B. Blackman, and J. W. Tukey, The measurement of power spectra from the point of view of communications engineering: part II, *Bell System Technical Journal*, 37(2) 1958, 485–569.

100. J. W. Tukey, Modern techniques of power spectrum estimation, *IEEE Transactions on Information Theory*, 15(2) 1967, 56–66.

101. W. A. Garden, Measurement of spectral correlation, *IEEE Transactions on Acoustics, Speech and Signal Processing*, 34(5) 1986, 1111–1123.

102. A. Papoulis, and S. U. Pillai, *Probability, Random Variables, and Stochastic Processes*, 3rd ed., McGraw-Hill, New York, 2002.

103. R. E. Mortensen, *Random Signals and Systems*, John Wiley & Sons, New York, 1987.

104. M. Li, A method for requiring block size for spectrum measurement of ocean surface waves, *IEEE Transactions on Instrumentation and Measurement*, 55(6) 2006, 2207–2215.

105. W. Q. Zhu, *Random Vibrations*, Science Press, Beijing, 1992. (In Chinese)

106. H. H. Wylton, J. H. Rabbahtuxin, and B. A. Loskov, Translated from Russian by J. C. Pan, *Probability and Statistical Analysis of Wind-Waves*, Ocean Press, Beijing, 1984. (In Chinese)

107. D. Slepian, Fluctuations of random noise power, *Bell System Technical Journal*, 37(1) 1958, 163–184.

108. J. W. Tukey, Discussion, emphasizing the connection between analysis of variance and spectrum analysis, *Technometrics*, 3(2) 1961, 191–219.

109. J. W. Tukey, The future of data analysis, *The Annals of Mathematical Statistics*, 33(1) 1962, 1–67.

110. J. W. Tukey, Data analysis and the frontiers of geophysics, *Science*, 148(3675) 1965, 1283–1289.

111. D. R. Brillinger, and W. John Tukey's work on time series and spectrum analysis, *The Annals of Statistics*, 30(6) 2002, 1595–1618.

112. A. Schuster, The periodogram and its optical analogy, *Proceedings of the Royal Society*, 77(515) 1906, 136–140.

113. M. S. Bartlett, Smoothing periodograms from time-series with continuous spectra, *Nature*, 161(4096) 1948, 686–687.

114. M. S. Bartlett, Correlation or spectral analysis? *The Statistician*, 27(3–4) 1978, 147–158.

115. M. S. Bartlett, Periodogram analysis and continuous spectra, *Biometrika*, 37(1–2) 1950, 1–16.

116. M. S. Bartlett, Statistical estimation of density functions, *Sankhyā: The Indian Journal of Statistics, Series A (1961–2002)*, 25(3) 1963, 245–254.

117. Z. A. Lomnicki, and S. K. Zaremba, On estimating the spectral density function of a stochastic process, *Journal of the Royal Statistical Society B*, 19(1) 1957, 13–37.

118. H. E. Daniels, The estimation of spectral densities, *Journal of the Royal Statistical Society B*, 24(1) 1962, 185–198.

119. R. Dahlhaus, On a spectral density estimate obtained by averaging priodograms, *Journal of Applied Probability*, 22(3) 1985, 598–610.

120. P. Welch, The use of fast Fourier transform for the estimation of power spectra, *IEEE Trans. Audio and Electroacoustics*, 15(2) 1970, 70–73.

121. P. Hall, and M. Li, Using the periodogram to estimate period in nonparametric regression, *Biometrika*, 93(2) 2006, 411–424.

122. C. M. Harris, *Shock and Vibration Handbook*, 5th ed., McGraw-Hill, New York, 2002.

123. Schlumberger Electronics Ltd., *1200 Real Time Signal Processor Operating Manual*, UK, 1983.

124. B. W. Lindgren, and G. W. McElrath, *Introduction to Probability and Statistics*, The Macmillan Company, New York, 1959.

125. S. L. Meyer, *Data Analysis for Scientists and Engineers*, John Wiley & Sons, New York, 1975.

126. J. A. Rice, *Mathematical Statistics and Data Analysis*, 2nd ed., Wadsworth Inc., New York, 1995.

127. M. Li, *Applications of Spectrum to Random Fatigue Testing*, Master Thesis, China Ship Scientific Research Center, Wuxi, 1990. (In Chinese)

128. F. Raspall, Efficient packet sampling for accurate traffic measurements, *Computer Networks*, 56(6) 2012, 1667–1684.

129. V. Carela-Español, P. Barlet-Ros, A. Cabellos-Aparicio, and J. Solé-Pareta, Analysis of the impact of sampling on NetFlow traffic classification, *Computer Networks*, 55(5) 2011, 1083–1099.

130. H. Elbiaze, M. F. Zhani, O. Cherkaoui, and F. Kamoun, A new structure-preserving method of sampling for predicting self-similar traffic, *Telecommunication Systems*, 43(3-4) 2010, 265–277.

131. Y. Chabchoub, C. Fricker, F. Guillemin, and P. Robert, On the statistical characterization of flows in Internet traffic with application to sampling, *Computer Communications*, 33(1) 2010, 103–112.

132. S. Fernandes, C. Kamienski, J. Kelner, D. Mariz, and D. Sadok, A stratified traffic sampling methodology for seeing the big picture, *Computer Networks*, 52(14) 2008, 2677–2689.

133. B.-Y. Choi and Z.-L. Zhang, Adaptive random sampling for traffic volume measurement, *Telecommunication Systems*, 34(1-1) 2007, 71–80.

134. N. Duffield, Sampling for passive Internet measurement: A review, *Statistical Science*, 19(3) 2004, 472–498.

135. G. He, and J. C. Hou, On sampling self-similar Internet traffic, *Computer Networks*, 50(16) 2006, 2919–2936.

136. A. Scherrer, N. Larrieu, P. Owezarski, P. Borgnat, and P. Abry, Non-Gaussian and long memory statistical characterisations for internet traffic with anomalies, *IEEE Transactions on Dependable and Secure Computing*, 4(1) 2007, 56–70.

137. A. Feldmann, A. C. Gilbert, W. Willinger, and T. G. Kurtz, The changing nature of network traffic: Scaling phenomena, *ACM SIGCOMM Computer Communication Review*, 28(2) 1998, 5–29.

138. T. A. Dahl, and T. R. Willemain, The effect of long-memory arrivals on queue performance, *Operations Research Letters*, 29(3) 2001, 123–127.

139. R. D. Smith, The dynamics of Internet traffic: Self-similarity, self-organization, and complex phenomena, *Advances in Complex Systems*, 14(9) 2011, 905–949.

140. S. Ledesma, and D. Liu, Synthesis of fractional Gaussian noise using linear approximation for generating self-similar network traffic, *ACM SIGCOMM Computer Communication Review*, 30(2) 2000, 4–17.

141. J. C. López-Ardao, C. López-García, A. Suárez-González, M. Fernández-Veiga, and R. Rodríguez-Rubio, On the use of self-similar processes in network simulation, *ACM Trans. Modeling and Computer Simulation*, 10(2) 2000, 125–151.

142. S. Robert, and J.-Y. Le Boudec, New models for pseudo self-similar traffic, *Performance Evaluation*, 30(1-2) 1997, 57–68.

143. V. Paxson, Fast, approximate synthesis of fractional Gaussian noise for generating self-similar network traffic, *ACM SIGCOMM Computer Communication Review*, 27(5) 1997, 5–18.

144. M. Li, and S. C. Lim, Modeling network traffic using generalized Cauchy process, *Physica A*, 387(11) 2008, 2584–2594.

145. M. Li, Modeling autocorrelation functions of long-range dependent teletraffic series based on optimal approximation in Hilbert space-a further study, *Applied Mathematical Modelling*, 31(3) 2007, 625–631.

146. T. Karagiannis, M. Molle, and M. Faloutsos, Long-range dependence ten years of Internet traffic modeling, *IEEE Internet Computing*, 8(5) 2004, 57–64.

147. P. Borgnat, G. Dewaele, K. Fukuda, P. Abry, and K. Cho, Seven years and one day: Sketching the evolution of Internet traffic, *INFOCOM2009*, Rio de Janeiro, Brazil, April 2009.

148. L. Isserlis, On a formula for the product-moment coefficient of any order of a normal frequency distribution in any number of variables, *Biometrika*, 12(1–2) 1918, 134–139.

149. L. J. Herbst, Periodogram analysis and variance fluctuations, *Journal of the Royal Statistical Society B*, 25(2) 1963, 442–450.

150. J. H. Laning, and R. H. Battin, *Random Processes in Automatic Control*, McGraw Hill, New York, 1956.

151. B. B. Mandelbrot, Fast fractional Gaussian noise generator, *Water Resources Research*, 7(3) 1971, 543–553.

152. M. Li, and C. H. Chi, A correlation-based computational method for simulating long-range dependent data, *Journal of the Franklin Institute*, 340(6–7) 2003, 503–514.

153. M Li, Fractal time series — A tutorial review, *Mathematical Problems in Engineering*, 2010, 2010, Article ID 157264 (26 pp).

154. M. Li, and S. C. Lim, Power spectrum of generalized Cauchy process, *Telecommunication Systems*, 43(3–4) 2010, 219–222.

155. S. C. Lim, and M. Li, A generalized Cauchy process and its application to relaxation phenomena, *Journal of Physics A: Mathematical and General*, 39(12) 2006, 2935–2951.

156. S. C. Lim, and L. P. Teo, Gaussian fields and Gaussian sheets with generalized Cauchy covariance structure, *Stochastic Processes and Their Applications*, 119(4) 2009, 1325–1356.

157. P. Danzig, J. Mogul, V. Paxson, and M. Schwartz, The internet traffic archive, 2000. ftp://ita.ee.lbl.gov/traces/. [dataset].

158. M. Li, Multi-fractional generalized Cauchy process and its application to teletraffic, *Physica A*, 550, 2020, 123982 (14 pp).

159. M. Li, Long-range dependence and self-similarity of teletraffic with different protocols at the large time scale of day in the duration of 12 years: Autocorrelation modeling, *Physica Scripta*, 95(4) 2020, 065222 (15 pp).

160. M. Li, Generalized fractional Gaussian noise and its application to traffic modeling, *Physica A*, 579, 2021, 1236137 (22 pp).

161. P. M. Robinson, Log-periodogram regression of time series with long range dependence, *The Annals of Statistics*, 23(3) 1995, 1048–1072.

162. G. Chan, P. Hall, and D. S. Poskitt, Periodogram-based estimators of fractal properties, *The Annals of Statistics*, 23(5) 1995, 1684–1711.

163. I. Lobato, and P. M. Robinson, Averaged periodogram estimation of long memory, *Journal of Econometrics*, 73(1) 1996, 303–324.

164. R. J. Sela, and C. M. Hurvich, The averaged periodogram estimator for a power law in coherency, *Journal of Time Series Analysis*, 33(2) 2012, 340–363.

165. V. A. Reisen, E. Moulines, P. Soulier, and G. C. Franco, On the properties of the periodogram of a stationary long-memory process over different epochs with applications, *Journal of Time Series Analysis*, 31(1) 2010, 20–36.

166. M. Sergides, and E. Paparoditis, Bootstrapping the local periodogram of locally stationary processes, *Journal of Time Series Analysis*, 29(2) 2008, 264–299.

167. C. Velasco, The periodogram of fractional processes, *Journal of Time Series Analysis*, 28(4) 2007, 600–627.

168. G. M. Raymond, D. B. Percival, and J. B. Bassingthwaighte, The spectra and periodograms of anti-correlated discrete fractional Gaussian noise, *Physica A*, 322(2003), 169–179.

169. M. Henry, Averaged periodogram spectral estimation with long-memory conditional heteroscedasticity, *Journal of Time Series Analysis*, 22(4) 2001, 431–459.

170. P. M. Robinson, and D. Marinucci, The averaged periodogram for nonstationary vector time series, *Statistical Inference for Stochastic Processes*, 3(1–2) 2000, 149–160.

171. E. Moulines, and P. Soulier, Broadband log-periodogram regression of time series with long-range dependence, *The Annals of Statistics*, 27(4) 1999, 1415–1439.

172. P. Kokoszka, and T. Mikosch, The integrated periodogram for long-memory processes with finite or infinite variance, *Stochastic Processes and Their Applications*, 66(1) 1997, 55–78.

173. I. N. Lobato, Consistency of the averaged cross-periodogram in long memory series, *Journal of Time Series Analysis*, 18(2) 1997, 137–155.

174. R. J. Bhansali, Robustness of the autoregressive spectral estimate for linear processes with infinite variance, *Journal of Time Series Analysis*, 18(3) 1997, 213–229.

175. V. A. Reisen, Estimation of the fractional difference parameter in the ARIMA(p, d, q) model using the smoothed periodogram, *Journal of Time Series Analysis*, 15(3) 1994, 335–350.

176. C. M. Hurvich, and K. I. Beltrao, Asymptotics for the low-frequency ordinates of the periodogram of a long-memory time series, *Journal of Time Series Analysis*, 14(5) 1993, 455–472.

177. M. Li, and W. Zhao, Smoothing the sample autocorrelation of long-range dependent traffic, *Mathematical Problems in Engineering*, 2013, 2013, Article ID 631498 (10 pp).

178. M. Li, and W. Zhao, Convergence of sample autocorrelation of long-range dependent traffic, *Mathematical Problems in Engineering*, 2013 2013, Article ID 725730 (7 pp).

179. T. Gneiting, and M. Schlather, Stochastic models that separate fractal dimension and the Hurst effect, *SIAM Review*, 46(2) 2004, 269–282.

III

Multi-fractal Models of Traffic

Multi-Fractional Generalized Cauchy Process and Its Application to Traffic

The contributions given in this chapter are in two aspects. The first is to introduce a novel random function which we call the multi-fractional generalized Cauchy (mGC) process. The second is to dissertate its application to network traffic for studying the multi-fractal behavior of traffic on a point-by-point basis. The introduced mGC process is with the time-varying fractal dimension $D(t)$ and the time-varying Hurst parameter $H(t)$. The representations of the autocorrelation function (ACF) and the power spectrum density (PSD) of the mGC process are proposed. Besides, the asymptotic expressions of the ACF and PSD of the mGC process are presented. The computation formula of $D(t)$ is given. The mGC model may be a new tool to describe the multi-fractal behavior of traffic. Precisely, it may be used to reveal the local irregularity or local self-similarity (LSS) which is a small time scale behavior of traffic, and global long-term persistence or long-range dependence (LRD) which is a large time scale behavior of traffic, on a point-by-point basis. The cast study with real traffic traces exhibits that the variance of $D(t)$ is much greater than that of $H(t)$. Thus, the present mGC model may provide a novel way to explain the fact that traffic has highly local irregularity while its LRD is robust.

9.1 INTRODUCTION

Fractal traffic discussed in previous chapters is in the sense of mono-fractal. By mono-fractal we mean that the fractal dimension and the Hurst parameter of a traffic series are constant. In this chapter, we introduce a multi-fractal or multi-fractional model of traffic. It is called the multi-fractional generalized Cauchy process. By multi-fractal or multi-fractional we mean that either the fractal dimension or the Hurst parameter of traffic is time-varying.

Arrival traffic modeling plays a role in communication systems (Gibson [1]). On circuit switched communication systems, such as traditional telephone networks, Erlang [2]

DOI: 10.1201/9781003354987-12

gave the pioneering work on traffic modeling with the Poisson distribution (also see Brockmeyer et al. [3], Gall [4], Lin et al. [5], Manfield and Downs [6], Reiser [7], Akimaru and Kawashima [8], Bojkovic et al. [9]). The traffic model of the Poisson distribution type is simple and successfully applied to circuit switched communication networks (Gibson [1], Cooper [10]). However, things change in packet switched communications networks.

In packet communications networks, such as the Internet, it was found that the traffic is non-Poisson in the late 1970s (Tobagi et al. [11, p. 1427]). In the 1980s, traffic property of non-Poisson was clearly noticed and the burstiness, which is a local property of traffic, was observed (Jain and Routhier [12]).

In the 1990s, people began investigating traffic from the point of view of fractal time series. The fractal properties of traffic, namely, self-similarity (SS) and LRD, were paid attention to (Michiel and Laevens [13], Li and Borgnat [14]). The focus of this research is on representing multi-fractal behavior of traffic.

Let $x(t)$ be traffic series. Denote its ACF by $r(\tau)$ that is given by

$$r(\tau) = E[x(t)x(t+\tau)], \tag{9.1}$$

where τ is time lag and E is the mean operator. Then, it was found the $r(\tau)$ decays so slowly such that (Paxson and Floyd [15], Beran et al. [16])

$$\int_{-\infty}^{\infty} r(\tau)d\tau = \infty. \tag{9.2}$$

In mathematics, $x(t)$ is said to be LRD or have long memory if its ACF satisfies (9.2) (Beran [17, 18]). Eq. (9.2) can be equivalently expressed by

$$r(\tau) \sim c\tau^{2H-2} \ (\tau \to \infty), \tag{9.3}$$

where $c > 0$ is a constant and $0.5 < H < 1$ is the Hurst parameter.

Let $S(f)$ be the PSD of traffic. Then, Csabai [19] reported that

$$S(f) \sim \frac{1}{f} \ (f \to 0). \tag{9.4}$$

The above says that traffic is a type of $1/f$ noise, which reflects its LRD property in the frequency domain.

When a random function $X(t)$ satisfies

$$X(at) =_d a^H X(t), a > 0, \tag{9.5}$$

where the equality $=_d$ denotes the same probability distribution on both sides, we say that $X(t)$ is an exact self-similar (SS) process with the SS measure of H. Traffic asymptotically follows (9.5) at large time scales (Paxson and Floyd [15], Leland et al. [20]).

A stationary process that exactly follows (9.3), (9.4), and (9.5) is the fractional Gaussian noise (fGn) of the Wely type, which was introduced by Mandelbrot and van Ness [21]. Its ACF is given by

$$C_{fGn}(\tau) = \frac{\sigma^2 \varepsilon^{2H-2}}{2} \left[\left(\frac{|\tau|}{\varepsilon} + 1 \right)^{2H} + \left| \frac{|\tau|}{\varepsilon} - 1 \right|^{2H} - 2 \left| \frac{\tau}{\varepsilon} \right|^{2H} \right], \tag{9.6}$$

where $\sigma^2 = (H\pi)^{-1} \Gamma(1-2H) \cos(H\pi)$ stands for the intensity of fGn, $\varepsilon > 0$ is a parameter used for regularizing fractional Brownian motion (fBm) so that the regularized fBm is differentiable [21, p. 427, 428]. FGn is LRD if $0.5 < H < 1$.

The model of fGn is usually used to describe the main properties of traffic, such as LRD, $1/f$ noise, and SS (see Michiel and Laevens [13, p. 2018], Beran et al. [16, Eq. (3)], Beran [17, 18], Leland et al. [20, p. 6], Tsybakov and Georganas [22, Eq. (2.1)], Adas [23, p. 88], Abry et al. [24, Eq. (4)], Karagiannis et al. [25, Eq. (3)], Gong et al. [26, p. 379], Lee and Fapojuwo [27, Eq. (8)], Stoev et al. [28, p. 429], Bregni and Jmoda [29], Lokshina [30], Lee et al. [31], Pinchas [32], just to mention a few). However, scientists noticed its limitation in traffic modeling. For instance, Paxson and Floyd stated that "it might be difficult to characterize the correlations over the entire trace with a single Hurst parameter" [15, Last sentence, Paragraph 4, § 7.4]. Tsybakov and Georganas said that "the class of exactly self-similar processes is too narrow for modeling actual network traffic" [22, Paragraph 1, Section II]. The exact SS process Tsybakov and Georganas stated is the fGn of the Weyl type.

Note that the SS and LRD are two different properties of a random function $X(t)$. The former measures its local property, namely, local irregularity or LSS, while the latter measures its global property, say, LRD (Mandelbrot [33], Li [34], Gneiting and Schlather [35]). When fGn is used as a traffic model, however, either SS/LSS or LRD happens to be characterized by the same single parameter H (Gneiting and Schlather [35], Li and Lim [36]).

In the 2000s, Li and Lim introduced a traffic model with two parameters, namely, the Hurst parameter $0 < H < 1$ and the fractal dimension $1 \leq D < 2$, see Li and Lim [36]. It is called the generalized Cauchy (GC) process (Lim and Li [37]). The normalized ACF of the GC process is in the form

$$C(\tau) = \left(1 + |\tau|^{4-2D} \right)^{-\frac{1-H}{2-D}}, 1 \leq D < 2, 0 < H < 1. \tag{9.7}$$

The ACF $C(\tau)$ is positive-definite for the above ranges of H and D. It is a completely monotone for $1 \leq D < 2, 0 \leq H < 1$. When $D = 1$ and $H = 0$, $C(\tau)$ reduces to the usual Cauchy process. The condition for the GC process to be of LRD is $0.5 < H < 1$, referring Li [38], Li and Li [39] for the brief of the evolution of the GC process. Note that traffic has a particular property that it has high local irregularity or high LSS and robust LRD (Willinger et al. [40, 41], Feldmann et al. [42]). With the GC model, Li and Lim explained that property of traffic in the multi-scale sense on an interval-by-interval basis [36].

An interesting and important property of traffic is its multi-fractal behaviour (see Abry et al. [24], Willinger et al. [40, 41], Feldmann et al. [42], Veitch et al. [43], Stênico and Ling [44],

Xu and Feng [45], Ostrowsky et al. [46], Vieira et al. [47], Rocha et al. [48], Budhiraja and Liu [49], Vieira and Lee [50], Masugi and Takuma [51], Fontugne et al. [52]). The challenging issue in this regard is to find a proper way to explain the highly local irregularity and robust LRD of traffic from the point of view of multi-fractals (Willinger et al. [40, 41], Feldmann et al. [42], Fontugne et al. [52], Ribeiro et al. [53]). However, the reports of describing the multi-fractal behavior of traffic on a point-by-point basis are rarely seen. This chapter aims at introducing the multi-fractional GC (mGC) process and applying it to describing the multi-fractal behavior of traffic on a point-by-point basis. By mGC process we mean that the constants D and H in (9.7) are replaced by $1 \leq D(t) < 2$ and $H(t) \in (0.5, 1)$, respectively. In this way, $D(t)$ may serve as a tool to describe the local irregularity or LSS of traffic on a point-by-point basis while $H(t)$ may be used to characterize the global long-term persistence or LRD of traffic on a point-by-point basis.

The rest of chapter is organized as follows: In Section 9.2, we propose the ACF of the mGC process and its asymptotic expressions for $\tau \to \infty$ and $\tau \to 0$. In Section 9.3, we present the PSD of the mGC process and its asymptotic expressions for $\omega \to 0$ and $\omega \to \infty$. The computation formula of $D(t)$ is given in Section 9.4. Case study with the demonstrations of $D(t)$ and $H(t)$ of real traffic is shown in Section 9.5. Discussions are in Section 9.6 followed by summary.

9.2 THE MGC PROCESS

9.2.1 ACF of the mGC Process

Theorem 9.1. Let

$$C(\tau,t) = \left(1 + |\tau|^{4-2D(t)}\right)^{-\frac{1-H(t)}{2-D(t)}}, \tag{9.8}$$

where $H : [0,\infty) \to [0,1)$, $D : [0,\infty) \to [1,2)$. Then, $C(\tau, t)$ is an ACF.

Proof. The function $C(\tau, t)$ is an even function in terms of τ since

$$C(\tau, t) = C(-\tau, t). \tag{9.9}$$

It is positive-definite for $1 \leq D(t) < 2$ and $0 \leq H(t) < 1$. That is,

$$C(\tau, t) \geq 0. \tag{9.10}$$

Besides,

$$C(0, t) \geq C(\tau, t). \tag{9.11}$$

In addition,

$$\lim_{\tau \to \infty} C(\tau, t) = 0. \tag{9.12}$$

Thus, according to the theory of random processes (Nigam [54], Bendat and Piersol [55], Priestley [56]), $C(\tau, t)$ is an ACF. □

Definition 9.1. A random function $X(t)$ is called mGC process if its ACF is in the form

$$C(\tau,t)=\psi^2\left(1+|\tau|^{4-2D(t)}\right)^{-\frac{1-H(t)}{2-D(t)}}, \tag{9.13}$$

where $H:[0,\infty)\to[0,1)$, $D:[0,\infty)\to[1,2)$, and ψ^2 is the intensity of $X(t)$. $\square$

For the purpose of traffic modeling and the description of $D(t)$ and $H(t)$, we may use the random function $\frac{X(t)}{\psi}$ without the generality losing. In subsequent text, unless otherwise stated, we only consider the normalized ACF in the form expressed by (9.8).

Note 9.1. $C(\tau, t)$ reduces to the ACF of the conventional GC process if both $D(t)$ and $H(t)$ are constants. $\square$

For facilitating discussions, we denote

$$\alpha(t)=4-2D(t), \tag{9.14}$$

and

$$\beta(t)=2-2H(t). \tag{9.15}$$

Using $\alpha(t)$ and $\beta(t)$, $C(\tau, t)$ is written as

$$C(\tau,t)=\left(1+|\tau|^{\alpha(t)}\right)^{-\frac{\beta(t)}{\alpha(t)}}, \tag{9.16}$$

where $0<\alpha(t)\le 2$, $0<\beta(t)\le 2$.

Because $C(\tau, t)$ is sufficient smooth on $(0, \infty)$ and for $\tau\to 0$

$$C(0,t)-C(\tau,t)\sim c|\tau|^{\alpha(t)}, \tag{9.17}$$

according to Hall and Joy [57], Chan et al. [58], Kent and Wood [59], Adler [60], we have the time-varying fractal dimension given by

$$D(t)=2-\frac{\alpha(t)}{2}. \tag{9.18}$$

Thus, $\alpha(t)$ is a time-varying fractal index.
Since

$$\lim_{\tau\to\infty}C(\tau,t)=|\tau|^{-\beta(t)}, \tag{9.19}$$

the condition for the mGC process $X(t)$ to be of LRD is $0<\beta(t)<1$, which corresponds to $0.5<H(t)<1$. The quantity $\beta(t)$ is a time-varying LRD index. For $1<\beta(t)\le 2$, $X(t)$ is of short-range dependence (SRD). Figure 9.1 indicates the plots of $C(\tau, t)$.

9.2.2 Asymptotic Expressions
Asymptotically,

$$\lim_{\tau\to\infty}C(\tau,t)=|\tau|^{-\beta(t)}=|\tau|^{2-2H(t)}. \tag{9.20}$$

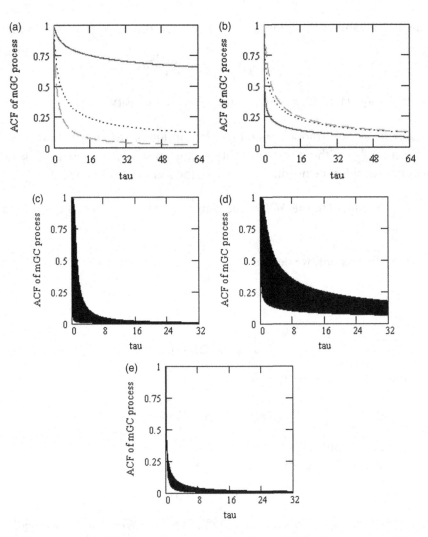

FIGURE 9.1 Plots of the ACF of mGC process. (a) Solid line: $D(t) = 1.5$. $H(t) = 0.95$. Dot line: $D(t) = 1.5$. $H(t) = 0.75$. Dash line: $D(t) = 1.5$. $H(t) = 0.55$. (b) Solid line: $H(t) = 0.75$. $D(t) = 1.85$. Dot line: $H(t) = 0.75$. $D(t) = 1.65$. Dash line: $H(t) = 0.75$. $D(t) = 1.45$. (c) $D(t) = 1.9|\cos(0.05t)|$. $H(t) = 0.5|\cos(0.05t)|$. (d) $D(t) = 1.9|\cos(0.05t)|$. $H(t) = 0.75$. (e) $D(t) = 1.75$. $H(t) = 0.5|\cos(0.05t)|$.

Thus, for large τ, we have the asymptotical expression of $C(\tau, t)$ in the form

$$C(\tau,t) \approx |\tau|^{-\beta(t)} = |\tau|^{2-2H(t)}. \tag{9.21}$$

For $\tau \geq 10$, $C(\tau, t)$ is well approximated by $|\tau|^{2-2H(t)}$.

The above exhibits that for large τ, $C(\tau, t)$ is solely associated with its LRD measure $H(t)$ and the measure of local irregularity, $D(t)$, may be neglected.

On the other hand, if $\tau \to 0$, we have

$$C(\tau,t) = 1 - |\tau|^{\alpha(t)} = 1 - |\tau|^{4-2D(t)}. \tag{9.22}$$

In fact, for small τ,

$$C(\tau,t) \approx 1 - |\tau|^{\alpha(t)} = 1 - |\tau|^{4-2D(t)}. \tag{9.23}$$

The above says that the LRD measure $H(t)$ may be ignored and $C(\tau, t)$ simply relates to the LSS measure $D(t)$ for small τ.

9.3 PSD OF THE MGC PROCESS

9.3.1 PSD

Denote the PSD of the mGC process by $S(\omega, t)$. Then (Nigam [54, Eq. (4.111)], Bendat and Piersol [55, Eq. (12.130)], Priestley [56], Priestley and Tong [61, Eq. (2.4)]), we have

$$S(\omega,t) = F[C(\tau,t)], \tag{9.24}$$

where F is the operator of Fourier transform. The above implies

$$S(\omega,t) = \int_{-\infty}^{\infty} \left(1+|\tau|^{\alpha(t)}\right)^{-\frac{\beta(t)}{\alpha(t)}} e^{-i\omega\tau} \, d\tau. \tag{9.25}$$

For the mGC process with LRD, we have

$$S(0,t) = \int_{-\infty}^{\infty} \left(1+|\tau|^{\alpha(t)}\right)^{-\frac{\beta(t)}{\alpha(t)}} d\tau = \infty. \tag{9.26}$$

Thus, the computation of (9.25) should be done in the domain of generalized functions over the Schwartz space of test functions.

Theorem 9.2. The PSD of the mGC process is expressed by

$$S(\omega,t) = \sum_{\infty}^{k=0} \frac{(-1)^k \Gamma\left\{\left[\frac{\beta(t)}{\alpha(t)}\right]+k\right\}}{\Gamma\left[\frac{\beta(t)}{\alpha(t)}\right]\Gamma(1+k)} I_1(\omega) * \mathrm{Sa}(\omega)$$

$$+ \sum_{\infty}^{k=0} \frac{(-1)^k \Gamma\left\{\left[\frac{\beta(t)}{\alpha(t)}\right]+k\right\}}{\Gamma\left[\frac{\beta(t)}{\alpha(t)}\right]\Gamma(1+k)} [\pi I_2(\omega) - I_2(\omega) * \mathrm{Sa}(\omega)], \tag{9.27}$$

where * is the convolution operation,

$$I_1(\omega,t) = F\left[|\tau|^{\alpha(t)k}\right] = \int_{\infty}^{-\infty} |\tau|^{\alpha(t)k} e^{-i\omega\tau} \, d\tau, \tag{9.28}$$

$$I_2(\omega,t) = F\left[|\tau|^{-[\beta(t)+\alpha(t)k]}\right] = \int_{-\infty}^{\infty} |\tau|^{-[\beta(t)+\alpha(t)k]} e^{-i\omega\tau} \, d\tau, \tag{9.29}$$

and $Sa(\omega) = \dfrac{\sin \omega}{\omega}$.

Proof. Let $u(\tau)$ be the unit step function. Then, with the binomial series, $C(\tau, t)$ can be expanded to be

$$C(\tau,t) = \left\{ \sum_{k=0}^{\infty} \frac{(-1)^k \Gamma\left\{\left[\dfrac{\beta(t)}{\alpha(t)}\right]+k\right\}}{\Gamma\left[\dfrac{\beta(t)}{\alpha(t)}\right]\Gamma(1+k)} |\tau|^{\alpha k} \right\} [u(\tau+1)-u(\tau-1)]$$

$$+ \left\{ \sum_{\infty}^{k=0} \frac{(-1)^k \Gamma\left\{\left[\dfrac{\beta(t)}{\alpha(t)}\right]+k\right\}}{\Gamma\left[\dfrac{\beta(t)}{\alpha(t)}\right]\Gamma(1+k)} |\tau|^{-(\beta+\alpha k)} \right\} [u(\tau-1)+u(-\tau-1)]. \tag{9.30}$$

The first item on the right side of the above equation is the binomial expansion of $C(\tau, t)$ when $|\tau| < 1$ while the second is for $|\tau| > 1$ (Li and Lim [62, 63]). □

Note that $F\left(|\tau|^{\lambda}\right) = -2\sin(\lambda\pi/2)\Gamma(\lambda+1)|\omega|^{-\lambda-1}$ for $\lambda \neq -1, -3, \dots$ (Gelfand and Vilenkin [64]). Thus, for $\alpha k \neq -1, -3, \dots$, we have

$$F\left[|\tau|^{\alpha(t)k}\right] = -2\sin[\alpha(t)k\pi/2]\Gamma[\alpha(t)k+1]|\omega|^{-\alpha(t)k-1}. \tag{9.31}$$

Similarly, for $-(\beta+\alpha k) \neq -1, -3, \dots$, we have

$$F\left[|\tau|^{-[\beta(t)+\alpha(t)k]}\right] = 2\sin\{[\beta(t)+\alpha(t)k]\pi/2\}\Gamma\{1-[\beta(t)+\alpha(t)k]\}|\omega|^{[\beta(t)+\alpha(t)k]-1}. \tag{9.32}$$

Since $F[u(\tau+1)-u(\tau-1)] = 2Sa(\omega)$, the Fourier transform of $C(\tau, t)$ is given by (9.27). The proof completes. □

9.3.2 Asymptotic Expressions of PSD

For $\omega \rightarrow 0$, we have

$$S(\omega,t) = \int_{\infty}^{-\infty} |\tau|^{-\beta(t)} e^{-i\omega\tau} \, d\tau = 2\sin[\beta(t)\pi/2]\Gamma[1-\beta(t)]|\omega|^{\beta(t)-1}$$

$$= 2\sin\{[H(t)-1]\pi\}\Gamma[3-2H(t)]|\omega|^{2H(t)-3}, \quad \omega \rightarrow 0. \tag{9.33}$$

Note 9.2. If $X(t)$ is LRD, $0.5 < H(t) < 1$ or $0 < \beta(t) < 1$. In that case, $S(\omega, t)$ is a type of $1/f$ noise which reflects the LRD property of the mGC process in the frequency domain. □

Note 9.3. The LRD property of the mGC process only relies on $H(t)$ or $\beta(t)$. □

On the other side, for $\omega \to \infty$, we have

$$S(\omega,t) = \int_{-\infty}^{\infty} \left[1 - |\tau|^{\alpha(t)}\right] e^{-i\omega\tau} d\tau = 2\pi\delta(\omega) - \int_{-\infty}^{\infty} |\tau|^{\alpha(t)} e^{-i\omega\tau} d\tau$$

$$= -\int_{-\infty}^{\infty} |\tau|^{\alpha(t)} e^{-i\omega\tau} d\tau = 2\sin\left[\alpha(t)\pi/2\right]\Gamma\left[\alpha(t)+1\right]|\omega|^{-\alpha(t)-1}$$

$$= 2\sin\left\{\left[2 - D(t)\right]\pi\right\}\Gamma\left[5 - 2D(t)\right]|\omega|^{2D(t)-5}, \quad \omega \to \infty. \tag{9.34}$$

Note that we consider $S(\omega, t)$ for $\omega \to \infty$. Thus, $\delta(\omega) = 0$ in the above. Since $1 \le D(t) < 2$, the PSD of the mGC process for $\omega \to \infty$ is also a kind of $1/f$ noise but it merely relates to $D(t)$ or $\alpha(t)$.

9.4 COMPUTATIONS OF $D(T)$ AND $H(T)$

9.4.1 Computing $D(t)$ of the mGC Process

Theorem 9.3. Denote by ε positive infinitesimal, that is, $\varepsilon \to 0+$. Then,

$$D(t) = \frac{4 - \left\{\ln\left[X(t+\varepsilon) - X(rt)\right] - \ln\left[X(t+\varepsilon) - X(t)\right]\right\}}{2\ln r}, \tag{9.35}$$

where the equality is in the sense that both sides have the same probability distribution.

Proof. The mGC process $X(t)$ is said to be locally SS of order v for $r > 0$ if (Lim and Li [37, p. 2937], Adler [60])

$$X(s) - X(rt) = r^v \left[X(s) - X(t)\right], \quad |t-s| \to 0, \tag{9.36}$$

where the equality means that the both sides have the same probability distribution. When $v = \alpha(t)$, the above becomes

$$X(s) - X(rt) = r^{\alpha(t)} \left[X(s) - X(t)\right], \quad |t-s| \to 0. \tag{9.37}$$

Let $s = t + \varepsilon$. Then, we have

$$X(t+\varepsilon) - X(rt) = r^{\alpha(t)} \left[X(t+\varepsilon) - X(t)\right]$$
$$= r^{4-2D(t)} \left[X(t+\varepsilon) - X(t)\right]. \tag{9.38}$$

Performing the logarithm operations on the both sides of (9.38) yields (9.35). This finishes the proof. □

Note 9.4. When ε is a small positive number, we have

$$D(t) \approx \frac{4 - \left\{\ln\left[X(t+\varepsilon) - X(rt)\right] - \ln\left[X(t+\varepsilon) - X(t)\right]\right\}}{2\ln r}. \tag{9.39}$$

9.4.2 Computing $H(t)$ of the mGC Process

Because $C(\tau,t) \approx |\tau|^{2-2H(t)}$ for large τ, the ACF of the mGC process is equivalent to that of the modified multi-fractional Gaussian noise (Li [65], also Chapter 10 in this book). Therefore, the computation of $H(t)$ of the mGC process may adopt that utilized in the multi-fractional Brownian motion (mfBm) introduced by Levy-Vehel [66], Peltier and Levy-Vehel [67, 68], also see Guevel and Levy-Vehel [69], Ayache et al. [70], Falconer and Levy-Vehel [71], Muniandy and Lim [72], Muniandy et al. [73]. To be precise,

$$H(t) = -\frac{\log(\sqrt{\pi/2} S_n(j))}{\log(N-1)}, \tag{9.40}$$

where

$$S_n(j) = \frac{m}{N-1} \sum_{j=0}^{j+n} |X(i+1) - X(i)|, \quad 1 < n < N, \tag{9.41}$$

where m is the largest integer not exceeding N/n and

$$t = j/(N-1). \tag{9.42}$$

9.5 CASE STUDY

The real traffic data provided by Bellcore (BC) in 1989 and Digital Equipment Corporation (DEC) in March 1995 were used for the work on traffic modeling and analysis from the point of view of fractals (see Paxson and Floyd [15], Leland et al. [20], Abry et al. [24], Li and Lim [36], Li [38], Willinger et al. [41], Feldmann et al. [42], Veitch et al. [43], Roughan et al. [74], Abry and Veitch [75], Li [76, 77], Li et al. [78]). The research by Fontugne et al. [52] exhibits that stochastic properties of traffic remain the same from the early data by BC in 1989 to the recent data by the MAWI (Measurement and Analysis on the WIDE Internet) Working Group Traffic Archive (Japan) in 2015, respectively. In this case study, I shall reveal that by using the mGC model, the property of $\text{Var}[D(t)] \gg \text{Var}[H(t)]$ holds for the traffic data by BC in 1989, DEC in 1995 to those by MAWI in 2019, providing new evidence to support the point of view stated by Fontugne et al. [52] as well as Borgnat et al. [79].

9.5.1 Data

The file names of the real traffic traces used in this work are BC-pAug89.TL, BC-pOct89. TL, BC-Oct89Ext.TL, and BC-Oct89Ext4.TL, which are Ethernet traffic collected at BC; DEC-PKT-1.TCP, DEC-PKT-2.TCP, DEC-PKT-3.TCP, and DEC-PKT-4.TCP; which are wide-area TCP traffic recorded at DEC; MAWI-pkt-1.TCP, MAWI-pkt-2.TCP, MAWI-pkt-3.TCP, and MAWI-pkt-4.TCP, which are wide-area TCP traffic measured by MAWI.

TABLE 9.1 Four Real Traffic Traces on the Ethernet Measured by BC

Trace Name	Starting Time of Measurement	Duration	Number of Packets
pAug.TL	11:25AM, 29 Aug 89	52 min	1 million
pOct.TL	11:00AM, 05 Oct 89	29 min	1 million
OctExt.TL	11:46PM, 03 Oct 89	34.111 h	1 million
OctExt4.TL	2:37PM, 10 Oct 89	21.095 h	1 million

TABLE 9.2 Four Wide-Area TCP Traces Recorded by DEC

Trace Name	Record Date	Duration	Number of Packets
DEC-pkt-1.TCP	08 Mar 95	10 PM–11 PM	3.3 million
DEC-pkt-2.TCP	09 Mar 95	2 AM–3 AM	3.9 million
DEC-pkt-3.TCP	09 Mar 95	10 AM–11 AM	4.3 million
DEC-pkt-4.TCP	09 Mar 95	2 PM–3 PM	5.7 million

TABLE 9.3 Four Wide-Area TCP Traces Recorded by MAWI

Trace Name	Starting Record Time	Duration	Number of Packets
MAWI-pkt-1.TCP	2:00 PM, 18 Apr 2019	6.77208 min	741,404
MAWI-pkt-2.TCP	2:00 PM, 19 Apr 2019	6.65965 min	742,638
MAWI-pkt-3.TCP	2:00 PM, 20 Apr 2019	12.55740 min	482,564
MAWI-pkt-4.TCP	2:00 PM, 21 Apr 2019	12.66541 min	576,495

The data at BC and DEC are available from ftp://ita.ee.lbl.gov/traces. Those at MAWI are available via http://mawi.wide.ad.jp/mawi/. The test data traces are listed in Tables 9.1–9.3.

9.5.2 Demonstrations

Let $x[t(i)]$ be the number of bytes in a packet at the time $t(i)$ ($i = 0, 1, 2, \ldots$), where $t(i)$ is the timestamp of the ith packet. In the case study, we consider traffic series $x(i)$ that represents the packet size or packet length of the ith packet. Then, $D(i)$ and $H(i)$ are denoted as the fractal dimension and the Hurst parameter of the ith packet, respectively.

Figure 9.2 shows the first 1024 points of 12 traffic traces, Figure 9.3 indicates their $D(i)$s to characterize their local irregularity on a point-by-point basis, and Figure 9.4 illustrates their $H(i)$s to demonstrate their global long-term persistence on a point-by-point basis.

We summarize the values of $\text{Var}[D(i)]$ and $\text{Var}[H(i)]$ of 12 traces in the captions of Figure 9.3, Figure 9.4 and also in Table 9.4. The results exhibit that $\text{Var}[D(i)]$ is much greater than $\text{Var}[H(i)]$, implying that the fluctuation of the local irregularity of traffic is considerably larger than that of its globally long-term persistence. Therefore, the present mGC model provides a new way to study the multi-fractal behavior of traffic and novel tool to investigate the local irregularity and LRD of traffic.

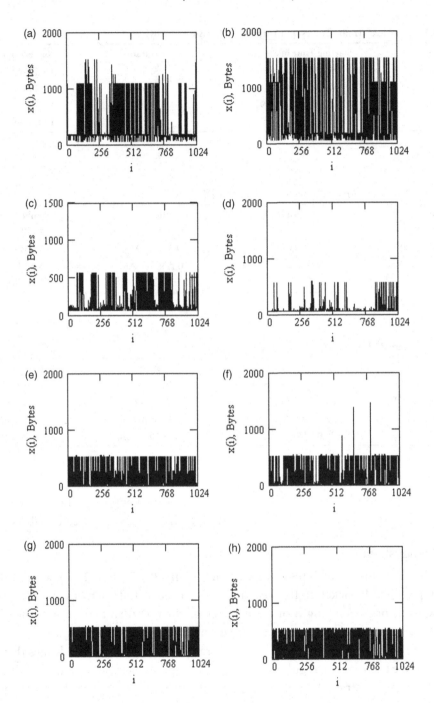

FIGURE 9.2 First 1024 data of traffic traces. (a) $x(i)$ of pAug.TL. (b) $x(i)$ of pOct.TL. (c) $x(i)$ of OctExt.TL. (d) $x(i)$ of OctExt4.TL. (e) $x(i)$ of DEC-pkt-1.TCP. (f) $x(i)$ of DEC-pkt-2.TCP. (g) $x(i)$ of DEC-pkt-3.TCP. (h) $x(i)$ of DEC-pkt-4.TCP. (i) $x(i)$ of MAWI-pkt-1.TCP. (j) $x(i)$ of MAWI-pkt-2. TCP. (k) $x(i)$ of MAWI-pkt-3.TCP. (l) $x(i)$ of MAWI-pkt-4.TCP. *(Continued)*

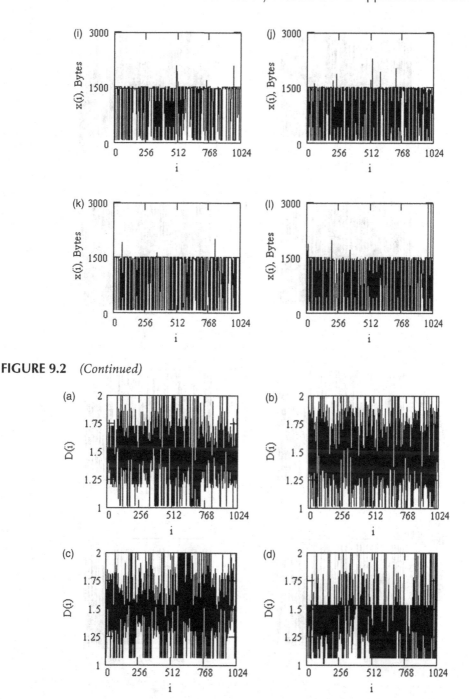

FIGURE 9.2 *(Continued)*

FIGURE 9.3 First 1024 data of $D(i)$ of traffic traces. (a) $D(i)$ of pAug.TL with Var[$D(i)$] = 0.093. (b) $D(i)$ of pOct.TL with Var[$D(i)$] = 0.125. (c) $D(i)$ of OctExt.TL with Var[$D(i)$] = 0.072. (d) $D(i)$ of OctExt4.TL with Var[$D(i)$] = 0.116. (e) $D(i)$ of DEC-pkt-1.TCP with Var[$D(i)$] = 0.104. (f) $D(i)$ of DEC-pkt-2.TCP with Var[$D(i)$] = 0.112. (g) $D(i)$ of DEC-pkt-3.TCP with Var[$D(i)$] = 0.130. (h) $D(i)$ of DEC-pkt-4.TCP with Var[$D(i)$] = 0.119. (i) $D(i)$ of MAWI-pkt-1.TCP with Var[$D(i)$] = 0.118. (j) $D(i)$ of MAWI-pkt-2.TCP with Var[$D(i)$] = 0.128. (k) $D(i)$ of MAWI-pkt-3.TCP with Var[$D(i)$] = 0.173. (l) $D(i)$ of MAWI-pkt-4.TCP with Var[$D(i)$] = 0.173. *(Continued)*

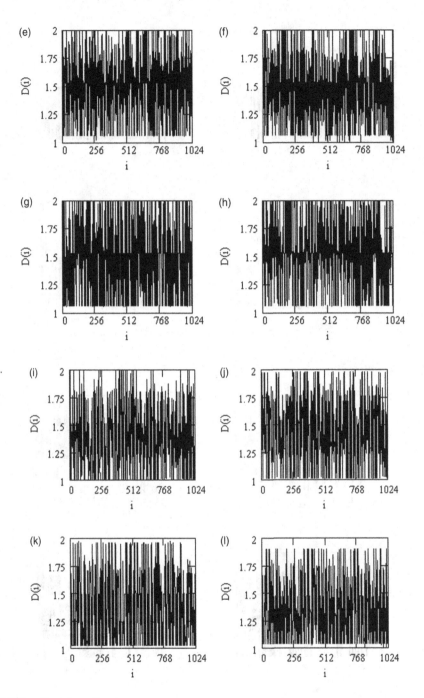

FIGURE 9.3 *(Continued)*

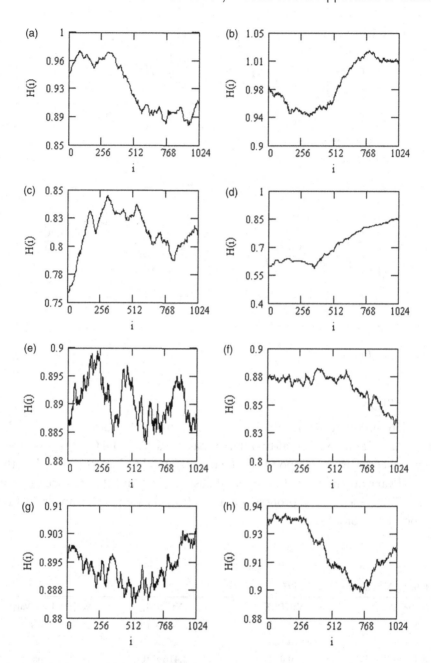

FIGURE 9.4 First 1024 data of $H(i)$ of traffic traces. (a) $H(i)$ of pAug.TL with $\mathrm{Var}[H(i)] = 1.143 \times 10^{-3}$. (b) $H(i)$ of pOct.TL with $\mathrm{Var}[H(i)] = 9.152 \times 10^{-4}$. (c) $H(i)$ of OctExt.TL with $\mathrm{Var}[H(i)] = 3.416 \times 10^{-4}$. (d) $H(i)$ of OctExt4.TL with $\mathrm{Var}[H(i)] = 7.953 \times 10^{-3}$. (e) $H(i)$ of DEC-pkt-1.TCP with $\mathrm{Var}[H(i)] = 1.352 \times 10^{-5}$. (f) $H(i)$ of DEC-pkt-2.TCP with $\mathrm{Var}[H(i)] = 1.550 \times 10^{-4}$. (g) $H(i)$ of DEC-pkt-3.TCP with $\mathrm{Var}[H(i)] = 2.203 \times 10^{-5}$. (h) $H(i)$ of DEC-pkt-4.TCP with $\mathrm{Var}[H(i)] = 1.768 \times 10^{-4}$. (i) $H(i)$ of MAWI-pkt-1.TCP with $\mathrm{Var}[H(i)] = 2.082 \times 10^{-5}$. (j) $H(i)$ of MAWI-pkt-2.TCP with $\mathrm{Var}[H(i)] = 1.619 \times 10^{-5}$. (k) $H(i)$ of MAWI-pkt-3.TCP with $\mathrm{Var}[H(i)] = 6.076 \times 10^{-6}$. (l). $H(i)$ of MAWI-pkt-4.TCP with $\mathrm{Var}[H(i)] = 9.232 \times 10^{-6}$. *(Continued)*

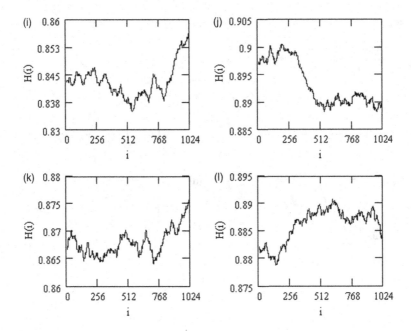

FIGURE 9.4 *(Continued)*

9.6 DISCUSSIONS

Li and Lim [36] proposed the results in traffic modeling with constants D and H, as well as varying D and H on an interval-by-interval basis based on the GC process. The case of D and H being constants corresponds to the mono-fractal GC process. On the other side, that D and H vary on an interval-by-interval basis corresponds to the case of multi-scale GC process. The present mGC process is a multi-fractal case which implies that D and H vary on a point-by-point basis.

TABLE 9.4 Variances of $D(i)$ and $H(i)$

Trace Name	Var[$D(i)$]	Var[$H(i)$]	Var[$D(i)$] >> Var[$H(i)$]
pAug.TL	0.093	1.143×10^{-3}	Yes
pOct.TL	0.125	9.152×10^{-4}	Yes
OctExt.TL	0.072	3.416×10^{-4}	Yes
OctExt4.TL	0.116	7.953×10^{-3}	Yes
DEC-pkt-1.TCP	0.104	1.352×10^{-5}	Yes
DEC-pkt-2.TCP	0.112	1.550×10^{-4}	Yes
DEC-pkt-3.TCP	0.130	2.203×10^{-5}	Yes
DEC-pkt-4.TCP	0.119	1.768×10^{-4}	Yes
MAWI-pkt-1.TCP	0.118	2.208×10^{-5}	Yes
MAWI-pkt-2.TCP	0.128	1.619×10^{-5}	Yes
MAWI-pkt-3.TCP	0.173	6.076×10^{-6}	Yes
MAWI-pkt-4.TCP	0.173	9.232×10^{-6}	Yes

Note that a challenging issue in traffic modeling and analysis is to explain its multi-fractal property in that traffic has highly local irregularity and robust LRD (Paxson and Floyd [15], Abry et al. [24], Willinger et al. [40, 41], Feldmann et al. [42], Veitch et al. [43], Willinger and Paxson [80]). The present mGC model has the time-varying fractal dimension $D(t)$ and the Hurst parameter $H(t)$. It may separately characterize the local irregularity and LRD of traffic on a point-by-point basis. Therefore, it may yet be a tool to describe the multi-fractal behavior of traffic.

Note that it needs $\varepsilon \to 0$ for the computation of $D(t)$, see Theorem 9.3. However, in practice, a traffic trace $x(i)$ $(i = 0, 1, 2, \ldots)$ is discrete. Thus, the minimum ε is 1 in the numerical computation of $D(i)$ when (9.35) is used. That may cause computation errors in a way with respect to $D(i)$. Out future work will be on a numeric computation about how to improve the computation accuracy of $D(i)$ of traffic in discrete time series.

9.7 SUMMARY

We have presented a traffic model of mGC process by introducing a time-varying ACF $C(\tau, t)$ in Theorem 9.1. We have put forward its asymptotic expressions for $\tau \to \infty$ and $\tau \to 0$ in (9.20) and (9.22), respectively. In addition, we have given the PSD of the mGC process in Theorem 9.2. Its asymptotic expressions for $\omega \to 0$ and $\omega \to \infty$ are given in (9.33) and (9.34), respectively. The mGC model contains separate representations of the time-varying fractal dimension $D(t)$ and the time-varying Hurst parameter $H(t)$ on a point-by-point basis. We have proposed the computation formula of $D(t)$ in Theorem 9.3. The case study has shown that $\text{Var}[D(t)] \gg \text{Var}[H(t)]$, providing a new view of multi-fractal behavior of traffic.

REFERENCES

1. J. D. Gibson (ed.), *The Communications Handbook*, IEEE Press, New York, 1997.
2. A. K. Erlang, Telefon-ventetider. et stykke sandsynlighedsregning, *Matematisk Tidsskrift*, B, 1920, 25–42.
3. E. Brockmeyer, H. L. Halstrom, and A. Jensen, The life of A. K. Erlang, *Transactions of the Danish Academy of Technical Sciences*, No. 2, 1948, 23–100.
4. F. Le Gall, One moment model for telephone traffic, *Applied Mathematical Modelling*, 6(6) 1982, 415–423.
5. P. Lin, B. Leon, and C. Stewart, Analysis of circuit-switched networks employing originating-office control with spill-forward, *IEEE Transactions on Communications*, 26(6) 1978, 754–766.
6. D. Manfield, and T. Downs, On the one-moment analysis of telephone traffic networks, *IEEE Transactions on Communications*, 27(8) 1979, 1169–1174.
7. M. Reiser, Performance evaluation of data communication systems, *Proceedings of the IEEE*, 70(2) 1982, 171–196.
8. H. Akimaru, and K. Kawashima, *Teletraffic: Theory and Applications*, Springer, Berlin, 1993.
9. Z. Bojkovic, M. Bakmaz, and B. Bakmaz, Originator of teletraffic theory, *Proceedings of the IEEE*, 98(1) 2010, 123–127.
10. R. B. Cooper, *Introduction to Queueing Theory*, Elsevier, 2nd ed., North Holland, 1981.
11. F. A. Tobagi, M. Gerla, R. W. Peebles, and E. G. Manning, Modeling and measurement techniques in packet communication networks, *Proceedings of the IEEE*, 66(11) 1978, 1423–1447.
12. R. Jain, and S. Routhier, Packet trains-measurements and a new model for computer network traffic, *IEEE Journal on Selected Areas in Communications*, 4(6) 1986, 986–995.
13. H. Michiel, and K. Laevens, Teletraffic engineering in a broad-band era, *Proceedings of the IEEE*, 85(12) 1997, 2007–2033.

14. M. Li, and P. Borgnat, Foreword to the special issue on traffic modeling, its computations and applications, *Telecommunication Systems*, 43(3–4) 2010, 145–146.
15. V. Paxson, and S. Floyd, Wide area traffic: The failure of Poisson modeling, *IEEE/ACM Transactions on Networking*, 3(3) 1995, 226–244.
16. J. Beran, R. Shernan, M. S. Taqqu, and W. Willinger, Long-range dependence in variable bit-rate video traffic, *IEEE Transactions on Communications*, 43(2–3–4) 1995, 1566–1579.
17. J. Beran, *Statistics for Long-Memory Processes*, Chapman & Hall, New York, 1994.
18. J. Beran, Statistical methods for data with long-range dependence, *Statistical Science*, 7(4) 1992, 404–416.
19. I. Csabai, 1/*f* noise in computer network traffic, *The Journal of Physics A: Mathematical and Theoretical*, 27(12) 1994, L417–L421.
20. W. E. Leland, M. S. Taqqu, W. Willinger, and D. V. Wilson, On the self-similar nature of ethernet traffic (extended version), *IEEE/ACM Transactions on Networking*, 2(1) 1994, 1–15.
21. B. B. Mandelbrot, and J. W. van Ness, Fractional Brownian motions, fractional noises and applications, *SIAM Review*, 10(4) 1968, 422–437.
22. B. Tsybakov, and N. D. Georganas, Self-similar processes in communications networks, *IEEE Transactions on Information Theory*, 44(5) 1998, 1713–1725.
23. A. Adas, Traffic models in broadband networks, *IEEE Communication Magazine*, 35(7) 1997, 82–89.
24. P. Abry, R. Baraniuk, P. Flandrin, R. Riedi, and D. Veitch, Multiscale nature of network traffic, *IEEE Signal Processing Magazine*, 19(3) 2002, 28–46.
25. T. Karagiannis, M. Molle, and M. Faloutsos, Long-range dependence: Ten years of internet traffic modeling, *IEEE Internet Comput.*, 8(5) 2004, 57–64.
26. W.-B. Gong, Y. Liu, V. Misra, and D. Towsley, Self-similarity and long range dependence on the internet: A second look at the evidence, origins and implications, *Computer Networks*, 48(3) 2005, 377–399.
27. I. W. C. Lee, and A. O. Fapojuwo, Stochastic processes for computer network traffic modeling, *Computer Communications*, 29(1) 2005, 1–23.
28. S. Stoev, M. S. Taqqu, C. Park, and J. S. Marron, On the wavelet spectrum diagnostic for Hurst parameter estimation in the analysis of internet traffic, *Computer Networks*, 48(3) 2005, 423–445.
29. S. Bregni, and L. Jmoda, Accurate estimation of the Hurst parameter of long-range dependent traffic using modified Allan and Hadamard variances, *IEEE Transactions on Communications*, 56(11) 2008, 1900–1906.
30. I. Lokshina, Study on estimating probabilities of buffer overflow in high-speed communication networks, *Telecommunication Systems*, 62(2) 2016, 269–302.
31. J.-S. R. Lee, S.-K. Ye, and H.-D. J. Jeong, ATMSim: An anomaly teletraffic detection measurement analysis simulator, *Simulation Modelling Practice and Theory*, 49, 2014, 98–109.
32. M. Pinchas, Cooperative multi PTP slaves for timing improvement in an fGn environment, *IEEE Communications Letters*, 22(7) 2018, 1366–1369.
33. B. B. Mandelbrot, *Gaussian Self-Affinity and Fractals*, Springer, New York, 2001.
34. M. Li, Fractal time series: A tutorial review, *Mathematical Problems in Engineering*, 2010, 2010, Article ID 157264 (26 pp).
35. T. Gneiting, and M. Schlather, Stochastic models that separate fractal dimension and Hurst effect, *SIAM Review*, 46(2) 2004, 269–282.
36. M. Li, and S. C. Lim, Modeling network traffic using generalized Cauchy process, *Physica A*, 387(11) 2008, 2584–2594.
37. S. C. Lim, and M. Li, Generalized Cauchy process and its application to relaxation phenomena, *The Journal of Physics A: Mathematical and Theoretical*, 39(12) 2006, 2935–2951.
38. M. Li, Record length requirement of long-range dependent teletraffic, *Physica A*, 472, 2017, 164–187.

39. M. Li, and J.-Y. Li, Generalized Cauchy model of sea level fluctuations with long-range dependence, *Physica A*, 484, 2017, 309–335.

40. W. Willinger, R. Govindan, S. Jamin, V. Paxson, and S. Shenker, Scaling phenomena in the internet critically, *Proceedings of the National Academy of Sciences of the United States of America*, 99(Suppl 1) 2002, 2573–2580.

41. W. Willinger, M. S. Taqqu, R. Sherman, and D. V. Wilson, Self-similarity through high-variability: Statistical analysis of ethernet LAN traffic at the source level, *IEEE/ACM Transactions on Networking*, 5(1) 1997, 71–86.

42. A. Feldmann, A. C. Gilbert, W. Willinger, and T. G. Kurtz, The changing nature of network traffic: Scaling phenomena, *ACM SIGCOMM Computer Communication Review*, 28(2) 1998, 5–29.

43. D. Veitch, N. Hohn, and P. Abry, Multifractality in TCP/IP traffic: The case against, *Computer Networks*, 48(3) 2005, 293–313.

44. J. W. G. Stênico, and L. L. Ling, General solution to the losses estimation for multifractal traffic, *Journal of the Franklin Institute*, 351(10) 2014, 4904–4922.

45. Y. Xu, and H. Feng, Revisiting multifractality of TCP traffic using multifractal detrended fluctuation analysis, *Journal of Statistical Mechanics: Theory and Experiment*, 2014, 2014, P02007.

46. L. O. Ostrowsky, N. L. S. da Fonseca, and C. A. V. Melo, A multiscaling traffic model for UDP streams, *Simulation Modelling Practice and Theory*, 26, 2012, 32–48.

47. F. H. T. Vieira, F. G. C. Rocha, and J. A. dos Santos, Loss probability estimation and control for OFDM/TDMA wireless systems considering multifractal traffic characteristics, *Computer Communications*, 35(2) 2012, 263–271.

48. E. Rocha, P. Salvador, and A. Nogueira, Can multiscale traffic analysis be used to differentiate internet applications? *Telecommunication Systems*, 48(1–2) 2011, 19–30.

49. A. Budhiraja, and X. Liu, Multiscale diffusion approximations for stochastic networks in heavy traffic, *Stochastic Processes and Their Applications*, 121(3) 2011, 630–656.

50. F. H. T. Vieira, and L. L. Lee, Adaptive wavelet-based multifractal model applied to the effective bandwidth estimation of network traffic flows, *IET Communications*, 3(6) 2009, 906–919.

51. M. Masugi, and T. Takuma, Multi-fractal analysis of IP-network traffic for assessing time variations in scaling properties, *Physica D*, 225(2) 2007, 119–126.

52. R. Fontugne, P. Abry, K. Fukuda, D. Veitch, K. Cho, P. Borgnat, and H. Wendt, Scaling in internet traffic: A 14 year and 3 day longitudinal study, with multiscale analyses and random projections, *IEEE/ACM Transactions on Networking*, 25(4) 2017, 2152–2165

53. V. J. Ribeiro, Z.-L. Zhang, S. Moon, and C. Diot, Small-time scaling behavior of internet backbone traffic, *Computer Networks*, 48(3) 2005, 315–334.

54. N. C. Nigam, *Introduction to Random Vibrations*, The MIT Press, New York, 1983.

55. J. S. Bendat, and A. G. Piersol, *Random Data: Analysis and Measurement Procedure*, 4th ed., John Wiley & Sons, New York, 2010.

56. M. B. Priestley, *Spectral Analysis and Time Series*, Academic Press, New York, 1981.

57. P. Hall, and R. Roy, On the relationship between fractal dimension and fractal index for stationary stochastic processes, *The Annals of Applied Probability*, 4(1) 1994, 241–253.

58. G. Chan, P. Hall, and D. S. Poskitt, Periodogram-based estimators of fractal properties, *The Annals of Statistics*, 23(5) 1995, 1684–1711.

59. J. T. Kent, and A. T. Wood, Estimating the fractal dimension of a locally self-similar Gaussian process by using increments, *Journal of the Royal Statistical Society B*, 59(3) 1997, 679–699.

60. A. J. Adler, *The Geometry of Random Fields*, John Wiley & Sons, New York, 1981.

61. M. B. Priestley, and H. Tong, On the analysis of bivariate non-stationary processes, *Journal of the Royal Statistical Society B*, 35(2) 1973, 153–166.

62. M. Li, and S. C. Lim, Power spectrum of generalized Cauchy process, *Telecommunication Systems*, 43(3–4) 2010, 219–222.

63. M. Li, and S. C. Lim, A rigorous derivation of power spectrum of fractional Gaussian noise, *Fluctuation and Noise Letters*, 6(4) 2006, C33–C36.
64. I. M. Gelfand, and K. Vilenkin, *Generalized Functions*, Vol. 1, Academic Press, New York, 1964.
65. M. Li, Modified multifractional Gaussian noise and its application, *Physica Scripta*, 96(12) 2021, 125002 (12 pp).
66. J. Levy-Vehel, Beyond multifractional Brownian motion: New stochastic models for geophysical modeling, *Nonlinear Processes in Geophysics*, 20(5) 2013, 643–655.
67. R. F. Peltier, and J. Levy-Vehel, *Multifractional Brownian Motion: Definition and Preliminaries Results*, INRIA RR 2645, France, 1995.
68. R. F. Peltier, and J. Levy-Vehel, *A New Method for Estimating the Parameter of Fractional Brownian Motion*, INRIA RR 2696, France, 1994.
69. R. Le Guevel, and J. Levy-Vehel, Incremental moments and Holder exponents of multifractional multistable processes, *ESAIM: Probability and Statistics*, 17, 2013, 135–178.
70. A. Ayache, S. Cohen, and J. Levy-Vehel, The covariance structure of multifractional Brownian motion, with application to long range dependence, *ICASSP, 2000 IEEE International Conference on Acoustics, Speech, and Signal Processing*, 6, 2000, 3810–3813.
71. K. J. Falconer, and J. Levy-Vehel, Multifractional, multistable, and other processes with prescribed local form, *Journal of Theoretical Probability*, 22(2) 2009, 375–401.
72. S. V. Muniandy, and S. C. Lim, On some possible generalizations of fractional Brownian motion, *Physics Letters A*, 266(2–3) 2000, 140–145.
73. S. V. Muniandy, S. C. Lim, and R. Murugan, Inhomogeneous scaling behaviors in Malaysia foreign currency exchange rates, *Physica A*, 301(1–4) 2001, 407–428.
74. M. Roughan, D. Veitch, and P. Abry, Real time estimation of the parameters of long-range dependence, *IEEE/ACM Transactions on Networking*, 8(4) 2000, 467–478.
75. P. Abry, and D. Veitch, Wavelet analysis of long-range dependent traffic, *IEEE Trans. Information Theory*, 44(1) 1998, 2–15.
76. M. Li, Modeling autocorrelation functions of long-range dependent teletraffic series based on optimal approximation in Hilbert space: A further study, *Applied Mathematical Modelling*, 31(3) 2007, 625–631.
77. M. Li, Generalized fractional Gaussian noise and its application to traffic modeling, *Physica A*, 579, 2021, 1236137 (22 pp).
78. M. Li, W. J. Jia, and W. Zhao, Correlation form of timestamp increment sequences of self-similar traffic on ethernet, *Electronics Letters*, 36(19) 2000, 1668–1669.
79. P. Borgnat, G. Dewaele, K. Fukuda, P. Abry, and K. Cho, Seven years and one day: Sketching the evolution of Internet traffic, *Proceedings of the 28th IEEE INFOCOM 2009*, Rio de Janeiro (Brazil), May 2009, 711–719.
80. W. Willinger, and V. Paxson, Where mathematics meets the internet, *Notices of the AMS*, 45(8) 1998, 961–970.

Modified Multi-fractional Gaussian Noise and Its Application to Traffic

In this chapter, we introduce a novel multi-fractional noise called modified multi-fractional Gaussian noise (mmfGn). The introduced mmfGn is equipped with the time-varying Hurst parameter $H(t)$ on a point-by-point basis based on the autocorrelation function (ACF) of the fGn. The representations of the ACF and the power spectrum density (PSD) of the mmfGn are presented in Theorems 10.1 and 10.4, respectively. Besides, we propose the approximation of the ACF of mmfGn in Theorem 10.2. We put forward the properties of the long-range dependence (LRD) and short-range dependence (SRD) of mmfGn in Theorem 10.3 and the condition of mmfGn to be $1/f$ noise in Theorem 10.5. The fractal dimension of mmfGn, which is also time-varying on a point-by-point basis, is proposed in Theorem 10.6. In addition, we propose the concepts that mmfGn is stationary for $H(t) <$ 0.70 irrelevant of function forms of $H(t)$ and it is conditional stationary when $H(t) > 0.70$. A set of stationary ranges regarding $H(t) > 0.70$ are given in Table 10.1. As an application case, we use the time-varying Hurst parameter $H(t)$ to describe the stationarity of real traffic and introduce the concept of the stationarity scale of traffic.

10.1 INTRODUCTION

We introduced a multi-fractal traffic model of multi-fractional generalized Cauchy (mGC) process in Chapter 9. In this chapter, we bring forward another multi-fractal model called modified multi-fractional Gaussian noise (mmfGn).

Fractional processes have attracted interests of researchers [1–8]. In this regard, multi-fractional ones are paid attention to (see refs [9–26]). Multi-fractional noise is a class of fractional processes with the time-varying Hurst parameter and/or time-varying fractal dimension. A kind of widely used multi-fractional noise is multi-fractional Brownian motion (mfBm). By mfBm one means that it is a fractional Brownian motion (fBm) with

DOI: 10.1201/9781003354987-13

TABLE 10.1 Stationary Ranges of mmfGn with LRD with the Rule of $\mathrm{Corr}[r(\tau; H(t_1)), r(\tau; H(t_2))] \geq 0.7$

Ordinal Number of Row	$H(t) \in (H_{\min}, H_{\max})$	$\mathrm{Corr}[r(\tau; H_{\min}), r(\tau; H_{\max})] \geq$	$\Delta = H_{\max} - H_{\min}$
1	$H(t) \in (0.680, 0.990)$	0.700	0.310
2	$H(t) \in (0.672, 0.980)$	0.701	0.308
3	$H(t) \in (0.663, 0.970)$	0.701	0.307
4	$H(t) \in (0.653, 0.960)$	0.700	0.307
5	$H(t) \in (0.643, 0.950)$	0.700	0.307
6	$H(t) \in (0.632, 0.940)$	0.700	0.308
7	$H(t) \in (0.621, 0.930)$	0.700	0.309
8	$H(t) \in (0.609, 0.920)$	0.700	0.311
9	$H(t) \in (0.596, 0.910)$	0.700	0.314
10	$H(t) \in (0.582, 0.900)$	0.700	0.318
11	$H(t) \in (0.568, 0.890)$	0.700	0.322
12	$H(t) \in (0.552, 0.880)$	0.700	0.328
13	$H(t) \in (0.535, 0.870)$	0.700	0.335
14	$H(t) \in (0.517, 0.860)$	0.700	0.343
15	$H(t) \in (0.498, 0.850)$	0.700	0.352
16	$H(t) \in (0.477, 0.840)$	0.700	0.363
17	$H(t) \in (0.455, 0.830)$	0.700	0.375
18	$H(t) \in (0.431, 0.820)$	0.700	0.389
19	$H(t) \in (0.405, 0.810)$	0.700	0.405
20	$H(t) \in (0.377, 0.800)$	0.700	0.423
21	$H(t) \in (0.347, 0.790)$	0.700	0.443
22	$H(t) \in (0.315, 0.780)$	0.700	0.465
23	$H(t) \in (0.282, 0.770)$	0.700	0.488
24	$H(t) \in (0.245, 0.760)$	0.700	0.515
25	$H(t) \in (0.207, 0.750)$	0.700	0.543
26	$H(t) \in (0.167, 0.740)$	0.700	0.573
27	$H(t) \in (0.127, 0.730)$	0.700	0.603
28	$H(t) \in (0.084, 0.720)$	0.700	0.636
29	$H(t) \in (0.040, 0.710)$	0.700	0.670
30	$H(t) \in (10^{-32}, 0.700)$	0.702	0.702
31	$H(t) \in (10^{-32}, 0.650)$	0.766	0.766
32	$H(t) \in (10^{-32}, 0.600)$	0.819	0.819
33	$H(t) \in (10^{-32}, 0.550)$	0.861	0.861
34	$H(t) \in (10^{-32}, 0.501)$	0.895	0.895

time-varying Hurst parameter (Lim [10], Levy-Vehel [12]). There are other types of multi-fractional noises. Examples include multi-fractional Ornstein–Uhlenbeck processes (Lim and Teo [11]), tempered multi-fractional stable motion (Fan and Levy-Vehel [13]), multi-fractional Brownian motion (mfBm) (Ayache et al. [14]), Lévy multistable processes (Guével and Levy-Vehel [15]), and the author brought forward a novel multi-fractional noise called multi-fractional generalized Cauchy process, see Chapter 9 and [26]. In this chapter, I introduce another new multi-fractional noise called modified multi-fractional Gaussian noise (mmfGn).

A Gaussian process is an increment process of the Brownian motion(Bm). Denote the Bm by $B(t)$. Let $g(t)$ be a Gaussian process. Then, for $a > 0$, $g(t) = B(t + a) - B(t)$. Denote the operator of the Weyl fractional derivative of order $v > 0$ by $_tW_{-\infty}^v$. Then (Miller and Ross [27], Lavoie et al. [28, p. 245], [29–33]),

$$_tW_{-\infty}^v f(t) = \frac{1}{\Gamma(-v)} \int_{-\infty}^{t} \frac{f(u)}{(t-u)^{v+1}} du. \tag{10.1}$$

There are other types of fractional integrals and derivatives [27–33]. Different types of fractional integrals/derivatives may yield different results for a specific function (see Lim [10], Lim and Teo [11], Miller and Ross [27], and Lavoie et al. [28]). Note that the integral lower limit of $W_{-\infty}^v f(t)$ is $-\infty$. An application of the Weyl fractional derivative is in Fourier transform [27, 28]. With the Weyl fractional derivative and the Riemann–Liouville one, Mandelbrot and van Ness introduced two types of fBms and fGn in ref [34] (also see Lim [10], Li [35]). Let $B_H(t)$ be the fBm with $0 < H < 1$, where H is the Hurst parameter. Using the Weyl fractional derivative [34, p. 424], Mandelbrot and van Ness defined a type of fBm in the form ([34, Definition 2.1, Eq. (2.1)])

$$B_H(t) - B_H(0) = \frac{1}{\Gamma(H+1/2)} \left\{ \int_{-\infty}^{0} [(t-u)^{H-0.5} - (-u)^{H-0.5}] dB(u) + \int_{0}^{t} (t-u)^{H-0.5} dB(u) \right\}, \tag{10.2}$$

where $dB(u)$ is done in the domain of generalized functions. The above is the fBm of the Weyl type. It is widely used (see Zeinali and Pourdarvish [36], Yu [37]).

Let $g_1(t)$ be the fGn. Then, $g_1(t) = B_H(t + a) - B_H(t)$. Denote the ACF of $g_1(t)$ by $C_{fGn}(\tau)$. Then [34, p. 433],

$$C_{fGn}(\tau) = \frac{V_H \varepsilon^{2H-2}}{2} \left[\left(\frac{|\tau|}{\varepsilon} + 1 \right)^{2H} + \left| \frac{|\tau|}{\varepsilon} - 1 \right|^{2H} - 2 \left| \frac{\tau}{\varepsilon} \right|^{2H} \right], \tag{10.3}$$

where τ is lag and

$$V_H = \text{Var}[B_H(1)] = \Gamma(1-2H) \frac{\cos \pi H}{\pi H} \tag{10.4}$$

is the strength of $B_H(t)$ and/or $g_1(t)$. The quantity $\varepsilon > 0$ in (10.3) is used for smoothing the fBm so that the smoothed fBm is differentiable in the domain of generalized functions [34]. By letting $\varepsilon \to 0$, (10.3) reduces to the commonly used expression of $C_{fGn}(\tau)$ in the form

$$C_{fGn}(\tau) = \frac{V_H}{2} \left[(|\tau|+1)^{2H} + \|\tau|-1|^{2H} - 2|\tau|^{2H} \right]. \tag{10.5}$$

Note that fGn stands for three classes of processes [34–38]. Since $\int_{-\infty}^{\infty} C_{fGn}(\tau)d\tau = \infty$ for $0.5 < H < 1$, $g_1(t)$ for $0.5 < H < 1$ is of LRD according to the definition of LRD. The larger the value of H, the stronger the LRD for a time series. When $0 < H < 0.5$, one has $\int_{-\infty}^{\infty} C_{fGn}(\tau)d\tau = 0$. Thus, $g_1(t)$ is of SRD for $0 < H < 0.5$. If $H = 0.5$, $g_1(t)$ reduces to the white noise.

Let $S_{\text{fGn}}(\omega)$ be the PSD of $g_1(t)$. According our previous work [38], $S_{\text{fGn}}(\omega)$ is given by

$$S_{\text{fGn}}(\omega) = V_H \sin(H\pi)\Gamma(2H+1)|\omega|^{1-2H}. \tag{10.6}$$

Denote the mfBm by $X(t)$. Then, following Peltier and Levy-Vehel [39], $X(t)$ is given by

$$X(t) = \frac{1}{\Gamma(H(t)+1/2)} \left\{ \int_{-\infty}^{0} [(t-u)^{H(t)-0.5} - (-u)^{H(t)-0.5}] dB(u) + \int_{0}^{t} (t-u)^{H(t)-0.5} dB(u) \right\}, \tag{10.7}$$

where the function $H(t)$ satisfies $H : [0, \infty) \rightarrow (0, 1)$. The above implies that the mfBm is simply obtained by replacing the constant H in (10.2) with $H(t)$. Numerically, $H(t)$ at $t = j/(N-1)$ is given by

$$H(t) = -\frac{\log(\sqrt{\pi/2} S_k(j))}{\log(N-1)}, \tag{10.8}$$

where

$$S_k(j) = \frac{m}{N-1} \sum_{j=0}^{j+k} |X(i+1) - X(i)|, \quad 1 < k < N, \tag{10.9}$$

where m is the largest integer not exceeding N/k [39–41].

Let $G(t)$ be the multi-fractional Gaussian noise (mfGn). Then, it is given by

$$G(t) = X(t+a) - X(t), \quad a > 0. \tag{10.10}$$

However, how to represent the ACF of $G(t)$ or the PSD of $G(t)$ remains a challenging problem. In order to evade the problem in practical applications, for instance, in the field of teletraffic modeling, people attain the ACF on an interval-by-interval basis instead of point-by-point basis. To be precise, for $n = 0, 1, \ldots, C_{\text{fGn}}(\tau)$ on an interval-by-interval basis is given by

$$C_{\text{fGn}}(\tau) = \frac{V_{H(n)}}{2} \left[(|\tau|+1)^{2H(n)} + \||\tau|-1\|^{2H(n)} - 2|\tau|^{2H(n)} \right], \tag{10.11}$$

where $H(n)$ is the Hurst parameter of fGn $g_1(t)$ in the nth interval (see refs [42–45]). Note that $C_{\text{fGn}}(\tau)$ in (10.11) is still the ACF of fGn but on an interval-by-interval basis instead of the ACF of mfGn.

Our previous work [46] stated the stationarity issue of teletraffic with the ACF given by

$$C_{\text{mmfGn}}(\tau) = \frac{V_{H(t)}}{2} \left[(|\tau|+1)^{2H(t)} + \||\tau|-1\|^{2H(t)} - 2|\tau|^{2H(t)} \right]. \tag{10.12}$$

However, ref [46] lacks in the theory of the noise the ACF of which obeys (10.12). We call a random function $G_1(t)$ that obeys the above ACF of mmfGn, instead of mfGn, because

$C_{\mathrm{mmfGn}}(\tau)$ takes the function form as that of $C_{\mathrm{fGn}}(\tau)$. In this chapter, I aim at presenting a theory profile with respect to mmfGn. It consists of the expressions of ACF, PSD, the approximation of ACF, its properties of LRD and $1/f$ noise, and the fractal dimension of mmfGn. In addition, we address its stationarity issue and its application to studying the stationarity test of traffic.

The remaining chapter is organized as follows: In Section 10.2, we present the theoretic results of mmfGn. The results regarding the stationarity of mmfGn are discussed in Section 10.3. Application to the stationarity test of real traffic is given in Section 10.4, which is followed by summary.

10.2 MODIFIED MULTI-FRACTIONAL GUASSIAN NOISE

10.2.1 ACF of mmfGn

Theorem 10.1. Let $r(\tau)$ be a function in the form

$$r(\tau) = \frac{V_{H(t)}}{2}\left[\left(|\tau|+1\right)^{2H(t)} + \left\||\tau|-1\right\|^{2H(t)} - 2|\tau|^{2H(t)}\right], \tag{10.13}$$

where $0 < H(t) < 1$ for $t \in [0,\infty)$ and

$$V_{H(t)} = \Gamma(1-2H(t))\frac{\cos \pi H(t)}{\pi H(t)}. \tag{10.14}$$

Then, $r(\tau)$ is an ACF.

Proof. The function $r(\tau)$ has the properties $r(\tau) = r(-\tau)$ and $r(0) \geq r(\tau)$. Thus, $r(\tau)$ is an ACF according to the theory of random processes (Yaglom [47]). The proof finishes. $\square$

For facilitating the discussions, we denote $r(\tau)$ in (10.13) by $C_{\mathrm{mmfGn}}(\tau)$, implying that we call a process mmfGn when its ACF follows $C_{\mathrm{mmfGn}}(\tau)$. When $H(t) = \text{constant} \in (0, 1)$, $C_{\mathrm{mmfGn}}(\tau)$ reduces to $C_{\mathrm{fGn}}(\tau)$. Therefore, we say that the fGn is a special case of mmfGn. Figure 10.1 and Figure 10.2 show the plots of $C_{\mathrm{mmfGn}}(\tau)$ with the deterministic $H(t)$ and the random $H(t)$, respectively. Figure 10.3 shows the case of $C_{\mathrm{mmfGn}}(\tau) = C_{\mathrm{fGn}}(\tau)$ when $H(t)$ is a constant.

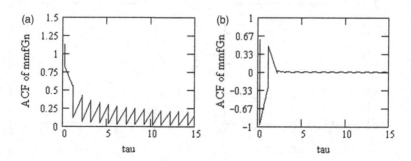

FIGURE 10.1 Plots of $C_{\mathrm{mmfGn}}(\tau)$ with deterministic $H(t)$. (a) $C_{\mathrm{mmfGn}}(\tau)$ with $H(t) = 0.2|\cos t| + 0.6$. (b) $C_{\mathrm{mmfGn}}(\tau)$ with $H(t) = 0.2|\cos t| - 0.6$.

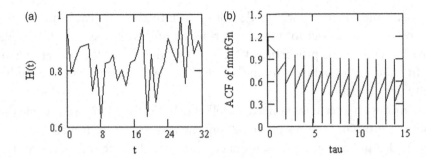

FIGURE 10.2 Plots of $H(t)$ and $C_{mmfGn}(\tau)$ with random $H(t)$. (a) $H(t)$ is a random sequence with the normal distribution with the mean 0.8 and the standard deviation 0.08. (b) $C_{mmfGn}(\tau)$ with $H(t)$ in (a).

The above figures exhibit that $C_{mmfGn}(\tau)$ decays slowly for $0.5 < H(t) < 1$ due to LRD. On the other hand, it decays fast when $0 < H(t) < 0.5$ owing to SRD. When $0.5 < H(t) < 1$, $C_{mmfGn}(\tau)$ is positive for all τ. When $0 < H(t) < 0.5$, $C_{mmfGn}(\tau)$ may be negative for some τ as can be seen from Figure 10.1 (b) and Figure 10.3 (b).

10.2.2 Approximation of ACF of mmfGn

The ACF of mmfGn has the approximation expressed by Theorem 10.2.

Theorem 10.2. An approximation of $C_{mmfGn}(\tau)$ is given by

$$C_{mmfGn}(\tau) \approx V_{H(t)}H(t)[2H(t)-1]|\tau|^{2H(t)-2}. \tag{10.15}$$

Proof. Since $\frac{V_{H(t)}}{2}\left[(|\tau|+1)^{2H(t)} + \big\|\tau|-1\big|^{2H(t)} - 2|\tau|^{2H(t)}\right]$ is the finite second-order difference of $0.5V_{H(t)}|\tau|^{2H(t)}$ in terms of τ, approximating it by the second-order difference produces

$$\frac{V_{H(t)}}{2}\left[(|\tau|+1)^{2H(t)} + \big\|\tau|-1\big|^{2H(t)} - 2|\tau|^{2H(t)}\right] \approx \frac{\partial^2 0.5V_{H(t)}|\tau|^{2H(t)}}{\partial \tau^2}$$

$$= V_{H(t)}H(t)[2H(t)-1]|\tau|^{2H(t)-2}. \tag{10.16}$$

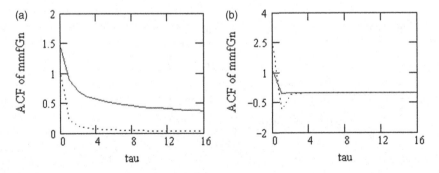

FIGURE 10.3 Plots of $C_{mmfGn}(\tau)$ that reduces to $C_{fGn}(\tau)$ when $H(t)$ is a constant. (a) $C_{mmfGn}(\tau)$ with $H(t) = 0.85$ (solid), 0.65 (dot). (b) $C_{mmfGn}(\tau)$ with $H(t) = 0.45$ (solid), 0.15 (dot).

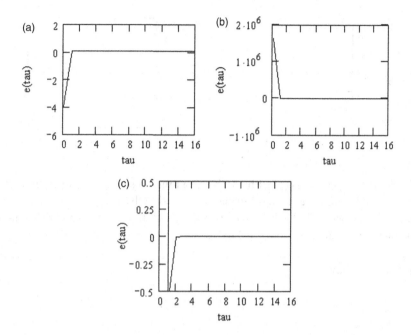

FIGURE 10.4 Plots of $e(\tau)$ with the constant $H(t)$. (a) $e(\tau)$ for $H(t) = 0.95$. (b) $e(\tau)$ for $H(t) = 0.35$. (c) Observing the zoom of $e(\tau)$ for $H(t) = 0.95$ with the ordinate axis from -0.5 to 0.5.

The proof completes. □

Denote the error by

$$e(\tau) = C_{\text{mmfGn}}(\tau) - V_{H(t)}H(t)[2H(t)-1]|\tau|^{2H(t)-2}. \tag{10.17}$$

We use Figures 10.4–10.6 to show the plots of $e(\tau)$ for indicating the approximation of $C_{\text{mmfGn}}(\tau)$. Figure 10.4 indicates the plots of $e(\tau)$ with the constant $H(t)$, Figure 10.5 gives the plot of $e(\tau)$ with the random $H(t)$, and Figure 10.6 shows the plots of $e(\tau)$ with the deterministic $H(t)$. For small τ, $e(\tau)$ may be positive or negative, relying on the value of $H(t)$, as can be seen from Figures 10.4–10.6. In practice, whether $e(\tau)$ is positive or negative does not matter because one concerns with $|e(\tau)|$. One important thing is that the approximation (10.15) is quite accurate for $\tau > 5$ as can be seen from the plots in Figures 10.4–10.6. Note that we use the ordinate axis from -0.5 to 0.5 to observe the zoom of $e(\tau)$ in Figure 10.4 (c), Figure 10.5 (b), and Figure 10.6 (c).

10.2.3 Statistical Dependences of mmfGn

Theorem 10.3. When $0.5 < H(t) < 1$, mmfGn is of LRD. It is of SRD if $0 < H(t) < 0.5$.

Proof. According to Theorem 10.2, we have

$$\int_0^\infty C_{\text{mmfGn}}(\tau)d\tau \approx V_{H(t)}H(t)[2H(t)-1]\int_0^\infty |\tau|^{2H(t)-2}\,d\tau. \tag{10.18}$$

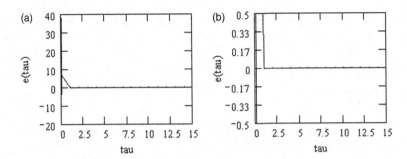

FIGURE 10.5 Plots of $e(\tau)$. (a) $e(\tau)$ with $H(t)$ being a random sequence with the normal distribution with the mean 0.8 and the standard deviation 0.08. (b) Observing the zoom of $e(\tau)$ with $H(t)$ being a random sequence with the normal distribution with the mean 0.8 and the standard deviation 0.08 using the ordinate axis from -0.5 to 0.5.

If $0.5 < H(t) < 1$, we have

$$\int_0^\infty |\tau|^{2H(t)-2} \, d\tau = \infty. \tag{10.19}$$

Hence, mmfGn is of LRD if $0.5 < H(t) < 1$. On the other side, if $0 < H(t) < 0.5$, we have

$$\int_0^\infty |\tau|^{2H(t)-2} \, d\tau < \infty. \tag{10.20}$$

Thus, mmfGn is of SRD when $0 < H(t) < 0.5$. This completes the proof. □

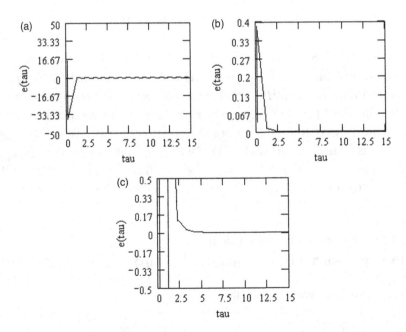

FIGURE 10.6 Plots of $e(\tau)$ with the deterministic $H(t)$. (a) $H(t) = 0.2|cost| - 0.6$. (b) $H(t) = 0.2|cost| + 0.6$. (c) Observing the zoom of $e(\tau)$ with $H(t) = 0.2|cost| - 0.6$ using the ordinate axis from -0.5 to 0.5.

Note that the statistical dependence, either LRD or SRD, is a property characterizing a global property of a time series (Li [6, 26, 35, 42], Li and Li [7], Li and Lim [44], Lim and Li [48], Gneiting and Schlather [49], Li and Wang [50], Li and Zhao [51], He et al. [52]). In fact, the LRD can be described by $C_{\mathrm{mmfGn}}(\tau) \sim |\tau|^{2H(t)-2}$ for $0.5 < H(t) < 1$ when $\tau \to \infty$. The SRD can be described by $C_{\mathrm{mmfGn}}(\tau) \sim |\tau|^{2H(t)-2}$ for $0 < H(t) < 0.5$ for $\tau \to \infty$. The condition of $\tau \to \infty$ implies that LRD or SRD is a global property of a time series.

10.2.4 PSD of mmfGn

Theorem 10.4. Denote the PSD of mmfGn by $S_{\mathrm{mmfGn}}(\omega)$. Then

$$S_{\mathrm{mmfGn}}(\omega) = V_{H(t)} \sin\left[H(t)\pi\right]\Gamma\left[2H(t)+1\right]|\omega|^{1-2H(t)}. \tag{10.21}$$

Proof. Let $S_{\mathrm{fGn}}(\omega)$ be the PSD of fGn. According to ref [38],

$$S_{\mathrm{fGn}}(\omega) = \int_{-\infty}^{\infty} C_{\mathrm{fGn}}(\tau)e^{-i\omega\tau}d\tau = V_H \sin(H\pi)\Gamma(2H+1)|\omega|^{1-2H}. \tag{10.22}$$

On the other hand,

$$S_{\mathrm{mmfGn}}(\omega) = \int_{-\infty}^{\infty} C_{\mathrm{mmfGn}}(\tau)e^{-i\omega\tau}d\tau. \tag{10.23}$$

Since the integral variable in the above integrand is τ, from (10.22), we have (10.21). The proof completes. □

10.2.5 Condition of mmfGn to be 1/f Noise

Let $S(\omega)$ be the PSD of $x(t)$. Following Chapter 2, if $S(0) = \infty$, $x(t)$ is called 1/f noise (also see Mandelbrot [53], Li and Zhao [54]). Denote the ACF of $x(t)$ by $r(\tau)$. Because LRD implies $S(0) = \int_{-\infty}^{\infty} r(\tau)d\tau = \infty$, so-called 1/f noise is actually a process with LRD but is described in the frequency domain.

Theorem 10.5. The condition of mmfGn to be 1/f noise is $0.5 < H(t) < 1$.

Proof. According to Theorem 10.4, we have

$$S_{\mathrm{mmfGn}}(0) = \begin{cases} \infty, & 0.5 < H(t) < 1 \\ 0, & 0 < H(t) < 0.5 \end{cases}. \tag{10.24}$$

Therefore, the condition of mmfGn to be 1/f noise is $0.5 < H(t) < 1$. The proof finishes. □

As a matter of fact, the property of 1/f noise reflects the LRD of mmfGn in the frequency domain. Note that $S_{\mathrm{mmfGn}}(0) = 0$ when $0 < H(t) < 0.5$. It means that $S_{\mathrm{mmfGn}}(0) = \int_{-\infty}^{\infty} C_{\mathrm{mmfGn}}(\tau)d\tau = 0$ for $0 < H(t) < 0.5$. Therefore, in the frequency domain, we can also obtain that the SRD condition of mmfGn is $0 < H(t) < 0.5$.

10.2.6 Fractal Dimension of mmfGn

Theorem 10.6. Let $D(t)$ be the fractal dimension of mmfGn. Then,

$$D(t) = 2 - H(t). \tag{10.25}$$

Proof. Note that for $\tau \to 0$, we have

$$C_{\mathrm{mmfGn}}(0) - C_{\mathrm{mmfGn}}(\tau) \sim c_1 |\tau|^{2H(t)}, \tag{10.26}$$

where c_1 is a constant in terms of τ and $2H(t)$ is the fractal index of mmfGn. Following Li [6, 26, 35, 42], Li and Li [7], Li and Lim [44], Lim and Li [48], Gneiting and Schlather [49], Li and Wang [50], Mandelbrot [55], Kent and Wood [56], Hall and Roy [57], Constantine and Hall [58], Davies and Hall [59], Chan et al. [60], Hall [61], Feuerverger et al. [62], Hall and Wood [63], Adler [64], with probability one, the relationship between the fractal index $2H(t)$ and the fractal dimension $D(t)$ of mmfGn is in the form

$$D(t) = 2 - \frac{2H(t)}{2}, \quad (\tau \to 0). \tag{10.27}$$

Therefore, (10.25) holds. This finishes the proof. $\square$

Note that the fractional dimension $D(t)$ is a parameter measuring a local property of $x(t)$, called local self-similarity or local irregularity or local roughness [56–64], because it is obtained under the condition of $\tau \to 0$. Although the fractal index $2H(t)$ plays the same role as that of the fractal dimension $D(t)$ for mmfGn, one usually utilizes the fractal dimension to measure local self-similarity or local roughness of $x(t)$, see refs [6, 7, 26, 35, 44, 48–64]. Note that different types of fractional processes may have different relationships between D and H. For example, for the GC process, D and H are independent of each other [6, 7, 26, 35, 44, 48–52]. However, for mmfGn, $D(t)$ linearly relates to $H(t)$ by (10.25). In other words, its LRD measure $H(t)$ is reflected in its local self-similarity measure $D(t)$ and vice versa.

10.3 ON STATIONARITY OF MMFGN

10.3.1 Problem Statements

It is well known that fBm is non-stationary while fGn, that is its increment process, is stationary. It is obvious that mBm is non-stationary. However, an interesting question is whether mmfGn is stationary or not.

We write the ACF of mmfGn as $r(\tau; H(t))$. Since the ACF is time-varying owing to $H(t)$, one might consider mmfGn as non-stationary according to the definition of stationarity. Nevertheless, that may be inappropriate in practice. Practically, we can only obtain an mmfGn series with finite record length. Assume that a series of mmfGn consists of N points of data. Suppose that this series is sectioned into B blocks and each block consists of L points of data such that $N = B \times L$. Denote the estimate of the ACF of that series in the lth block by $r_l(\tau; H(t))$ for $l = 1, 2, \ldots, B$. Then, in general,

$$r_l(\tau; H(t)) \neq r_m(\tau; H(t)) \text{ for } l, m = 1, 2, \ldots, B. \tag{10.28}$$

In addition to the above, there are usually errors in the estimate $r_l(\tau; H(t))$. Thus, the estimate $r_l(\tau; H(t))$ appears time-varying when l changes from 1 to B even if $H(t)$ is a constant. Thus, an mmfGn series with slightly time-varying $H(t)$ may not be taken as an evidence of non-stationarity of that series. As a matter of fact, in practical terms, the series is non-stationary if its ACF varies with time *significantly*, see Bendat and Piersol [65], Li et al. [66]. By significantly we mean that the distinction of the stationarity or non-stationarity of a series should not be affected by errors of proper estimates of ACFs of that series, referring Li [67, 68] for the interpretation of a proper estimate of ACF. With this point of view, we say that mmfGn may be stationary if the variations of $H(t)$ are non-significant.

10.3.2 Problem-Solving Ideas

Intuitively, $r(\tau; H(t))$ may not vary with time significantly if $H(t)$ just varies with time slightly. More precisely, denote the range of slight variations of $H(t)$ by (H_{min}, H_{max}). Then, we say that $r(\tau; H(t))$ does not vary with time significantly if $H(t) \in (H_{min}, H_{max})$ for the purpose of the evaluation of stationarity from an engineering view. We call (H_{min}, H_{max}) the stationary range in what follows unless otherwise stated.

10.3.3 Highlights

The contributions of this section in stationarity test are in two aspects. First, we shall exhibit that $H(t) = 0.70$ is the critical point for the stationarity evaluation of mmfGn. By critical point, we mean that mmfGn is stationary if $H(t) \leq 0.70$. For $H(t) > 0.70$, its stationarity depends on a variation range of $H(t)$. Therefore, mmfGn is conditionally stationary when $H(t) > 0.70$. A consequence from the stationarity of mmfGn under the condition of $H(t) \leq 0.70$ is that mmfGn with SRD is stationary. On the other hand, we will present a set of stationary ranges for $H(t) > 0.70$.

10.3.4 Stationary Ranges of mmfGn and the Critical Point

MmfGn is Gaussian. A Gaussian process is uniquely determined by its ACF. Although $r(\tau; H(t))$ is time-varying in general, as previously mentioned, it may be inappropriate to generally regard it as a non-stationary noise in engineering because variations of $H(t)$ may be non-significant. The question is what the ranges of slight variations of $H(t)$ are? We shall present the answer to this question below.

Denote the correlation coefficient between $r(\tau; H(t_1))$ and $r(\tau; H(t_2))$ by

$$\text{Corr}[r(\tau; H(t_1)), r(\tau; H(t_2))], t_1 \neq t_2. \tag{10.29}$$

It is known that $0 \leq |\text{Corr}[r(\tau; H(t_1)), r(\tau; H(t_2))]| \leq 1$. The larger the value of $\text{Corr}[r(\tau; H(t_1)), r(\tau; H(t_2))]$, the more the similarity between the pattern of $r(\tau; H(t_1))$ and that of $r(\tau; H(t_2))$. Mathematically, the case of

$$\text{Corr}[r(\tau; H(t_1)), r(\tau; H(t_2))] = 1 \tag{10.30}$$

implies that the pattern of $r(\tau; H(t_1))$ is exactly the same as that of $r(\tau; H(t_2))$. On the contrary,

$$\text{Corr}[r(\tau; H(t_1)), r(\tau; H(t_2))] = 0 \tag{10.31}$$

means that the pattern of $r(\tau, H(t_1))$ is totally different from that of $r(\tau, H(t_2))$. From the point of view of engineering, however, the extreme case of either $Corr[r(\tau, H(t_1)), r(\tau, H(t_2))] = 1$ or $Corr[r(\tau, H(t_1)), r(\tau, H(t_2))] = 0$ does not make much sense due to errors and uncertainties in data measurements and numerical computations. In practical terms, one uses a threshold for $Corr[r(\tau, H(t_1)), r(\tau, H(t_2))]$ to evaluate the similarity between the two. The concrete value of the threshold depends on the requirement designed by researchers but it is quite common to take 0.7 as the smallest value of the threshold for the pattern patching purpose, according to the correlation technology in statistical pattern matching (Fu [69], Maimon and Rokach [70], Li [71]). Suppose we consider 0.7 as the threshold value. Then, we say that the pattern of $r(\tau, H(t_1))$ is similar to that of $r(\tau, H(t_2))$ if $Corr[r(\tau, H(t_1)), r(\tau, H(t_2))] \geq$ 0.7 and dissimilar otherwise.

Note that if $Corr[r(\tau, H(t_1)), r(\tau, H(t_2))] \geq 0.7$, the ACFs of mmGn at t_1 and at t_2, respectively, are similar such that they may be taken as identical in the sense of statistical pattern matching. Thus, the meaning of (H_{min}, H_{max}) is $Corr[r(\tau, H(t_1)), r(\tau, H(t_2))] \geq 0.7$ if $H(t_1)$, $H(t_2) \in (H_{min}, H_{max})$.

By computations, we obtain the results in Table 10.1 and Table 10.2. Table 10.1 describes the stationary ranges of mmfGn. For instance, the mmfGn with $H(t) \in (0.680, 0.990)$ is stationary since $Corr[r(\tau, H(t_1)), r(\tau, H(t_2))] \geq 0.7$ for $0.680 < H(t) < 0.990$, see the first row in Table 10.1. If $\min[H(t)] = 0.6$ and $\max[H(t)] = 0.99$, the mmfGn with $H(t) \in (0.600, 0.990)$ is non-stationary since $Corr[r(\tau, \min[H(t)]), r(\tau, \max[H(t)])] = 0.579$.

Remark 10.1. The stationarity of mmfGn is dependent on ranges of $H(t)$ if $H(t) > 0.70$. That is, mmfGn with $H(t) > 0.70$ is conditionally stationary. $\square$

In engineering for numerical computations, one may take $\delta = 10^{-32}$ as infinitesimal. Thus, from Table 10.1 and Table 10.2, we see that mmfGn is stationary when $H(t) \in (\delta, H_{max})$ for $H_{max} \leq 0.700$ since $Corr[r(\tau, \delta), r(\tau, H_{max})] \geq 0.702$. In other words, mmfGn with $H(t) \leq 0.70$ is stationary. Therefore, we reveal important phenomena described by Remark 10.2 and Remark 10.3.

TABLE 10.2 Describing Stationarity of mmfGn with SRD

$H(t) \in (10^{-32}, H_{max})$	$Corr[r(\tau, 10^{-32}), r(\tau, H_{max})] \geq$
$H(t) \in (10^{-32}, 0.500)$	0.895
$H(t) \in (10^{-32}, 0.490)$	0.901
$H(t) \in (10^{-32}, 0.450)$	0.922
$H(t) \in (10^{-32}, 0.400)$	0.943
$H(t) \in (10^{-32}, 0.350)$	0.960
$H(t) \in (10^{-32}, 0.300)$	0.973
$H(t) \in (10^{-32}, 0.250)$	0.982
$H(t) \in (10^{-32}, 0.200)$	0.990
$H(t) \in (10^{-32}, 0.150)$	0.995
$H(t) \in (10^{-32}, 0.100)$	0.998
$H(t) \in (10^{-32}, 0.050)$	0.999

Remark 10.2. The stationarity of mmfGn is independent of $H(t)$ if $H(t) \leq 0.70$, implying that $H = 0.70$ is the critical point for determining whether mmfGn is unconditionally stationary or conditionally stationary. □

Remark 10.3. MmfGn with SRD is stationary because $\mathrm{Corr}[r(\tau; \delta), r(\tau; H_{max})] \geq 0.895$ if $\delta < H(t) < 0.500$, see Table 10.2. □

10.4 APPLICATION TO STATIONARITY TEST OF TRAFFIC

Teletraffic time series (traffic for short) is multi-fractal. Due to the presence of LRD in traffic, testing the stationarity of traffic becomes a tough issue (Abry and Veitch [45, first paragraph, Section III.A]). In this case study, we use a well-known real traffic trace named BC-OctExt89.TL, which was recorded at the Bellcore, Morristown research and engineering facility at 11:46 PM on Oct. 3, 1989 (Paxson and Floyd [72]). It consists of 10^6 packets. Denote the size of the ith packet ($i = 0, \ldots$) by $x(i)$. Note that without considering the Ethernet preamble, header, or cyclic redundancy check (CRC), the Ethernet protocol forces all packets to have at least a minimum size of 64 bytes and at most the maximum size of 1518 bytes. The fixed limit 1518 bytes is specified by IEEE standard without technical reason (Stalling [73]). Thus, traffic has the behavior of "burstiness," which simply implies that there would be no packets transmitted for a while, then flurry of transmission, no transmission for another long time, and so forth if one observes traffic over a long period of time. This also means that traffic has intermittency (Tobagi et al. [74], McDysan [75], Jiang and Dovrolis [76]). Due to this, traffic may also be described by ON/OFF model (Willinger et al. [77], Heath et al. [78]).

There are several types of data traffics, such as data on a packet-by-packet basis, total bytes or packet count of accumulated traffic within an interval on an interval-by-interval basis, and interarrival times. Their statistics, however, are, in general, identical (Borgnat et al. [79]). The traffic series used in the case study is in packet size on packet-by-packet basis with the ordinate of "bytes." In most case, the size of packets during ON periods is similar as indicated in Figure 10.7 (a) though not exactly the same in general as demonstrated in Figure 10.7 (b). It may have larger peaks but the size of all packets is limited to 1518 bytes, see Figure 10.7 (c).

We now set different time scales as shown in Table 10.3. Figure 10.8 demonstrates $H(t)$ s of BC-OctExt89.TL at the different time scales of 1024, 2048, 4096, and 8192. At small time scale, for instance, $t = 0, 1, \ldots, 1024$, for BC-OctExt89.TL, $\max[H(t)] - \min[H(t)] = 0.08$. Thus, according to Table 10.1 and Table 10.3, OctExt89.TL should be considered as stationary at this time scale. On the other side, at large time scale, e.g., $t = 0, 1, \ldots, 4096$, $\max[H(t)] - \min[H(t)] = 0.70$. Consequently, this traffic trace should be taken as non-stationary in this case based on Table 10.1 and Table 10.3. One thing in common in this chapter and Chapter 7 is that traffic is stationary at small time scales and non-stationary at large time scales. In other words, the stationarity of traffic is observation scale dependent.

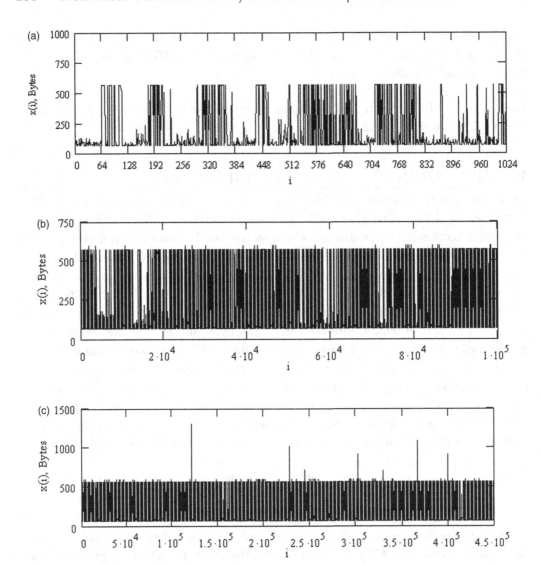

FIGURE 10.7 Real traffic BC-OctExt89.TL in packet size. (a) First 1024 data. (b) First 100,000 data. (c) First 450,000 data.

TABLE 10.3   Stationarity Observations at Different Time Scales

Time Scales	max[H(t)]	min[H(t)]	max[H(t)] − min[H(t)]	Corr[r(τ; min[H(t)]), r(τ; max[H(t)])]	Stationary
$t = 0, \ldots, 1024$	0.874	0.758	0.08	0.974	Yes
$t = 0, \ldots, 2048$	0.848	0.563	0.283	0.767	Yes
$t = 0, \ldots, 4096$	0.845	0.149	0.700	0.512	No
$t = 0, \ldots, 8192$	0.889	0.127	0.700	0.429	No

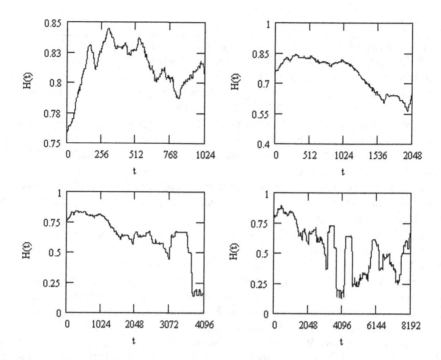

FIGURE 10.8 Local Hölder exponents of BC-OctExt89.TL at different time scales.

10.5 SUMMARY

In this chapter, we have presented a kind of multi-fractional noise which is termed mmfGn. Its ACF takes the form as that of the fGn but it is equipped with the time-varying Hurst parameter $H(t)$ on a point-by-point basis. The computation formulas with respect to the PSD, the approximation of the ACF, and the fractal dimension of mmfGn have been proposed. The condition of mmfGn to be of LRD or $1/f$ noise has been brought forward.

The computation formula of $H(t)$ of mmfGn is the same as that of the mfBm. However, using $H(t)$ of mmfGn to investigate the time-varying LRD substantially differs from using $H(t)$ of mfBm. In fact, the LRD condition of mmfGn is $0.5 < H(t) < 1$ while that of mfBm is $0 < H(t) < 1$. In addition, using $H(t)$ of mfBm implies that the investigated data series is non-stationary at any time scales because mmfBm is non-stationary at any time scales while using $H(t)$ of mmfGn does not imply so. As a whole, using $H(t)$ of mfBm implies that the investigated data series follows mfBm while using $H(t)$ of mmfGn has not that implication. Therefore, the mmfGn model may yet be a new tool to describe the time varying LRD behavior of time series.

In Section 10.3, we have discussed that there are two states of the stationarity of mmfGn. One is conditionally stationary, which occurs when $H(t) > 0.70$. The other is stationary irrelevant of $H(t)$ if $H(t) \leq 0.70$. The critical point is $H(t) = 0.70$. We have presented a set of ranges of $H(t)$ within which mmfGn is stationary under the condition of $H(t) > 0.70$. We have explained that a natural consequence of the unconditional stationarity is that the mmfGn with SRD is always stationary. We have shown a case for applying the present results to the stationarity test of a real traffic trace.

REFERENCES

1. J. Levy-Vehel, and E. Lutton, *Fractals in Engineering*, Springer, Berlin, 2005.
2. M. M. Meerschaert, and F. Sabzikar, Tempered fractional Brownian motion, *Statistics & Probability Letters*, 83(10) 2013, 2269–2275.
3. M. Pinchas, Cooperative multi PTP slaves for timing improvement in an fGn environment, *IEEE Communications Letters*, 22(7) 2018, 1366–1369.
4. H. Liu, W. Q. Song, M. Li, A. Kudreyko, and E. Zio, Fractional Levy stable motion: Finite difference iterative forecasting model, *Chaos, Solitons and Fractals*, 133, 2020, 109632.
5. M. Li, Generalized fractional Gaussian noise and its application to traffic modeling, *Physica A*, 579, 2021, 1236137 (22 pp).
6. M. Li, Long-range dependence and self-similarity of teletraffic with different protocols at the large time scale of day in the duration of 12 years: Autocorrelation modeling, *Physica Scripta*, 95(4) 2020, 065222 (15 pp).
7. M. Li, and J.-Y. Li, Generalized Cauchy model of sea level fluctuations with long-range dependence, *Physica A*, 484, 2017, 309–335.
8. M. Li, X. Sun, and X. Xiao, Revisiting fractional Gaussian noise, *Physica A*, 514, 2019, 56–62.
9. Y. Chen, X. Wang, and W. Deng, Localization and ballistic diffusion for the tempered fractional Brownia–Langevin motion, *Journal of Statistical Physics*, 169(1) 2017, 18–37.
10. S. C. Lim, Fractional Brownian motion and multifractional Brownian motion of Riemann-Liouville type, *Journal of Physics A: Mathematics & General*, 34(7) 2001, 1301–1310.
11. S. C. Lim, and L. P. Teo, Weyl and Riemann-Liouville multifractional Ornstein-Uhlenbeck processes, *Journal of Physics A: Mathematical and Theoretical*, 40(23) 2007, 6035–6060.
12. J. Levy-Vehel, Beyond multifractional Brownian motion: New stochastic models for geophysical modeling, *Nonlinear Processes in Geophysics*, 20(5) 2013, 643–655.
13. X. Fan, and J. Levy-Vehel, Tempered fractional multistable motion and tempered multifractional stable motion, *ESAIM: Probability and Statistics*, 23, 2019, 37–67.
14. A. Ayache, N.-R. Shieh, and Y. Xiao, Multiparameter multifractional Brownian motion: Local nondeterminism and joint continuity of the local times, *Annales de l'Institut Henri Poincaré, Probabilités et Statistiques*, 47(4) 2011, 1029–1054.
15. R. Le Guével, and J. Levy-Vehel, Hausdorff, large deviation and Legendre multifractal spectra of Lévy multistable processes, *Stochastic Processes and their Applications*, 130(4), 2020, 2032–2057.
16. D. Molina-García, T. Sandev, H. Safdari, G. Pagnini, G. Pagnini, A. V. Chechkin, and R. Metzler, Crossover from anomalous to normal diffusion: Truncated power-law noise correlations and applications to dynamics in lipid bilayers, *New Journal of Physics*, 20(10) 2018, 103027.
17. K. V. Ralchenko, and G. Shevchenko, Path properties of multifractal Brownian motion, *Theory of Probability and Mathematical Statistics*, 80, 2010, 119–130.
18. K. C. Lee, Characterization of turbulence stability through the identification of multifractional Brownian motions, *Nonlinear Processes in Geophysics*, 20(1) 2013, 97–106.
19. J. Ryvkina, Fractional Brownian motion with variable Hurst parameter: Definition and properties, *Journal of Theoretical Probability*, 28(3) 2015, 866–891.
20. A. Benassi, S. Cohen, and J. Istas, Identifying the multifractional function of a Gaussian process, *Statistics and Probability Letters*, 39(4) 1998, 337–345.
21. G. Chan, and A. T. Wood, Simulation of multifractional Brownian motion, *COMPSTAT (Proceedings in Computational Statistics 13th Symposium held in Bristol, Great Britain, 1998)*, Springer, Berlin/Heidelberg, 1998, pp. 233–238.
22. S. A. Stoev, and M. S. Taqqu, How rich is the class of multifractional Brownian motions, *Stochastic Processes and Their Applications*, 116(2) 2006, 200–221.
23. M. Balcerek, and K. Burnecki, Testing of multifractional Brownian motion, *Entropy*, 22(12) 2020, 1403.

24. Y. Karaca, M. Moonis, and D. Baleanu, Fractal and multifractional-based predictive optimization model for stroke subtypes' classification, *Chaos Solitons & Fractals*, 136, 2020, 109820.

25. Y. Karaca, and D. Baleanu, A novel R/S fractal analysis and wavelet entropy characterization approach for robust forecasting based on self-similar time series modeling, *Fractals*, 28(8) 2020, 2040032.

26. M. Li, Multi-fractional generalized Cauchy process and its application to teletraffic, *Physica A*, 550, 2020, 123982 (14 pp).

27. K. S. Miller, and B. Ross, *An Introduction to the Fractional Calculus and Fractional Differential Equations*, John Wiley & Sons, New York, 1993.

28. J. L. Lavoie, T. J. Osler, and R. Tremblay, Fractional derivatives and special functions, *SIAM Review*, 18(2) 1976, 240–268.

29. J. Klafter, S. C. Lim, and R. Metzler, *Fractional Dynamics: Recent Advances*, World Scientific, Singapore, 2012.

30. M. Li, *Theory of Fractional Engineering Vibrations*, De Gruyter, Berlin/Boston, 2021.

31. C. P. Li, and M. Cai, *Theory and Numerical Approximations of Fractional Integrals and Derivatives*, SIAM, Philadelphia, 2019.

32. M. D. Ortigueira, and D. Valério, *Fractional Signals and Systems*, De Gruyter, Berlin/Boston, 2020.

33. T. M. Atanackovic, S. Pilipovic, B. Stankovic, and D. Zorica, *Fractional Calculus with Applications in Mechanics*, John Wiley & Sons, Croydon, 2014.

34. B. B. Mandelbrot, and J. W. van Ness, Fractional Brownian motions, fractional noises and applications, *SIAM Review*, 10(4) 1968, 422–437.

35. M. Li, Fractal time series: A tutorial review, *Mathematical Problems in Engineering*, 2010, 2010, Article ID 157264 (26 pp).

36. N. Zeinali, and A. Pourdarvish, An entropy-based estimator of the Hurst exponent in fractional Brownian motion, *Physica A*, 591, 2022, 126690.

37. Q. Yu, Higher-order derivative of self-intersection local time for fractional Brownian motion, *Journal of Theoretical Probability*, 34(4) 2021, 1749–1774.

38. M. Li, and S. C. Lim, A rigorous derivation of power spectrum of fractional Gaussian noise, *Fluctuation and Noise Letters*, 6(4) 2006, C33–C36.

39. R. F. Peltier, and J. Levy-Vehel, *Multifractional Brownian Motion: Definition and Preliminaries Results*, INRIA RR 2645, France, 1995.

40. S. C. Lim, and S. V. Muniandy, On some possible generalizations of fractional Brownian motion, *Physics Letters A*, 266(2–3) 2000, 140–145.

41. S. V. Muniandy, S. C. Lim, and R. Murugan, Inhomogeneous scaling behaviors in Malaysia foreign currency exchange rates, *Physica A*, 301(1–4) 2001, 407–428.

42. M. Li, Generation of teletraffic of generalized Cauchy type, *Physica Scripta*, 81(2) 2010, 025007 (10 pp).

43. M. Li, Change trend of averaged Hurst parameter of traffic under DDOS flood attacks, *Computers & Security*, 25(3) 2006, 213–220.

44. M. Li, and S. C. Lim, Modeling network traffic using generalized Cauchy process, *Physica A*, 387(11) 2008, 2584–2594.

45. P. Abry, and D. Veitch, Wavelet analysis of long-range dependent traffic, *IEEE Transactions on Information Theory*, 44(1) 1998, 2–15.

46. M. Li, and W. Zhao, Quantitatively investigating locally weak stationarity of modified multi-fractional Gaussian noise, *Physica A*, 391(24) 2012, 6268–6278.

47. A. M. Yaglom, *Correlation Theory of Stationary and Related Random Functions*, Vol. I, Springer, New York, 1987.

48. S. C. Lim, and M. Li, A generalized Cauchy process and its application to relaxation phenomena, *Journal of Physics A: Mathematical and General*, 39(12) 2006, 2935–2951.

49. T. Gneiting, and M. Schlather, Stochastic models that separate fractal dimension and the Hurst effect, *SIAM Review*, 46(2) 2004, 269–282.
50. M. Li, and A. Wang, Fractal teletraffic delay bounds in computer networks, *Physica A*, 557, 2020, 124903 (13 pp).
51. M. Li, and W. Zhao, Representation of a stochastic traffic bound, *IEEE Transactions on Parallel and Distributed Systems*, 21(9) 2010, 1368–1372.
52. J. He, G. Christakos, J. Wu, M. Li, and J. Leng, Spatiotemporal BME characterization and mapping of sea surface chlorophyll in Chesapeake Bay (USA) using auxiliary sea surface temperature data, *Science of the Total Environment*, 794, 2021, 148670.
53. B. B. Mandelbrot, *Multifractals and 1/f Noise*, Springer, New York, 1998.
54. M. Li, and W. Zhao, On 1/f noise, *Mathematical Problems in Engineering*, 2012, 2012, Article ID 673648 (22 pp).
55. B. B. Mandelbrot, *The Fractal Geometry of Nature*, W. H. Freeman, New York, 1982.
56. J. T. Kent, and A. T. Wood, Estimating the fractal dimension of a locally self-similar Gaussian process by using increments, *Journal of the Royal Statistical Society-Series B*, 59(3) 1997, 679–699.
57. P. Hall, and R. Roy, On the relationship between fractal dimension and fractal index for stationary stochastic processes, *The Annals of Applied Probability*, 4(1) 1994, 241–253.
58. A. G. Constantine, and P. Hall, Characterizing surface smoothness via estimation of effective fractal dimension, *Journal of the Royal Statistical Society-Series B*, 56(1) 1994, 97–113.
59. S. Davies, and P. Hall, Fractal analysis of surface roughness by using spatial data, *Journal of the Royal Statistical Society Series B*, 61(1) 1999, 3–37.
60. G. Chan, P. Hall, and D. S. Poskitt, Periodogram-based estimators of fractal properties, *The Annals of Statistics*, 23(5) 1995, 1684–1711.
61. P. Hall, On the effect of measuring a self-similar process, *SIAM Journal on Applied Mathematics*, 55(3) 1995, 800–808.
62. A. Feuerverger, P. Hall, and A. T. A. Wood, Estimation of fractal index and fractal dimension of a Gaussian process by counting the number of level crossings, *Journal of Time Series Analysis*, 15(6) 1994, 587–606.
63. P. Hall, and A. Wood, On the performance of box-counting estimators of fractal dimension, *Biometrika*, 80(1) 1993, 246–252.
64. R. J. Adler, *The Geometry of Random Fields*, John Wiley & Sons, New York, 1981.
65. J. S. Bendat, and A. G. Piersol, *Random Data: Analysis and Measurement Procedure*, 3rd ed., John Wiley & Sons, New York, 2000.
66. M. Li, W.-S. Chen, and L. Han, Correlation matching method of the weak stationarity test of LRD traffic, *Telecommunication Systems*, 43(3–4) 2010, 181–195.
67. M. Li, A method for requiring block size for spectrum measurement of ocean surface waves, *IEEE Transactions on Instrumentation and Measurement*, 55(6) 2006, 2207–2215.
68. M. Li, Record length requirement of long-range dependent teletraffic, *Physica A*, 472, 2017, 164–187.
69. K. S. Fu (ed.), *Digital Pattern Recognition*, Springer, New York, 1976.
70. O. Maimon, and L. Rokach, *Data Mining and Knowledge Discovery Handbook*, Springer, New York, 2010.
71. M. Li, An iteration method to adjusting random loading for a laboratory fatigue test, *International Journal of Fatigue*, 27(7) 2005, 783–789.
72. V. Paxson, and S. Floyd, Wide area traffic: The failure of Poisson modeling, *IEEE/ACM Transactions on Networking*, 3(3) 1995, 226–244.
73. W. Stalling, *Data and Computer Communications*, 7th ed., Pearson Education, New York, 2004.
74. F. A. Tobagi, M. Gerla, R. W. Peebles, and E. G. Manning, Modeling and measurement techniques in packet communication networks, *Proceedings of the IEEE*, 66(11) 1978, 1423–1447.

75. D. McDysan, *QoS & Traffic Management in IP & ATM Networks*, McGraw-Hill, New York, 2000.
76. H. Jiang, and C. Dovrolis, Why is the internet traffic bursty in short time scales, *Performance Evaluation Review*, 33(1) 2005, 241–252.
77. W. Willinger, V. Paxson, and M. S. Taqqu, Self-similarity and heavy tails: Structural modeling of network traffic, In *A Practical Guide to Heavy Tails: Statistical Techniques and Applications*, R. Adler, R. Feldman, and M. S. Taqqu (eds.), Birkhauser, Boston, 1998.
78. D. Heath, I. S. Resnick, and G. Samorodnitsky, Heavy tails and long range dependence in on-off processes and associated fluid models, *Mathematics of Operations Research*, 23(1) 1998, 145–165.
79. P. Borgnat, G. Dewaele, K. Fukuda, P. Abry, and K. Cho, Seven years and one day: Sketching the evolution of Internet traffic, *Proc. the 28th IEEE INFOCOM 2009*, Rio de Janeiro (Brazil), May 2009, pp. 711–719.

CHAPTER 11

Traffic Simulation

Generation of long-range-dependent (LRD) traffic is crucial to networking, for example, simulating the Internet. In this aspect, it is desired to generate an LRD traffic series according to a given correlation structure that may well reflect the statistics of real traffic. Research in traffic modeling exhibits that the LRD traffic is well modeled by the generalized Cauchy (GC) process indexed by two parameters that separately characterize the self-similarity (SS), which is a local property described by the fractal dimension D, and long-range dependence (LRD), which is a global property measured by the Hurst parameter H. In this chapter, we present a computational method to generate the LRD traffic based on the correlation form of the GC process in both the uni-fractal and multi-scale cases. It may yet be used as a way to flexibly simulate the realizations that reflect the fractal phenomena of traffic for both short-term lags and long-term ones.

11.1 INTRODUCTION

Teletraffic (traffic for short) is a type of fractal series with both (local) self-similarity SS and LRD. It is one of the common cases to exhibit the fractal phenomena of time series (see Li [1–5], Mandelbrot [6], Beran [7], and references therein). Its simulation is desired in the Internet communications (Paxson and Floyd [8]). Since fractal time series has the statistical properties that are substantially different from those of conventional time series, such as SS, LRD, and very big variance, the simulation of the LRD traffic differs, in methodology, from those used in the generation of conventional series.

There are two basic characteristics of conventional random processes, namely, probability density function (PDF) and power spectrum density (PSD) function (Papoulis and Pillai [9]). Consequently, there are two categories of methods in random data generation. One is to synthesize a random function $x(t)$ such that it follows a predetermined PDF $p(x)$ (see Press et al. [10]). The other is to generate a random function $x(t)$ according to a given PSD $S_x(f)$ (see Chakrabarti [11], Li [12, 13], Li et al. [14], Smozuka [15], and Stanislaw [16]).

The two have their applications to practical issues. For instance, queuing theory needs random data that are synthesized based on a predetermined PDF (Gibson [17]). Investigating the frequency response of structures desires random data that are generated

DOI: 10.1201/9781003354987-14

according to a given PSD (see Chakrabarti [11], Li [12, 13], Li et al. [14], Smozuka [15], Stanislaw [16], and Mann [18]). Methods in both categories are quite mature. However, both encounter difficulties in the LRD traffic generation. Consequently, synthesizing the LRD traffic becomes an issue worth studying in computer science (see Paxson and Floyd [8, First sentence, Subparagraph 4, Paragraph 2, Section 6.1]).

Various methods of simulating the LRD traffic have been reported (see Garrett and Willinger [19], Jeong et al. [20], Lau et al. [21], Ledesma and Liu [22], Li and Chi [23], Paxson [24], and Rroughan [25]). Those are based on the model of the fractional Gaussian noise (fGn) with a single parameter introduced by Mandelbrot and van Ness [26]. The realizations by using those methods, therefore, only have the statistical properties of fGn with a single parameter, which characterizes SS and LRD by the linear relationship $D = 2 - H$ (Li [1–5], Li and Lim [27, 28]). Nevertheless, the limitation of the single-parameter fGn in traffic modeling was noticed as can be seen from the last sentence in Paragraph 4, §7.4 in Paxson and Floyd [29], which stated that "it might be difficult to characterize the correlations over the entire traffic traces with a single Hurst parameter". As a matter of fact, Kaplan and Kuo [30] also mentioned the shortcoming that the single-parameter fGn may not well fit series for short-term lags.

We studied a class of stationary Gaussian processes indexed by two parameters which separately characterize SS and LRD of fractal time series (Li [1, 3–5], Li and Lim [27, 28], Lim and Li [31]). Our research exhibits that the GC model is in far more agreement with the real traffic in comparison with fGn. Nevertheless, reports regarding the simulation of the CG type traffic, the multi-scale LRD traffic of the GC type in particular, are rarely seen. In this chapter, we aim at providing a method for synthesizing the LRD traffic of the GC type.

The rest of chapter is organized as follows: In Section 11.2, we shall respectively brief the PDF-based method and PSD-based method for the simulation of conventional random functions and point out their limitations in synthesizing the LRD traffic. Besides, we will brief the autocorrelation function (ACF)-based method of the random data generation. The simulation of the LRD traffic based on a given ACF of the GC process in the multi-scale case is presented in Section 11.3. Discussions are given in Section 11.4, which is followed by summary.

11.2 SIMULATIONS BASED ON GIVEN PDF/PSD/ACF

11.2.1 PDF-Based Simulation Method

Let x be a random series to be generated and $x \in [x_{min}, x_{max}]$. Let $p(x)$ be the PDF of x. Then, the PDF-based generation method can be used to generate a random series x whose PDF $p(x)$ is predetermined. A commonly used way, which is called the inverse transform method in Press et al. [10], is explained as follows.

Let U be a uniformly distributed random number between 0 and 1. That is, $U \sim U(0, 1)$. Let $P(x)$ be the cumulative distribution function of x. Then,

$$P(x) = \int_{-\infty}^{x} p(t)dt. \tag{11.1}$$

Because the PDF of x is given, the cumulative distribution function of x is known. The PDF-based method, that is, the inverse transform method, says that if

$$X = P^{-1}(U), \tag{11.2}$$

then x has the PDF $p(x)$, where P^{-1} is the inverse of P.

Recall that the LRD traffic is a class of two-order random processes (see Beran [7], Paxson and Floyd [29], Adas [32], Beran et al. [33], Lee and Fapojuwo [34], Leland et al. [35], Li [36], Michiel and Laevens [37], and Tsybakov and Georganas [38]). It usually has very big variance. To interpret this, we use a real traffic trace from the Internet Traffic Archive (ftp://ita.ee.lbl.gov/traces/). Its file name is pAug89, which was measured on an Ethernet at the Bellcore (BC) Morristown Research and Engineering facility. Let $x(t_i)$ be a traffic series, indicating the number of bytes of a packet at time t_i. Then, $x(i)$ implies the number of bytes in the ith packet ($i = 0, 1, \ldots$). Figure 11.1 gives the plot of the first 1024 points of that trace.

Note that the LRD traffic at large time scales is Gaussian as can be seen from the work by Scherrer et al. [39]. Let μ_ξ and σ_ξ be the expectation and the standard deviation of a random variable ξ that follows the Gaussian distribution, respectively. Then,

$$\xi \sim \frac{1}{\sqrt{2\pi}\sigma_\xi} e^{\frac{(\xi - \mu_\xi)^2}{2\sigma_\xi^2}}.$$

Denote the mean and the standard deviation of $x(i)$ with the sample size n by (11.3) and (11.4), respectively. Figure 11.2 shows its standard deviation plot and Figure 11.3 shows the mean plot.

$$E[x(i); n] = \frac{\sum\limits_{i=0}^{n} x(i)}{n-1}, \tag{11.3}$$

$$Std[x(i); n] = \sqrt{\frac{\sum\limits_{i=0}^{n} \{x(i) - E[x(i)]\}^2}{n-1}}. \tag{11.4}$$

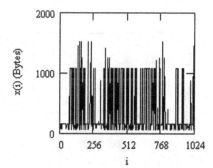

FIGURE 11.1 Real traffic trace pAug89.

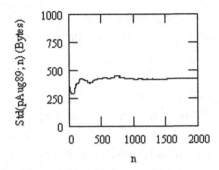

FIGURE 11.2 Standard deviation plot of pAug89.

By numeric computation, we have

$$\text{Std}(\text{pAug89}) = \sigma_{\text{pAug89}} = 484.974 \ (\text{Bytes}), \tag{11.5}$$

$$E(\text{pAug89}) = \mu_{\text{pAug89}} = 434.292 \ (\text{Bytes}). \tag{11.6}$$

Without considering the Ethernet preamble, header, or cyclic redundancy check (CRC), the Ethernet protocol forces all packets to have at least a minimum size of 64 bytes and at most the maximum size of 1518 bytes. The fixed limit 1518 bytes is specified by IEEE standard without technical reason. On the other side, $x(i) \geq 0$. Therefore, the very big variance makes the Gaussian PDF inappropriate to be used to synthesize the LRD traffic using the PDF-based method. As a matter of fact, by using the PDF-based method, one should first set the range $x \in [x_{\min}, x_{\max}]$. In practical terms, however, $x_{\max}$ is forced to be 1518 bytes while $x_{\min}$ must be equal to or greater than 0. Taking the traffic trace pAug89 as an example, however, we have

$$\mu_{\text{pAug89}} + 3\sigma_{\text{pAug89}} = 1889 \ (\text{Bytes}),$$

and

$$\mu_{\text{pAug89}} - 3\sigma_{\text{pAug89}} = -1021 \ (\text{Bytes}).$$

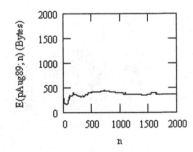

FIGURE 11.3 Mean plot of pAug89.

Thus, if the PDF-based method were used with the Gaussian PDF, the size of the synthesized data would be in contradiction to the packet size the real traffic requires. Hence, the limitation of the PDF-based method for the LRD traffic simulation.

11.2.2 PSD-Based Simulation Method

Let $x(t)$ be a random function with the frequency bandwidth $(0, f_{\max})$, where $f_{\max}$ is the maximum effective frequency of $x(t)$. Let $S_x(f)$ be its PSD. The discrete $S_x(f)$ is given by $S_x(n\Delta f)$, which implies that its PSD is divided into N equal increments of width Δf between 0 and $f_{\max}$. According to the analysis in Chakrabarti [11] and Stanislaw [16],

$$x(t) = \sum_{n=1}^{N} a_n \cos[2\pi n\Delta f t + \vartheta(n\Delta f)],\ 0 \le t \le T_r, \tag{11.7}$$

where $\vartheta(n\Delta f)$ is a uniformly distributed random phase between $-\pi$ and π. The coefficients a_n are given by

$$a_n = a(n\Delta f) = \sqrt{2S_x(n\Delta f)\Delta f}. \tag{11.8}$$

Eq. (11.7) produces the random data whose PSD $S_x(f)$ is predetermined. The random data generated by using (11.7) contain periodicity. Let T_r be the period of generated random data. Then,

$$T_r = \frac{1}{\Delta f}. \tag{11.9}$$

In the case of the finite Fourier transform,

$$N\Delta f = f_{max}, \tag{11.10}$$

$$\frac{N}{T_r} = f_{max}. \tag{11.11}$$

Within $[0, T_r]$, the generated random data are not repeated. Refer to Li [11, 12] and Li et al. [13] for the application cases of this method.

Recall that the PSD of the LRD traffic has the property of $1/f$ noise (Li [1], Csabai [40]). A $1/f$ noise is divergent when $f \to 0$. For instance, the PSD of fGn is given by (Li and Lim [41], Li [42])

$$S(f) = \sigma^2 \sin(H\pi)\Gamma(2H+1)|2\pi f|^{1-2H}, \tag{11.12}$$

where $0 < H < 1$ is the Hurst parameter. This makes the proper computation of (11.8) difficult owing to $\lim_{f \to 0} S(f) = \infty$ if $0.5 < H < 1$.

The previous discussions exhibit that either the PDF-based or the PSD-based simulation scheme has its limitation in synthesizing the LRD traffic. Though the PSD of the LRD traffic is divergent at $f \to 0$, its ACF is an ordinary function. Therefore, the ACF-based simulation was discussed in our previous work (Li and Chi [23]).

11.2.3 ACF-Based Simulation

Let $r_x(\tau)$ be the ACF of $x(t)$. Then,

$$r_x(\tau) = E[x(t)x(t+\tau)], \tag{11.13}$$

where E is the expectation operator. Let $w(t)$, $W(f)$, and $S_w(f)$ be the white noise function, its spectrum, and its PSD, respectively. Then,

$$W(f) = F[w(t)], \tag{11.14}$$

$$S_w(f) = WW^* = \text{Constant}, \tag{11.15}$$

where W^* represents the complex conjugation of W and F the operator of Fourier transform. Let r_w be the ACF of white noise. Then,

$$r_w(\tau) = E[w(t)w(t+\tau)] = \delta(\tau), \tag{11.16}$$

where $\delta(\tau)$ is the delta function.

Denote $h(t)$ and $H(f)$ the impulse function and the system function of a linear filter, respectively, see Figure 11.4.

Note that

$$y = w * F^{-1}\left\{[F(r_x)]^{0.5}\right\}. \tag{11.17}$$

where * means the convolution operation. Let

$$r_y = r_x, \tag{11.18}$$

where r_y is the ACF of y. If h is such that $r_y = r_x$, we say that y is statistically identical to x in the sense of $r_y = r_x$. In fact, under the excitation of white noise, one has

$$S_y(f) = |H(f)|^2. \tag{11.19}$$

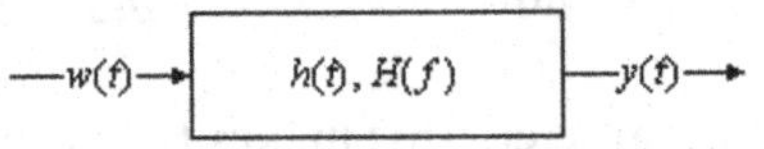

FIGURE 11.4 Linear system under the excitation of white noise.

Consequently, we have

$$S_y(f) = S_x(f). \tag{11.20}$$

Let $\psi(f)$ be the phase function of $H(f)$. Then,

$$H(f) = \sqrt{S_x(f)}e^{-j\psi(f)}. \tag{11.21}$$

Without the loss of generality, let

$$\psi(f) = 2n\pi \, (n = 0, 1, \ldots). \tag{11.22}$$

Then,

$$H(f) = [S_x(f)]^{0.5}. \tag{11.23}$$

Denote F^{-1} as the inverse of F. Then, we have

$$h = F^{-1}\left\{[F(r_x)]^{0.5}\right\}. \tag{11.24}$$

Therefore, for a given ACF r_x, we have the synthesized random data expressed as

$$y = w * F^{-1}\left\{[F(r_x)]^{0.5}\right\}. \tag{11.25}$$

It is worth noting that Eq. (11.25) has no limitation as $1/f$ noise has at $f = 0$ because $F^{-1}\{[F(r_x)]^{0.5}\}$ is an item in time domain. Consequently, Eq. (11.25) represents a method in time domain.

11.3 GENERATION OF LRD TRAFFIC OF GC TYPE

Let $X(t)$ be a traffic series. Then, $X(t)$ being of SS with the SS index κ implies

$$X(at) =_d a^\kappa X(t), a > 0, \tag{11.26}$$

where $=_d$ denotes equality in finite joint finite distribution. On the other hand, $X(t)$ is LRD if its ACF follows the power law given by

$$r(\tau) \sim c\tau^{-\beta}(\tau \to \infty), c > 0, 0 < \beta < 1. \tag{11.27}$$

In principle, the fractal dimension D and the Hurst parameter H of a series can be measured independently. The former characterizes SS while the later LRD. The ACF of fGn is given by

$$Rfgn(\tau) = \frac{\sigma^2 \varepsilon^{2H-2}}{2}\left[\left(\frac{|\tau|}{\varepsilon}+1\right)^{2H} + \left|\frac{|\tau|}{\varepsilon}-1\right|^{2H} - 2\left|\frac{\tau}{\varepsilon}\right|^{2H}\right], \tau \in \mathbb{R}, \tag{11.28}$$

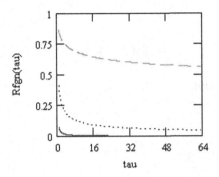

FIGURE 11.5 ACF of fGn. Solid line: $H = 0.55$, dot line: $H = 0.75$, dash line: $H = 0.95$.

where $\sigma^2 = (H\pi)^{-1}\Gamma(1-2H)\cos(H\pi)$ is the intensity of fGn and $\varepsilon > 0$ is used by smoothing the fractional Brownian motion (fBm) so that the smoothed fBm is differentiable (Mandelbrot and van Ness [26, p. 427–428], Li and Lim [41]). The case $0.5 < H < 1$ corresponds to LRD while $0 < H < 0.5$ implies short-range dependence (SRD). FGn reduces to standard white noise when $H = 0.5$. FGn with a single parameter characterizes both SS and LRD of traffic by H, or equivalently D since there is a linear relationship between them, that is, $D = 2 - H$, which is its limitation in the LRD traffic modeling (Li and Lim [27]). Figure 11.5 indicates the plots of the ACF of fGn with different values of H.

11.3.1 Concept of GC Process

$X(t)$ is called the GC process if it is a stationary Gaussian centered process with the following ACF:

$$r(\tau) = E\left[X(t+\tau)X(t)\right] = \left(1+\tau^{\alpha}\right)^{-\frac{\beta}{\alpha}}, \tau > 0, \tag{11.29}$$

where $0 < \alpha < 2$, and $\beta > 0$.

Note that $r(\tau)$ is positive-definite for the above ranges of α and β, and it is a completely monotone for $0 < \alpha \leq 1$, $\beta > 0$. When $\alpha = 2$ and $\beta = 1$, one gets the usual Cauchy process. The GC process satisfies the LRD property for $\beta < 1$ since

$$\int_0^{\infty} r(\tau)d\tau = \int_0^{\infty} \left(1+|\tau|^{\alpha}\right)^{-\frac{\beta}{\alpha}} d\tau = \infty \text{ if } \beta < 1. \tag{11.30}$$

Note that the GC process is locally self-similar. In fact, it is a Gaussian stationary process with the local SS of order α as its ACF satisfies for $\tau \rightarrow 0$,

$$r(\tau) = 1 - \frac{\beta}{\alpha}|\tau|^{\alpha}\left\{1 + O\left(|\tau|^{\gamma}\right)\right\}, \ \gamma > 0. \tag{11.31}$$

The above expression is equivalent to the following more commonly used definition of the local SS of a Gaussian process

$$X(t) - X(a\tau) =_d a^\kappa [X(t) - X(\tau)], \quad \tau \to 0. \tag{11.32}$$

The equivalence can be shown by noting that for τ_1 and $\tau_2 \to 0$, (11.31) gives for $\kappa = \alpha/2$

$$\mathrm{E}\big[(X(t+b\tau_1) - X(t))(X(t+b\tau_2) - X(t))\big] = \frac{\beta}{\alpha}\Big[|b\tau_1|^\alpha + |b\tau_2|^\alpha - |b(\tau_1 - \tau_2)|^\alpha\Big]$$

$$= \frac{\beta b^\alpha}{\alpha}\big(|\tau_1|^\alpha + |\tau_2|^\alpha - |\tau_1 - \tau_2|^\alpha\big) = \mathrm{E}\Big\{\big[b^{\alpha/2}(X(t+\tau_1) - X(t))\big]\big[b^{\alpha/2}(X(t+\tau_2) - X(t))\big]\Big\}.$$

In order to determine D of the graph of $X(t)$, we consider the local property of traffic. The fractal dimension D of a locally self-similar traffic of order α is given by (Kent and Wood [43], Constantine and Hall [44], and Taylor and Taylor [45])

$$D = 2 - \frac{\alpha}{2}. \tag{11.33}$$

From (11.27), one has

$$H = 1 - \frac{\beta}{2}. \tag{11.34}$$

From (11.33) and (11.34), we see that D and H may vary independently. $X(t)$ is LRD for $0 < \beta < 1$ and SRD for $\beta > 1$. Thus, we have the fractal index α which determines D and the index β that characterizes H. Figure 11.6 gives the plots of the ACF of the GC process with the constant H but different values of D, which clearly exhibits the advantage of the GC model in fitting a series for short-term lags. In Figure 11.7, on the other side, the ACF of the GC process is plotted with the constant D but different values of H. Both figures evidently interpret the flexibility of the GC process in modeling the fractal type traffic.

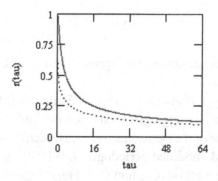

FIGURE 11.6 ACF of GC process with $H = 0.75$. Solid: $D = 1.2$, dot: $D = 1.8$.

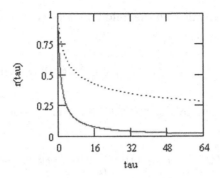

FIGURE 11.7 ACF of GC process with $D = 1.5$. Solid: $H = 0.55$, dot: $H = 0.85$.

Note that

$$\left(1+|\tau|^{\alpha}\right)^{-\beta/\alpha} \sim |\tau|^{-\beta} \ (\tau \to \infty).$$

Thus, the GC process approaches fGn at large time scale, more precisely in mathematics, for $\tau \to \infty$.

In the end of this section, we mention the estimates of D and H of the GC process for the purpose of completing the description of the GC process though the focus of this chapter does not relate to their estimators. There are some techniques popularly used to estimate D, such as box-counting, spectral, and variogram-based methods (see Mandelbrot [6], Dubuc et al. [46], Hall and Wood [47]). Nevertheless, some of the more popular methods, such as box-counting method, suffer from biases (Taylor and Taylor [45]). The method worth noting is called the variogram estimator explained by Constantine and Hall [44]. Denote by $\gamma(d)$ the observed mean of the square of the difference between two values of the series at points that are distance d apart. Then, for $d \to 0$, this estimator has the scaling law given by

$$\log \gamma(d) = \text{constant} + \alpha \log d + \text{error}. \tag{11.35}$$

The above scaling law is suitable for stationary processes satisfying

$$1 - r(\tau) \sim |\tau|^{-\alpha}$$

for $\tau \to 0$. The variogram estimator of D is expressed by $\hat{D} = 2 - \hat{\alpha}/2$, where $\hat{\alpha}$ is the slope in a log-log plot of $\gamma(d)$ versus d.

The reported estimators of H are rich, such as R/S analysis (Mandelbrot [6]), maximum likelihood method (Beran [7]), aggregated variance method, the method of differencing the variance, absolute value of the aggregated series, Higuchi's method, residuals of regression, periodogram method, modified periodogram, Whittle estimator (Taqqu et al. [48]), scaled windowed variance method (Cannon et al. [49]), dispersional method (Caccia et al. [50]), wavelet method (Abry and Veitch [51]), ACF regression method (Li [1, 3, 4, 36], Li and

Lim [27]), and fractional Fourier transform method (Chen et al. [52]). By using the method called the detrended fluctuation analysis discussed in Kantelhardt et al. [53] and Peng et al. [54], a series is partitioned into blocks of size m. Within each block, the least square fitting is used. Denote $v(m)$ the average of the sample variances. Then, the detrended fluctuation analysis is based on the following scaling law:

$$\log v(m) = \text{constant} + (2 - \beta)\log m + \text{error} \quad \text{for } m \to \infty. \tag{11.36}$$

The above is applicable for stationary processes satisfying (11.27) (see Taqqu et al. [48] for details). The estimate $\hat{H}$ of H is half the slope in a log-log plot of $v(m)$ versus m.

11.3.2 Simulating LRD Traffic of GC Type

Denote the GC process by X. Then, $H(f) = \sqrt{S_X(f)}e^{-j\psi(f)}$, where $\psi(f)$ satisfies (11.22). Note that

$$S_X(f) = F\left[\left(1 + |t|^\alpha\right)^{-\beta/\alpha}\right]. \tag{11.37}$$

According to (11.24), the impulse function of the simulator to generate traffic following the GC process under the excitation of white noise is

$$h(t) = F^{-1}\left\{\left[F\left(1 + |t|^\alpha\right)^{-\frac{\beta}{\alpha}}\right]^{0.5}\right\}. \tag{11.38}$$

Consequently, the synthesized traffic of the GC type is given by

$$X(t) = w(t) * F^{-1}\left\{\left[F\left(1 + |t|^\alpha\right)^{-\frac{\beta}{\alpha}}\right]^{0.5}\right\}. \tag{11.39}$$

Respectively expressing α and β by D and H, we have

$$X(t) = w(t) * F^{-1}\left\{\left[F\left(1 + |t|^{4-2D}\right)^{-\frac{1-H}{2-D}}\right]^{0.5}\right\}. \tag{11.40}$$

Let D and H be time varying on an interval-by-interval basis. Then, we have the values of D and H in the nth interval expressed by

$$D(n) = 2 - \frac{\alpha(n)}{2}, \tag{11.41}$$

$$H(n) = 1 - \frac{\beta(n)}{2}, \tag{11.42}$$

where n is the interval index. Therefore, we have the ACF of the multi-scale GC process on an interval-by-interval basis

$$r(\tau;n) = \left(1+\tau^{\alpha(n)}\right)^{-\frac{\beta(n)}{\alpha(n)}}, \quad \tau > 0. \tag{11.43}$$

To synthesize the multi-scale traffic of the GC type on an interval-by-interval basis, we use the following computational model:

$$X(t) = w(t) * \mathrm{F}^{-1}\left\{\left[\mathrm{F}\left(1+|t|^{4-2D(n)}\right)^{-\frac{1-H(n)}{2-D(n)}}\right]^{0.5}\right\}. \tag{11.44}$$

In the discrete case, the white noise is $w(i) = \mathrm{IFFT}\{W(f)\}$, where IFFT represents the inverse of FFT (fast Fourier transform). The synthesized multi-scale traffic of the GC type in the discrete case on an interval-by-interval basis is given by

$$X(i) = w(i) * \mathrm{IFFT}\left\{\left[\mathrm{FFT}\left(1+|i|^{4-2D(n)}\right)^{-\frac{1-H(n)}{2-D(n)}}\right]^{0.5}\right\}. \tag{11.45}$$

11.3.3 Case Study
11.3.3.1 Simulation of White Noise
White noise plays a role of building block in the ACF-based simulation of traffic in this chapter, see (11.44) or (11.45). We brief the simulation of white noise in the case study.

As early as 1958, Cox and Muller [55] presented a method of the generation of the Gaussian white noise. It is called the Box–Muller method (Press et al. [10] and Harger [56]). According to the Box-Muller method,

$$w(i) = C\left\{\mu + \sigma[-2\ln(\alpha(i))]^{0.5}\cos(2\pi\alpha(i))\right\}, \tag{11.46}$$

where $\alpha(i)$ is a random number uniformly distributed between 0 and 1, $C > 0$ is a constant, μ is the mean and σ the standard deviation.

Note that a practical signal is band-limited. Therefore, one may use

$$W(f) = \begin{cases} e^{j\phi(f)}, & |f| \leq f_c \\ 0, & \text{otherwise} \end{cases}, \tag{11.47}$$

where $j = \sqrt{-1}$ and $\phi(f)$ is a real random function with arbitrary distribution. The cutoff frequency f_c is such that it completely covers the band of the random sequence of interest. In this way,

$$w(t) = \mathrm{F}^{-1}\left[W(f)\right] \tag{11.48}$$

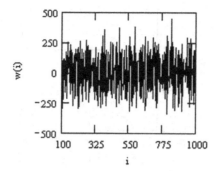

FIGURE 11.8 Synthesized white noise.

is a white noise function in the engineering sense. In fact, in $[0, f_c]$, the PSD of $w(t)$ is constant because $\left|e^{j\phi(f)}\right| = 1$. Figure 11.8 gives the plot of a white noise series. The computation for the PSD of $w(i)$ yields its corresponding PSD in the normalized case as indicated in Figure 11.9.

Nevertheless, the method given by (11.48) has a pitfall that $w(t)$ may be a complex function since $W(f)$ is complex. In this chapter, therefore, we utilize the following method.

Taking into account the method described by (11.7), we let $S_x(f) = 1$ for $f \in [0, f_c]$. In this way, we immediately obtain a white noise expressed by

$$w(t) = \sum_{n=1}^{N} \sqrt{2\Delta f} \, \cos[2\pi n\Delta ft + \vartheta(n\Delta f)], \, 0 \le t \le T_r. \tag{11.49}$$

The advantage of the method given in (11.49) is that it always synthesizes real white noise instead of complex one. We give a constant PSD in Figure 11.10. By using (11.49), we have a generated white noise series shown in Figure 11.11.

11.3.3.2 Simulation in the Mono-Fractal Case

Simulated realizations are shown in Figure 11.12 and Figure 11.13. Due to the advantage of the separated characterizations of D and H by using the GC correlation model, we can observe the distinct effects of D and H. In Figure 11.12, H is constant 0.7 but D decreases ($D = 1.9$ and 1.2). In Figure 11.13, D is constant 1.5 while H decreases ($H = 0.95$ and 0.55).

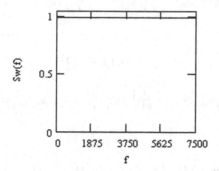

FIGURE 11.9 PSD of $w(i)$ in Figure 11.8.

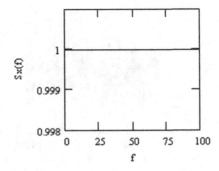

FIGURE 11.10 Given PSD.

11.3.3.3 Simulation in the Multi-Scale Case

There are six categories of multi-scale traffic of the GC type according to the function forms of $D(n)$ and $H(n)$. They are as follows:

- C1: $D(n)$ and $H(n)$ are the same deterministic functions.

- C2: $D(n)$ and $H(n)$ are different deterministic functions.

- C3: $D(n)$ is a deterministic function but $H(n)$ is random.

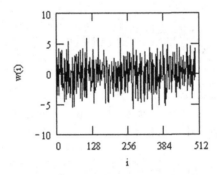

FIGURE 11.11 Generated white noise with the given PSD in Figure 11.10.

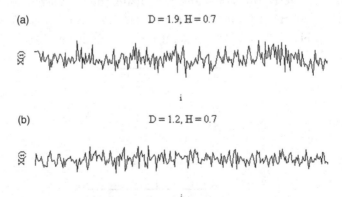

FIGURE 11.12 Realizations. (a) $D = 1.9$, $H = 0.7$. (b) $D = 1.2$, $H = 0.7$.

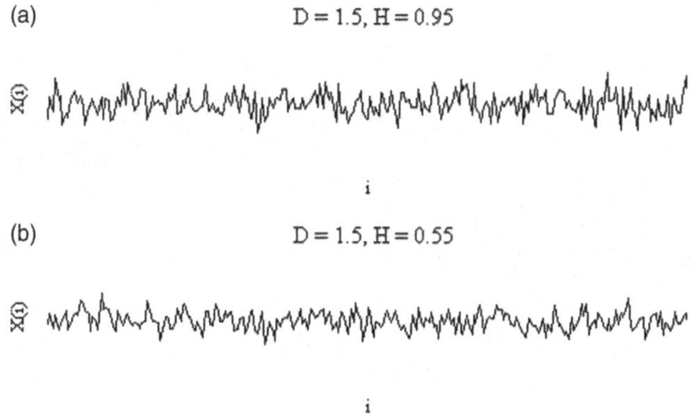

(a) $D = 1.5, H = 0.95$

i

(b) $D = 1.5, H = 0.55$

i

FIGURE 11.13   Realizations. (a) $H = 0.95$, $D = 1.5$. (b) $H = 0.55$, $D = 1.5$.

- C4: $D(n)$ is random but $H(n)$ is a deterministic function.
- C5: $D(n)$ and $H(n)$ are the same random functions.
- C6: $D(n)$ and $H(n)$ are different random functions.

Suppose the interval length is 64. For $n = 1, ..., 5$, we demonstrate the following four cases of realizations.

Case 1: $D(n) = 1.5$, $H(n) = 0.55 + 0.1(n - 1)$.

Case 2: $D(n) = 0.15n + 1$, $H(n) = 0.55$.

Case 3: $D(n) = \exp(0.12n)$, $H(n) = \exp(-0.1n)$.

Case 4: $D(n) = 1.786, 1.688, 1.749, 1.459, 1.946$; $H(n) = 0.577, 0.518, 0.684, 0.972, 0.861$.

In the first case, the synthesized series has the LRD linearly varying in terms of n from $H(1) = 0.55$ to $H(5) = 0.95$ but keeps a constant local irregularity as $D(n) = 1.5$, see Figure 11.14 (a). In order to demonstrate five sections with five different values of H, we plot Figure 11.14 (a) by separating five sections downward as indicated in Figure 11.14 (b).

In the second case, $H(n) = 0.55 = $ constant. Hence, it is a traffic series with the LRD invarying. However, its local irregularity linearly changes in terms of n from $D(1) = 1.15$ to $D(5) = 1.75$, see Figure 11.15.

Figure 11.16 shows the third case of realization. Its LRD changes in the form of $H(n) = \exp(-0.1n)$ from $H(1) = \exp(-0.1) = 0.905$ to $H(5) = \exp(-0.5) = 0.607$ and its local irregularity varies in the form of $D(n) = \exp(0.12n)$ from $D(1) = \exp(0.12) = 1.127$ to $D(5) = \exp(0.6) = 1.822$.

In the fourth case, both $D(n)$ and $H(n)$ are random variables following uniform distribution in the ranges of $1 < D(n) < 2$ and $0.5 < H(n) < 1$, see Figure 11.17.

In short, the present method can be used to synthesize the multi-scale LRD traffic of the GC type with arbitrary combination of local irregularity and statistical dependence.

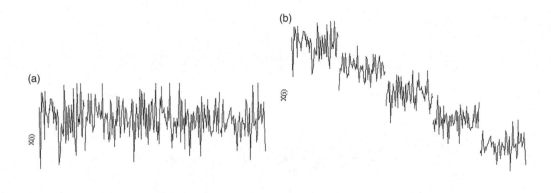

FIGURE 11.14 Realization for $D(n) = 1.5$ and $H(n) = 0.55 + 0.1(n - 1)$. (a) Five sections of series. (b) Top section: $H = 0.55$. The section below the top: $H = 0.65$. The middle section: $H = 0.75$. The section above the bottom: $H = 0.85$. The bottom section: $H = 0.95$.

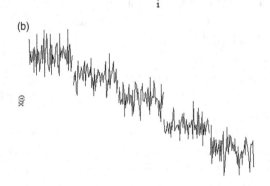

FIGURE 11.15 Realization for $D(n) = 0.15n + 1$ and $H(n) = 0.55$. (a) Five sections of series. (b) Top section: $D = 1.15$. The section below the top: $D = 1.3$. The middle section: $D = 1.45$. The section above the bottom: $D = 1.6$. The bottom section, $D = 1.75$.

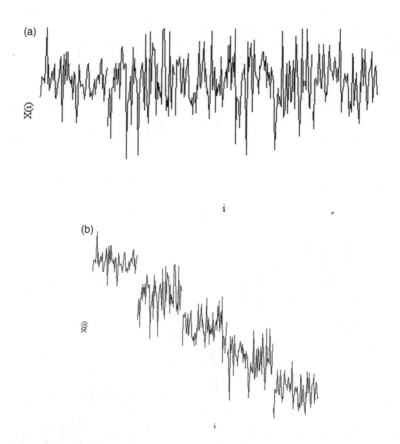

FIGURE 11.16 Realization for $D(n) = \exp(0.12n)$ and $H(n) = \exp(-0.1n)$. (a) Five sections of series. (b) Top section: $D = \exp(0.12)$ and $H = \exp(-0.1)$. The section below the top: $D = \exp(0.24)$ and $H = \exp(-0.2)$. The middle section: $D = \exp(0.36)$ and $H = \exp(-0.3)$. The section above the bottom: $D = \exp(0.48)$ and $H = \exp(-0.4)$. The bottom section: $D = \exp(0.6)$ and $H = \exp(-0.6)$.

11.4 DISCUSSIONS

Eq. (11.26) defines the global SS of $X(t)$ with the SS index κ. By global SS we mean that (11.26) holds for $0 < t < \infty$ with unique a and κ. This is a mono-fractal character. FGn is an only stationary increment process with SS (Samorodnitsky and Taqqu [57]). In general, however, it may be too restrictive for many practical applications, including traffic modeling. Note that a series, which is not self-similar, may be locally self-similar (Adler [58]). Expressions (11.31) and (11.32) are two definitions for the local SS. GC process is locally self-similar (Lim and Li [31]). On the other side, the property of LRD is defined by (11.27). Thus, the concept of SS differs from that of LRD (see Li [1, 59, 60] for details). Fractal dimension characterizes local SS, see, Mandelbrot [61, pp. 373] and while the Hurst parameter measures LRD (Li [1, 59, 60]). The fractal dimension can be computed by (11.33) while the Hurst parameter is computed by (11.34). They are different concepts (Mandelbrot [61]).

Note that the asymptotic expression (11.27) for $r(\tau)$ ($\tau \to \infty$) is used to represent the property of LRD in the time domain. The Fourier transform of $r(\tau)$ does not exist in the

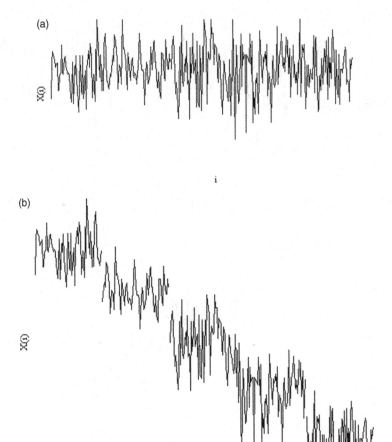

FIGURE 11.17 Realization for the random D and H with $D(n) = 1.786, 1.688, 1.749, 1.459, 1.946$ and $H(n) = 0.577, 0.518, 0.684, 0.972, 0.861$. (a) Five sections of series. (b) Top section: $D = 1.786$ and $H = 0.557$. The section below the top: $D = 1.688$ and $H = 0.518$. The middle section: $D = 1.749$ and $H = 0.684$. The section above the bottom: $D = 1.459$ and $H = 0.972$. The bottom section: $D = 1.946$ and $H = 0.861$.

domain of ordinary functions because $r(\tau)$ is non-integrable. However, in the domain of generalized functions over the Schwartz space of test functions, we have the property of LRD expressed in the frequency domain as follows:

$$F[r(\tau)] \sim |\omega|^{\beta-1} \quad (0 < \beta < 1, |\omega| \to 0), \tag{11.50}$$

where ω is the angular frequency. Therefore, the PSD of an LRD series is divergent at $\omega = 0$, which is a basic property of LRD series expressed in the frequency domain, see Chapters 1–3.

Denote $\mathrm{Sa}(\omega) = \frac{\sin(\omega)}{\omega}$. Denote the PSD of the GC process by $S_{\mathrm{GC}}(\omega)$. Then (Li and Lim [28]),

$$S_{\mathrm{GC}}(\omega) = \sum_{k=0}^{\infty} \frac{(-1)^k \Gamma[(\beta/\alpha)+k]}{\pi \Gamma(\beta/\alpha)\Gamma(1+k)} I_1(\omega) * \mathrm{Sa}(\omega)$$

$$+ \sum_{k=0}^{\infty} \frac{(-1)^k \Gamma[(\beta/\alpha)+k]}{\pi \Gamma(\beta/\alpha)\Gamma(1+k)} [\pi I_2(\omega) - I_2(\omega) * \mathrm{Sa}(\omega)], \qquad (11.51)$$

where

$$I_1(\omega) = -2\sin(\alpha k \pi/2)\Gamma(\alpha k+1)|\omega|^{-\alpha k-1}, \qquad (11.52)$$

$$I_2(\omega) = 2\sin[(\beta+\alpha k)\pi/2]\Gamma[1-(\beta+\alpha k)]|\omega|^{(\beta+\alpha k)-1}. \qquad (11.53)$$

The asymptotic expressions of $S_{\mathrm{GC}}(\omega)$ for $\omega \to 0$ is given by

$$S_{\mathrm{GC}}(\omega) \sim \frac{1}{\Gamma(\beta)\cos(\beta\pi/2)}|\omega|^{\beta-1}, \quad \omega \to 0. \qquad (11.54)$$

That is,

$$S_{\mathrm{GC}}(f) \sim \frac{|2\pi|^{\beta-1}}{\Gamma(\beta)\cos(\beta\pi/2)}|f|^{\beta-1}, \quad f \to 0. \qquad (11.55)$$

Recall that $\lim_{f\to 0} S(f) \to \infty$ is essential for a series with LRD. Otherwise, it is short-range-dependent. In practice, one usually expects T_r in (11.9) to be large enough. This equivalently requires that Δf should be small enough. Besides, the property of LRD needs that Δf has to be small enough such that $S(\Delta f)$ approaches infinite. However, this may produce the difficulty in computing a too large value of $S(\Delta f)$ for an too small Δf in addition to aliasing effects of spectrum. This is particularly true if the computation is run on an ordinary personal computer (Zhang [62]).

Note that we shall pay caution to the finite size effect when estimating the ACF of the simulation $X(t)$ with a finite sample size, say N. More precisely, let $\mathrm{Var}[r(k)]$ be the variance of the estimated ACF $r(k)$ of the discrete series $X(i)$ with the sample size N ($i = 1, 2, ..., N$). Then, a reference guideline to obtain the bound of $\mathrm{Var}[r(k)]$ is given by (Li [63] or Chapter 8)

$$\mathrm{Var}(r) \le \frac{4|\Gamma(1-2H)|^2 \cos^2(H\pi)(2H-1)^2}{\pi^2 N} \sum_{k=1}^{N} (k)^{4H-4}. \qquad (11.56)$$

Eq. (11.56) is derived based on the assumption that $X(i)$ is of fGn. Because for $\tau \to \infty$ (11.29) becomes

$$r(\tau) \sim \tau^{-\beta} \quad (\tau > 0, \tau \to \infty), \qquad (11.57)$$

the GC process is an approximation of fGn for large lags. Hence, (11.56) may yet be a reference to require the sample size of the simulation of the GC type traffic. Taking into the finite size effect account, we say that the ACF estimate of the simulation $X(i)$ ($i = 1, 2, ..., N$) may in general not exactly equal to the given ACF since there is estimate error. The variance bound of the ACF estimate may take (11.56) as a reference. An ACF estimate of synthesized series with too small sample size, say, e.g., 64 as the section size in Figures 11.14–11.17, may be unsatisfactory. The section size set to be 64 in Figures 11.14–11.17 is simply for the purpose of qualitatively demonstrate multi-scale simulations based on the computational model of (11.45) within a limited page space.

11.5 SUMMARY

In addition to the difficulties for the PDF-based method and the PSD-based one with respect to the simulation of the LRD traffic as explained in Section 11.2, we note that in traffic engineering, it is expected to accurately synthesize data series according to a predetermined correlation structure. This is because not only the ACF of the LRD traffic is an ordinary function while its PSD is a generalized function but also the ACF of arrival traffic greatly impacts the performances of queuing systems (Livny et al. [64]). Thus, performance analysis desires to accurately know how one packet arriving at time t statistically correlates to the other arriving at $t + \tau$ apart in future as remarked in Paxson and Floyd [8, First sentence, Subparagraph 4, Paragraph 2, Section 6.1]. Therefore, this chapter focuses on correlation-based computational method.

We have discussed a computational model to generate the multi-scale LRD traffic of the GC type. Since the ACF of the GC process can separately characterize the fractal dimension and the Hurst parameter of traffic, the present method can be used to flexibly synthesize realizations of the six categories of the multi-scale LRD traffic of the GC type on an interval-by-interval basis.

REFERENCES

1. M. Li, *Fractal Teletraffic Modeling and Delay Bounds in Computer Communications*, 1st ed., CRC Press, Boca Raton, 2022.
2. M. Li, Modified multifractional Gaussian noise and its application, *Physica Scripta*, 96(12) 2021, 125002 (12 pp).
3. M. Li, Generalized fractional Gaussian noise and its application to traffic modeling, *Physica A*, 579, 2021, 1236137 (22 pp).
4. M. Li, Long-range dependence and self-similarity of teletraffic with different protocols at the large time scale of day in the duration of 12 years: Autocorrelation modeling, *Physica Scripta*, 95(4) 2020, 065222 (15 pp).
5. M. Li, Multi-fractional generalized Cauchy process and its application to teletraffic, *Physica A*, 550, 2020, 123982 (14 pp).
6. B. B. Mandelbrot, *Gaussian Self-Affinity and Fractals*, Springer, New York, 2001.
7. J. Beran, *Statistics for Long-Memory Processes*, Chapman & Hall, New York, 1994.
8. V. Paxson, and S. Floyd, Why we don't know how to simulate the internet, *Proc., Winter Simulation Conf.*, USA, 1997, 1037–1044.
9. A. Papoulis, and S. U. Pillai, *Probability, In Random Variables and Stochastic Processes*, McGraw-Hill, New York, 1997.

10. W. H. Press, S. A. Teukolsky, W. T. Vetterling, and B. P. Flannery, *Numerical Recipes in C: the Art of Scientific Computing*, 2nd Edition, Cambridge University Press, Cambridge, 1992.

11. S. K. Chakrabarti, *Offshore Structure Modeling*, World Scientific, Singapore, 1994.

12. M. Li, An iteration method to adjusting random loading for a laboratory fatigue test, *International Journal of Fatigue*, 27(7) 2005, 783–789.

13. M. Li, An optimal controller of an irregular wave maker, *Applied Mathematical Modelling*, 29(1) 2005, 55–63.

14. M. Li, B.-H. Xu, and Y.-S. Wu, An H_2-optimal control of random loading for a laboratory fatigue test, *Journal of Testing and Evaluation*, 26(6) 1998, 619–625.

15. M. Smozuka, Simulation of multivariate and multidimensional random processes, *The Journal of the Acoustical Society of America*, 49(1B) 1971, 357–368.

16. R. M. Stanislaw, *Ocean Surface Waves: Their Physics and Prediction*, World Scientific, Singapore, 1996.

17. J. D. Gibson (ed.), *The Communications Handbook*, IEEE Press, New York, 1997.

18. J. Mann, Wind field simulation, *Probabilistic Engineering Mechanics*, 13(4) 1998, 269–282.

19. M. W. Garrett, and W. Willinger, Analysis, modeling and generation of self-similar VBR traffic, *Proceedings of ACM SigComm'94*, London, 1994, 269–280.

20. H.-D. J. Jeong, J.-S. R. Lee, D. McNickle, and P. Pawlikowski, Distributed steady-state simulation of telecommunication networks with self-similar teletraffic, *Simulation, Modelling Practice and Theory*, 13(3) 2005, 233–256.

21. W. C. Lau, A. Erramilli, J. L. Wang, and W. Willinger, Self-similar traffic generation: The random midpoint displacement algorithm and its properties, *IEEE International Conference on Communications, "gateway to Globalization,"* vol. 1, 1995, 466–472, Seattle, Washington, USA.

22. S. Ledesma, and D. Liu, Synthesis of fractional Gaussian noise using linear approximation for generating self-similar network traffic, *ACM SICOMM Computer Communication Review*, 30(2) 2000, 4–17.

23. M. Li, and C.-H. Chi, A correlation-based computational method for simulating long-range dependent data, *Journal of the Franklin Institute*, 340(6–7) 2003, 503–514.

24. V. Paxson, Fast approximate synthesis of fractional Gaussian noise for generating self-similar network traffic, *ACM SICOMM Computer Communication Review*, 27(5) 1997, 5–18.

25. M. Rroughan, Simplifying the synthesis of internet traffic matrices, *ACM SICOMM Computer Communication Review*, 35(5) 2005, 93–96.

26. B. B. Mandelbrot, and J. W. van Ness, Fractional Brownian motions, fractional noises and applications, *SIAM Review*, 10(4) 1968, 422–437.

27. M. Li, and S. C. Lim, Modeling network traffic using generalized Cauchy process, *Physica A*, 387(11) 2008, 2584–2594.

28. M. Li, and S. C. Lim, Power spectrum of generalized Cauchy process, *Telecommunication Systems*, 43(3–4) 2010, 219–222.

29. V. Paxson, and S. Floyd, Wide area traffic: The failure of Poisson modeling, *IEEE/ACM Transactions on Networking*, 3(3) 1995, 226–244.

30. L. M. Kaplan, and C.-C. J. Kuo, Extended self-similarity for fractional Brownian motion, *IEEE Transactions on Signal Processing*, 42(11) 1994, 3526–3530.

31. S. C. Lim, and M. Li, Generalized Cauchy process and its application to relaxation phenomena, *Journal of Physics A: Mathematical and General*, 39(12) 2006, 2935–2951.

32. A. Adas, Traffic models in broadband networks, *IEEE Communications Magazine*, 35(7) 1997, 82–89.

33. J. Beran, R. Shernan, M. S. Taqqu, and W. Willinger, Long-range dependence in variable bitrate video traffic, *IEEE Transactions on Communications*, 43(2–3–4) 1995, 1566–1579.

34. I. W. C. Lee, and A. O. Fapojuwo, Stochastic processes for computer network traffic modeling, *Computer Communications*, 29(1) 2005, 1–23.

35. W. E. Leland, M. S. Taqqu, W. Willinger, and D. V. Wilson, On the self-similar nature of ethernet traffic (extended version), *IEEE/ACM Transactions on Networking*, 2(1) 1994, 1–15.

36. M. Li, Modeling autocorrelation functions of long-range dependent teletraffic series based on optimal approximation in Hilbert space: A further study, *Applied Mathematical Modelling*, 31(3) 2007, 625–631.

37. H. Michiel, and K. Laevens, Teletraffic engineering in a broad-band era, *Proceedings of the IEEE*, 85(12) 1997, 2007–2033.

38. B. Tsybakov, and N. D. Georganas, Self-similar processes in communications networks, *IEEE Transactions on Information Theory*, 44(5) 1998, 1713–1725.

39. A. Scherrer, N. Larrieu, P. Owezarski, P. Borgnat, and P. Abry, Non-Gaussian and long memory statistical characterisations for internet traffic with anomalies, *IEEE Transactions on Dependable and Secure Computing*, 4(1) 2007, 56–70.

40. I. Csabai, $1/f$ noise in computer network traffic, *The Journal of Physics A: Mathematical and Theoretical*, 27(12) 1994, L417–L421.

41. M. Li, and S. C. Lim, A rigorous derivation of power spectrum of fractional Gaussian noise, *Fluctuation and Noise Letters*, 6(4) 2006, C33–C36.

42. M. Li, Power spectrum of generalized fractional Gaussian noise, *Advances in Mathematical Physics*, 2013, 2013, Article ID 315979 (3 pp).

43. J. T. Kent, and A. T. A. Wood, Estimating the fractal dimension of a locally self-similar Gaussian process by using increments, *Journal of the Royal Statistical Society, B*, 59(3) 1997, 579–599.

44. A. G. Constantine, and P. Hall, Characterizing surface smoothness via estimation of effective fractal dimension, *Journal of the Royal Statistical Society, B*, 56(1) 1994, 97–113.

45. C. C. Taylor, and S. J. Taylor, Estimating the dimension of a fractal, *Journal of the Royal Statistical Society B*, 53(2) 1991, 353–364.

46. B. Dubuc, J. F. Quiniou, C. Roques-Carmes, C. Tricot, and S. W. Zucker, Evaluating the fractal dimension of profiles, *Physical Review A*, 39(3) 1989, 1500–1512.

47. P. Hall, and A. Wood, On the performance of box-counting estimators of fractal dimension, *Biometrika*, 80(1) 1993, 246–252.

48. M. S. Taqqu, V. Teverovsky, and W. Willinger, Estimators for long-range dependence: An empirical study, *Fractals*, 3(4) 1995, 785–798.

49. M. J. Cannon, D. B. Percival, D. C. Caccia, G. M. Raymond, and J. B. Bassingthwaighte, Evaluating scaled windowed variance methods for estimating the Hurst coefficient of time series, *Physica A*, 241(3–4) 1997, 606–626.

50. D. C. Caccia, D. B. Percival, M. J. Cannon, G. M. Raymond, and J. B. Bassingthwaighte, Analyzing exact fractal time series: Evaluating dispersional analysis and rescaled range methods, *Physica A*, 246(3–4) 1997, 609–632.

51. P. Abry, and D. Veitch, Wavelet analysis of long-range dependent traffic, *IEEE Transactions on Information Theory*, 44(1) 1998, 2–15.

52. Y.-Q. Chen, R. Sun, and A. Zhou, An improved Hurst parameter estimator based on fractional Fourier transform, *Telecommunication Systems*, 43(3–4) 2010, 197–206.

53. J. W. Kantelhardt, E. Koscielny-Bunde, H. H. A. Rego, S. Havlin, and A. Bunde, Detecting long-range correlations with detrended fluctuation analysis, *Physica A*, 295(3–4) 2001, 441–454.

54. C.-K. Peng, S. V. Buldyrev, S. Havlin, M. Simons, H. E. Stanley, and A. L. Goldberger, Mosaic organization of DNA nucleotides, *Physical Review E*, 49(2) 1994, 1685–1689.

55. G. E. P. Cox, and M. E. Muller, A note on the generation of random normal deviates, *The Annals of Mathematical Statistics*, 29(2) 1958, 610–611.

56. R. O. Harger, *An Introduction to Digital Signal Processing with MATHCAD*, PWS Publishing Company, New York, 1999, Chap. 24.

57. G. Samorodnitsky, and M. S. Taqqu, *Stable Non-Gaussian Random Processes: Stochastic Models with Infinite Variance*, Chapman and Hall, New York, 1994.

58. A. J. Adler, *The Geometry of Random Fields*, John Wiley & Sons, New York, 1981.
59. M. Li, Self-Similarity and Long-Range Dependence in Teletraffic, *Proc., the 9th WSEAS Int. Conf. on Multimedia Systems and Signal Processing*, Hangzhou, China, May 2009, 19–24.
60. M. Li, Fractal time series: A tutorial review, *Mathematical Problems in Engineering*, 2010, 2010, Article ID 157264 (26 pp).
61. B. B. Mandelbrot, *The Fractal Geometry of Nature*, W. H. Freeman, New York, 1982.
62. P. Zhang, *Design and Implementation for the Simulation System of Fractional Gaussian Noise*, Master Thesis, East China Normal University, Shanghai, 2007 (in Chinese).
63. M. Li, Record length requirement of long-range dependent teletraffic, *Physica A*, 472, 2017, 164–187.
64. M. Livny, B. Melamed, and A. K. Tsiolis, The impact of autocorrelation on queuing systems, *Management Science*, 39(3) 1993, 322–339.

IV

Anomaly Detection of Traffic

Reliably Identifying Signs of DDOS Flood Attacks Based on Traffic Pattern Recognition

In the aspect of intrusion detection, reliable detection remains a challenge issue. Hence, reliable detection of distributed denial-of-service (DDOS) attacks is worth studying. By reliable detection we mean that signs of attacks can be identified with pre-determined detection probability and false alarm probability. This chapter focuses on reliable detection of DDOS flood attacks by identifying pattern of traffic with long-range dependence (LRD). In this aspect, there are three fundamental issues in theory and practice:

- What is a statistical feature of traffic to be used for pattern recognition?

- How to represent distributions of identification probability, false alarm probability, and miss probability?

- How to assure a decision-making that has high identification probability, low false alarm probability, and low miss probability?

This chapter gives a statistical detection scheme based on identifying abnormal variations of LRD traffic time series. The representations of three probability distributions mentioned above are given and a decision-making region is explained. With this region, one can know what an identification (or false alarm or miss) probability is for capturing signs of DDOS flood attacks. The significance of a decision-making region is that it provides a guideline to set appropriate threshold value so as to assure pre-determined high identification probability, pre-desired low false alarm probability and pre-determined low miss probability. A case study is demonstrated.

DOI: 10.1201/9781003354987-16

12.1 BACKGROUND

In [1], Kemmerer and Vigna stated that "The challenge is to develop a system that detects close to 100 percent of attacks with minimal false positives. We are still far from achieving this goal." In fact, that is challenge [2]. Various methods have been discussed for intrusion detections (see refs [1–3]). However, analog to radar systems [4], a detection scheme is usually objective-dependent. Recall that a DDOS attacker sends flood packets to a victim with identities of attack packets hiding. Hence, a DDOS attacker may engage the power of a vast number of coordinated Internet hosts to devour some key resources, such as bandwidth, router processing capacity, or network stack resources, at a target site (victim) such that it denies services it usually offers to legitimate clients. Targets include large e-commerce sites, mid-sized business, government, universities, and end users [5–7].

There are methods discussing detection of DDOS attacks, e.g., [7–11]. However, reliably statistical detection is rarely seen. Note that no matter what type of tools is used for performing a DDOS attacking, a radical feature of DDOS flood attacks can be investigated in traffic pattern because a DDOS flood attack tool may not make effects without arrival flood packets. Consequently, according to the particularities of DDOS flood attacks, statistical detection may yet be a method for reliably identifying signs of DDOS flood attacks. Obviously, reliably identifying signs of DDOS flood attacks requires (1) a model that well statistically features different types of DDOS flood attacks, and (2) making an identification decision such that it has high identification probability and low false alarm probability. This chapter explains my results in these aspects.

Traffic engineering exhibits that a traffic series is generally of long-range dependence (LRD) [12–20]. Thus, the traffic series on a specific site has its statistical patterns. Autocorrelation function may serve as a statistical model of LRD traffic [12–25].

From the point of view of pattern recognition, identification of DDOS flood attacks may be an issue of statistical pattern recognition of traffic. Consequently, there is a probability issue with respect to identification, false alarm, and miss. For reliably identifying DDOS flood attacks, it is desired to find an approach to assure high identification probability and low false alarm probability for an identification decision. This chapter is based on my previous work [25] from a view of statistical detection system by introducing several parts that are essential for a detection system. They are (1) the detailed explanations why autocorrelation function may serve as a statistical model for featuring traffic; (2) the detailed discussions of correlation techniques used for statistical pattern recognition; (3) a detailed description of transforming a random variable with arbitrary distribution to one with normal distribution; and (4) availability description. As remarked by C. C. Wood, techniques involved in information security are multi-disciplinary in nature [26]. This chapter consists of the author's previous research in decision-making [25], correlation techniques [27], traffic modeling and simulation [12–17, 28, 29].

In this chapter, a traffic trace is called the normal traffic when the site is not attacked. When the site is intruded by DDOS flood attacks, the traffic pattern appears considerably different from the normal one [6] and we call it abnormal traffic.

The rest of chapter is organized as follows: Section 12.2 discusses feature extraction. Section 12.3 introduces three formulas for three probabilities and explains how to achieve high identification probability (P_i) with low false alarm probability (P_f) and low miss probability (P_m). Section 12.4 demonstrates a case study. Discussions and summary are given in Section 12.5.

12.2 FEATURE EXTRACTION

12.2.1 Featuring LRD Traffic Via Autocorrelations

For a random series x, if its autocorrelation function r_{xx} is summable, x is statistically short-range dependent (SRD). Otherwise, it is statistically LRD. Traffic is LRD. The power spectrum of x is the Fourier transform of r_{xx}. Thus, the power spectrum of an SRD series exists in the domain of ordinary functions (Dirichlet condition) while the existence of the power spectrum of an LRD series should be explained in the domain of generalized functions. This is a primary difference between SRD processes and LRD processes.

For an LRD traffic series x, for $\tau \geq 0$, its autocorrelation has the following asymptotical property

$$r_{xx}(\tau) \sim c\tau^{2H-2}, \quad \tau \to \infty, 0.5 < H < 1, \tag{12.1}$$

where $c > 0$ is a constant, ~ stands for asymptotically equivalence under the limit $\tau \to \infty$ and H is the Hurst parameter. Eq. (12.1) implies the LRD property of traffic: $\int_0^\infty r_{xx}(\tau)d\tau = \infty$. As r_{xx} is an even function, it can be expressed by

$$r_{xx}(\tau) \sim c|\tau|^{2H-2}, \quad 0.5 < H < 1. \tag{12.2}$$

Let $S_{xx}(\omega)$ be the power spectrum of x. Then, $S_{xx}(\omega) = F(r_{xx})$, where F is the operator of the Fourier transform. In the domain of generalized functions [30], one gets

$$F\left(|\tau|^{2-2H}\right) = 2\sin\left[\frac{\pi(2H-2)}{2}\right]\Gamma(3-2H)|\omega|^{-(3-2H)}. \tag{12.3}$$

When $\omega \to 0$,

$$S_{xx}(\omega) \sim \frac{1}{\omega}. \tag{12.4}$$

Eqs. (12.2) and (12.4) are asymptotical expressions of autocorrelation functions and power spectra of LRD traffic, respectively. They draw attention that S_{xx} is a generalized function while r_{xx} is an ordinary function. Consequently, autocorrelation function is a commonly used model to characterize LRD traffic [12–24]. In practice, autocorrelations are obtained numerically. For simplifying the explanation of our approach, we take fractional Gaussian noise (fGn) as an example model of traffic. The normalized autocorrelation of LRD fGn in the discrete case is given by

$$r(k) = \frac{1}{2}\left[\left(|k|+1\right)^{2H} + \||k|-1\|^{2H} - 2|k|^{2H}\right], \quad 0.5 < H < 1, \tag{12.5}$$

where k is integer. Since $r(k)$ is an even function, it is considered for $k \geq 0$ in what follows.

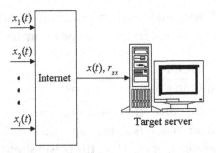

FIGURE 12.1 Normal traffic at input of a server.

12.2.2 Feature Analysis of Attack Traffic

Let $x(t)$ be the normal traffic, indicating the number of bytes in a packet arriving at a site at t on a packet-by-packet basis. Normally, the site serves $x(t)$ peacefully though it may sometimes be unpleasantly delayed because of the normal jam of traffic. Figure 12.1 shows an indication of traffic normally arriving at a target server input.

Assume that the site suffers from DDOS flood attacks at t_0. Generally, the site may not be overwhelmed immediately at the moment of t_0 [6]. Assume that intruders attack the site such that the site is overwhelmed to deny services at t_1. Then, the time interval (t_0, t_1) is called transition process of attacking. Let t_d be the time for making an identification decision. Then, if the protected site identifies attack signs at $t_d < (t_1 - t_0)$, operators on the site may have time to properly deal with abnormal traffic.

Let $y(t)$ be the abnormal traffic during transition process of attacking. Then, $y(t)$ can be abstractly expressed by

$$y(t) = x(t) + n(t), \tag{12.6}$$

where $n(t)$ is the component of attack traffic. Figure 12.2 indicates abnormal traffic arriving at a target server input. What we desire is how to attain an identification that has high P_i, low P_f and low P_m. Analog to radar system [4], three terms are explained as follows. Correctly recognizing an abnormal sign is termed *identification*. Failing to recognize it is called *miss*. Mistakenly recognizing a normal as abnormal is *false alarm*.

$$\|r_{xx} - r_{yy}\| = \xi > V, \text{ Identification}, \tag{12.7a}$$

$$\|r_{xx} - r_{xl}\| = \zeta > V, \text{ False alarm}, \tag{12.7b}$$

$$\xi < V, \text{ Miss}, \tag{12.7c}$$

where $\xi = \|r_{nn}\| = \|r_{xx} - r_{yy}\|$ is a distance between x and y, and $V > 0$ is the threshold. In (12.7b), r_{xl} stands for the autocorrelation, which is not used as the template but obtained when there is no attack.

Proposition 12.1. Let x and y be normal traffic and abnormal traffic, respectively. Let r_{xx} and r_{yy} be the autocorrelations of x and y, respectively. During the transition process of DDOS flood attacking, $\|r_{yy} - r_{xx}\|$ is noteworthy.

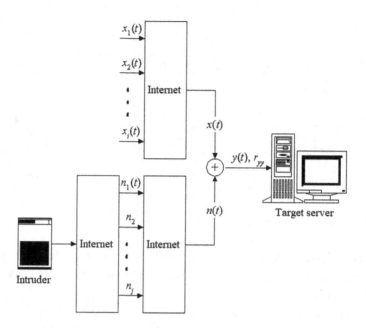

FIGURE 12.2 Illustration of abnormal traffic.

Proof. A network system is a queuing system. Arrival traffic x of a queuing system has its statistical pattern r_{xx} [31]. Suppose the site suffers from DDOS flood attacks. Suppose that $\|r_{yy} - r_{xx}\|$ is negligible in this case. Then, the site would be overwhelmed at its normal state even if there were no DDOS flood packets. This is an obvious contradiction. □

Proposition 12.2. Let x, y and n be normal traffic, abnormal traffic, and attack traffic, respectively. Let r_{xx}, r_{yy} and r_{nn} be autocorrelations of x, y, and n, respectively. Let x and n be statistically uncorrelated. Then,

$$r_{yy} = r_{xx} + r_{nn}. \tag{12.8}$$

Proof. Let $x(t)$ and $n(t)$ be two random functions of single history. Then, the function $r_{xn}(\tau) = \mathrm{E}[x(t)n(t + \tau)]$ is called cross-correlation function between $x(t)$ and $n(t)$. Similarly, the function $r_{nx}(\tau) = \mathrm{E}[n(t)x(t + \tau)]$ is called cross-correlation function between $n(t)$ and $x(t)$. The value of $r_{xn}(\tau)$ implies the correlation coefficient between $x(t)$ and $n(t)$ when $n(t)$ is shifted τ apart The value of $r_{nx}(\tau)$ implies the correlation coefficient between $n(t)$ and $x(t)$ when $x(t)$ is shifted τ apart. □

If $x(t)$ and $n(t)$ are uncorrelated, then $r_{xn}(\tau) = 0$ and $r_{nx}(\tau) = 0$. Since the autocorrelation function of $y(t)$ is given by $r_{yy}(\tau) = \mathrm{E}[y(t)y(t + \tau)] = \mathrm{E}\{[x(t) + n(t)][x(t + \tau) + n(t + \tau)]\}$, we have

$$r_{yy} = \mathrm{E}[x(t)x(t+\tau)] + \mathrm{E}[n(t)n(t+\tau)] + \mathrm{E}[x(t)n(t+\tau)] + \mathrm{E}[n(t)x(t+\tau)]$$

$$= r_{xx} + r_{nn} + r_{xn} + r_{nx}.$$

Because $x(t)$ and $n(t)$ are uncorrelated, $r_{xn}(\tau) = r_{nx}(\tau) = 0$. Therefore, Proposition 12.2 holds. □

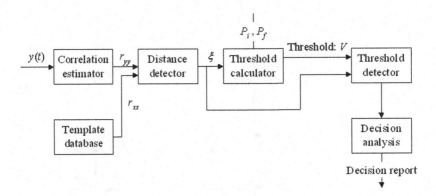

FIGURE 12.3 Diagram of an IDS.

According to (12.7a), $\xi > V$ means identification. Nevertheless, the question we are interested in is what the probability is when one makes a decision of $\xi > V$? We will discuss this in detail in Section 12.3.

Figure 12.3 shows the system diagram. The principle of the system is like this. The collected arrival traffic first passes through a correlation estimator. The result of correlation estimator outputs an online estimated autocorrelation r_{yy}. Both r_{xx} and r_{yy} are fed into the distance detector. The output of the distance detector is ξ. The input of the threshold calculator consists of ξ, pre-determined P_i and P_f. The threshold calculator outputs the threshold value V. The threshold V and ξ are compared in the threshold detector. The comparison result is used for decision-making under the conditions that P_i and P_f are given.

12.3 IDENTIFICATION DECISION

12.3.1 Identification Measure

Clearly, ξ and ζ are random variables. Mathematically, there are many measures available, e.g., [32, 33], but our experiments show that Itakura-Saito measure [33] works well:

$$r_{xx} - r_{yy} = \sum_k \left| \frac{r_{yy}}{r_{xx}} - \log \frac{r_{yy}}{r_{xx}} - 1 \right|. \tag{12.9}$$

12.3.2 Design of Probability Distributions

An autocorrelation sequence of arrival traffic is obtained on interval-by-interval basis. Denote $r_{yy,\ m}$ as the autocorrelation sequence in $[(m-1)L, mL]$, $(m = 1, 2, \ldots)$, where L is the size of sub-sub-interval, see below. Let ξ_m be the distance between r_{xx} and $r_{yy,\ m}$. Then,

$$\xi_m = \left\| r_{xx} - r_{yy,m} \right\|. \tag{12.10}$$

Usually, $r_{yy,\ m} \neq r_{yy,\ n}$ for $m \neq n$. Therefore, ξ_m usually is a random sequence in terms of the index m.

Let Q be the total length of an observation interval for one identification. Divide that interval into N non-overlapped sub-intervals. Each sub-interval is divided into M non-overlapped sub-sub-intervals. Each sub-sub-interval is of L length. Then,

$$Q = L \times M \times N. \tag{12.11}$$

Let $\bar{\xi}_n$ $(n = 1, 2, \ldots, N)$ be the average of ξ_m in each sub-interval. Then,

$$\bar{\xi}_n = \frac{1}{M} \sum_{j=(n-1)M}^{nM} r_{xx} - r_{yy}. \tag{12.12}$$

Obviously, $\bar{\xi}_n$ is a random sequence in terms of n. In practical terms, a normality assumption for the distribution of $\bar{\xi}_n$ becomes quite accurate in most cases for $N \geq 10$ [34]. In practice, it is not difficult to arrange enough sample such that $\bar{\xi}_n$ obeys Gaussian distribution. This also applies in ζ. For the simplicity, we denote $\bar{\xi}_n$ as ξ in the rest of chapter if no confusion is caused.

12.3.2.1 Identification Probability

Let μ_ξ and σ_ξ^2 be the expectation and the variance of ξ, respectively. Then,

$$\xi \sim \frac{1}{\sqrt{2\pi}\sigma_\xi} e^{-\frac{(\xi-\mu_\xi)^2}{2\sigma_\xi^2}}. \tag{12.13}$$

Let

$$\Phi(t) = \int_{-\infty}^{t} \frac{1}{\sqrt{2\pi}} e^{-\frac{t^2}{2}} dt. \tag{12.14}$$

Then, the identification probability is given by

$$P\{V < \xi < \infty\} = \int_{\frac{V-\mu_\xi}{\sigma_\xi}}^{\infty} \frac{1}{\sqrt{2\pi}} e^{-\frac{t^2}{2}} dt = 1 - \Phi\left(\frac{V-\mu_\xi}{\sigma_\xi}\right) = P_i. \tag{12.15}$$

12.3.2.2 False Alarm Probability

Let μ_ζ and σ_ζ^2 be the mean and the variance of ζ. Then, the false alarm probability is given by

$$P\{V < \varsigma < \infty\} = \int_{\frac{V-\mu_\varsigma}{\sigma_\varsigma}}^{\infty} \frac{1}{\sqrt{2\pi}} e^{-\frac{t^2}{2}} dt = 1 - \Phi\left(\frac{V-\mu_\varsigma}{\sigma_\varsigma}\right) = P_f. \tag{12.16}$$

12.3.2.3 Miss Probability

Since $P_i + P_m = 1$, the miss probability is given by

$$P\{-\infty < \xi < V\} = \int_{-\infty}^{\frac{V-\mu_\xi}{\sigma_\xi}} \frac{1}{\sqrt{2\pi}} e^{-\frac{t^2}{2}} dt = \Phi\left(\frac{V-\mu_\xi}{\sigma_\xi}\right) = P_m. \tag{12.17}$$

Generally, $\mu_\zeta = 0$. Besides, when the numeric computations in data processing are consistent, $\sigma_\zeta = \sigma_\xi = \sigma$. In this case, three probabilities are given by

$$P_i = \int_{\frac{V-\mu_\xi}{\sigma}}^{\infty} \frac{1}{\sqrt{2\pi}} e^{-\frac{t^2}{2}} dt = 1 - \Phi\left(\frac{V-\mu_\xi}{\sigma}\right), \tag{12.18}$$

$$P_f = \int_{\frac{V}{\sigma}}^{\infty} \frac{1}{\sqrt{2\pi}} e^{-\frac{t^2}{2}} dt = 1 - \Phi\left(\frac{V}{\sigma}\right), \tag{12.19}$$

$$P_m = \int_{-\infty}^{\frac{V-\mu_\xi}{\sigma}} \frac{1}{\sqrt{2\pi}} e^{-\frac{t^2}{2}} dt = \Phi\left(\frac{V-\mu_\xi}{\sigma}\right). \tag{12.20}$$

Figures 12.4–12.6 show the curves of three distributions, respectively.

12.3.3 Threshold, Threshold Area, and Decision-Making Region

12.3.3.1 Threshold

The selection of a threshold value is important for identification. The lower bound of the threshold can be obtained in this way. Given a false alarm probability f, we want to find the threshold V_f such that $P_f(V_f) \leq f$. Clearly,

$$V_f \geq -\sigma \Phi^{-1}(f). \tag{12.21}$$

In the case of $f = 0$ and the computation precision being 4, we obtain

$$V_f \geq -4\sigma. \tag{12.22}$$

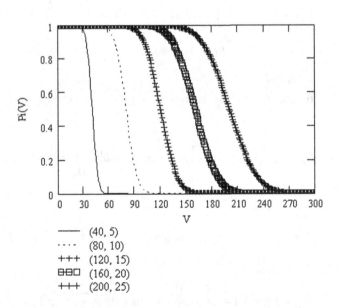

FIGURE 12.4 Identification probability.

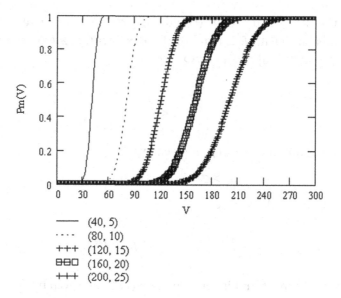

FIGURE 12.5 Miss probability.

The upper bound of the threshold can be obtained like this. Given an identification probability i, we want to find the threshold V_i such that $P_i(V_i) \geq i$. Clearly, when $\mu_\xi - \sigma \Phi^{-1}(i) > 0$, we have

$$V_i \leq \mu_\xi - \sigma \Phi^{-1}(i). \tag{12.23}$$

In the case of $i = 1$ and the computation precision being 4, when if $\mu_\xi - 4\sigma > 0$, we have

$$V_i \leq \mu_\xi - 4\sigma. \tag{12.24}$$

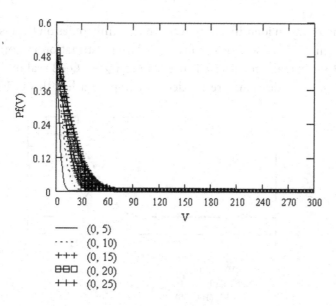

FIGURE 12.6 False alarm probability.

Therefore, when $-\sigma\Phi^{-1}(f)<\mu_\xi-\sigma\Phi^{-1}(i)$ and $V\in[-\sigma\Phi^{-1}(f),\mu_\xi-\sigma\Phi^{-1}(i)]$, the following identification probability with $P_i\geq i$ and the false one with $P_f\leq f$ are assured. That is, for $V\in[-\sigma\Phi^{-1}(f),\mu_\xi-\sigma\Phi^{-1}(i)]$ and $\mu_\xi-\sigma\Phi^{-1}(i)>0$,

$$\begin{cases} P_i\geq i, \\ P_f\leq f. \end{cases} \tag{12.25}$$

In (12.25), the constraint in terms of μ_ξ is given by

$$\mu_\xi>\sigma\Phi^{-1}(i)-\sigma\Phi^{-1}(f). \tag{12.26}$$

In the case of $i=1$ and $f=0$, we have

$$\begin{cases} P_i=1, \\ P_f=0, \end{cases} \quad V\in\left(4\sigma,\mu_\xi-4\sigma\right), \tag{12.27}$$

where $\mu_\xi-4\sigma>0$. The constraint in terms of μ_ξ in (12.27) is given by

$$\mu_\xi>8\sigma. \tag{12.28}$$

Eqs. (12.26) and (12.28) provide a guideline for setting V according to given P_i and P_f. Eq. (12.26) is for the general case while Eq. (12.28) is for the specific case of $i=1$ and $f=0$.

12.3.3.2 Decision-Making Region

Figure 12.7 demonstrates the region in the case of $\mu_\xi=100$ and $\sigma=10$. In fact, Figure 12.7 indicates the quality of a decision for a given value of V.

12.3.4 Applicability

To evaluate the applicability of the present approach, we investigate the time needed for an identification decision.

The time for making an identification decision t_d mainly relies on the fast Fourier transform (FFT). When FFT is used for the fast correlation estimation in correlation estimator and the split-radix real-valued FFT is used, $\frac{Q}{2}\log_2(Q)-\frac{3}{2}Q+2$ real multiplications and $\frac{3Q}{2}\log_2(Q)-\frac{5}{2}Q+4$ real additions are needed to compute a length Q FFT [35]. Suppose

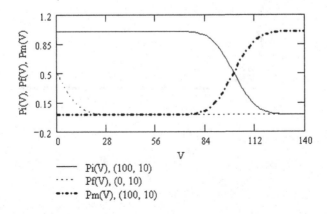

FIGURE 12.7 Decision-making region.

$L = 1024$, $M = 8$, and $N = 16$. Then, $Q = 2^{17}$. Thus, 8.52×10^5 real multiplications and 2.818×10^6 real additions are needed to compute a length $Q = 2^{17}$ FFT. In this case, an identification decision takes about 1 ms with the experiments on 133-MHz Pentium PC. This will be reduced if more powerful machine is used. Thus, the time for identification decision with the present approach is short enough to meet real time requirements in engineering.

12.4 CASE STUDY

Suppose the following function is the normalized template of a traffic trace as shown in Figure 12.8 (a):

$$r_{xx}(k) = \frac{1}{2}\left[(k+1)^{2H} + |k-1|^{2H} - 2k^{2H}\right]_{H=H_0=0.854}. \tag{12.29}$$

Figure 12.8 (b) illustrates an autocorrelation the H value of which is less than H_0.

Let the width of the confidence interval of autocorrelations be 0.104, corresponding to the confidence level 0.9999. Then, we assume y's $H \in [0.51, 0.75]$ during the transition process of attacking. Thus, the following is the normalized autocorrelation of a simulated abnormal traffic:

$$r_{yy}(k) = \frac{1}{2}\left[(k+1)^{2H} + |k-1|^{2H} - 2k^{2H}\right], \quad 0.51 \leq H \leq 0.75. \tag{12.30}$$

In the case study, 9999 points of Hs in $[0.51, 0.75]$ are randomly selected to simulate the autocorrelations of abnormal traffic since an autocorrelation function corresponds to a random series [18, 29] and a random series can be synthesized according to a given autocorrelation [29]. Figure 12.8 (c) indicates the distance ξ.

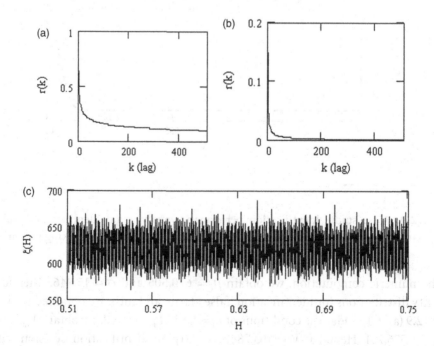

FIGURE 12.8 Case study: Identification distance. (a) Template autocorrelation $H_0 = 0.854$. (b). r_{yy} when $H = 0.60$. (c) Distance series ξ.

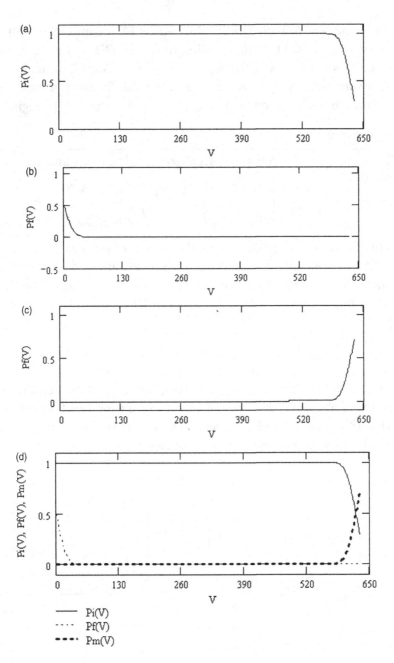

FIGURE 12.9 Case study: Identification performance analysis. (a) Identification probability. (b) False alarm probability. (c) Miss probability. (d) Identification decision-making region.

By the numeric computation, we obtain $\mu_\xi = 620.505$ and $\sigma = 15.946$. Therefore, the probability distributions for identification, false alarm, and miss are obtained as shown in Figures 12.9 (a)–(c). Under the conditions of $P_i = 1$ and $P_f = P_m = 0$, we obtain $V_{min} = 63.784$ and $V_{max} = 556.721$. Hence, if $V \in [63.784, 556.721]$, an identification decision will have $P_i = 1$ and $P_f = P_m = 0$, which can be easily observed from Figure 12.8 (c) or Figure 12.9 (d).

12.5 DISCUSSIONS AND SUMMARY

An approach for reliably identifying signs of DDOS flood attacks based on traffic pattern recognition has been presented and discussed. Representations of identification probability, false alarm probability, and miss probability have been proposed and explained. Following this approach, an identification regarding capturing signs of DDOS flood attacks from a view of detecting abnormal variations of traffic time series can be made with a given identification probability and a given false alarm probability.

In principle, this chapter exhibits a method for abnormal traffic detection by clustering the normal traffic. However, traffic time series has high variability due to reasons, such as new technologies and devices adopted frequently in the Internet [19, 24]. Consequently, a template established during a period of detection might not well match the one during another period of detection. To cope with such change or "drift," the study of the stationarity test, see Chapters 7 and 10, of traffic is crucial to the research issue with respect to adaptively and reliably identifying signs of DDOS attacks.

REFERENCES

1. R. A. Kemmerer, and G. Vigna, Intrusion detection: A brief history and overview, *Supplement to IEEE Computer, IEEE Security & Privacy*, 35(4) 2002, 27–30.
2. S.-J. Han, and S.-B. Cho, Detecting intrusion with rule-based integration of multiple models, *Computers & Security*, 22(7) 2003, 613–623.
3. S. H. Oh, and W. S. Lee, An anomaly intrusion detection method by clustering normal user behavior, *Computers & Security*, 22(7) 2003, 596–612.
4. B. R. Mahafza, *Introduction to Radar Analysis*, CRC Press, Boca Raton, 1998.
5. Z. Yang, X. Liu, T. Li, D. Wu, J. Wang, Y. Zhao, and H. Han, A systematic literature review of methods and datasets for anomaly-based network intrusion detection, *Computers & Security*, 116, 2022, 102675.
6. L. Garber, Denial-of-service attacks rip the internet, *IEEE Computer*, 33(4) 2000, 12–17.
7. M. McCormack, Denial of service, the future of dirty tricks, *Computer Fraud & Security*, 1996(11) 1996, 3.
8. A. Alsadhan, A. Hussain, P. Liatsis, M. Alani, H. Tawfik, P. Kendrick, and H. Francis, Locally weighted classifiers for detection of neighbor discovery protocol distributed denial-of-service and replayed attacks, *Transactions on Emerging Telecommunications Technologies*, 33(3) 2022, e3700.
9. V. Paxson, An analysis of using reflectors for distributed denial-of-service attacks, *ACM SIGCOMM Computer Communication Review*, 31(3) 2001, 38–47.
10. A. D. Keromytis, V. Misra, and D. Rubinstein, SOS: An architecture for mitigating DDoS attacks, *IEEE Journal on Selected Areas in Communications*, 22 (1) 2004, 176–188.
11. F. Musumeci, A. C. Fidanci, F. Paolucci, F. Cugini, and M. Tornatore, Machine-learning-enabled DDoS attacks detection in P4 programmable networks, *Journal of Network and Systems Management*, 30(1), Article number: 21, 2022.
12. M. Li, *Fractal Teletraffic Modeling and Delay Bounds in Computer Communications*, CRC Press, Boca Raton, 2022.
13. M. Li, and S. C. Lim, Modeling network traffic using generalized Cauchy process, *Physica A*, 387(11) 2008, 2584–2594.
14. M. Li, Long-range dependence and self-similarity of teletraffic with different protocols at the large time scale of day in the duration of 12 years: Autocorrelation modeling, *Physica Scripta*, 95(4) 2020, 065222 (15 pp).

15. M. Li, and A. Wang, Fractal teletraffic delay bounds in computer networks, *Physica A*, 557, 2020, 124903 (13 pp).
16. M. Li, Generalized fractional Gaussian noise and its application to traffic modeling, *Physica A*, 579, 2021, 1236137 (22 pp).
17. P. Abry, and D. Veitch, Wavelet analysis of long-range dependent traffic, *IEEE Transactions on Information Theory*, 44(1) 1998, 2–15.
18. J. Beran, *Statistics for Long-Memory Processes*, Chapman & Hall, New York, 1994.
19. V. Paxon, and S. Floyd, Why we don't know how to simulate the Internet, *Proc., Winter Simulation Conf.*, USA, 1997, 1037–1044.
20. M. Li, Multi-fractional generalized Cauchy process and its application to teletraffic, *Physica A*, 550, 2020, 123982 (14 pp).
21. B. B. Mandelbrot, *Gaussian Self-Affinity and Fractals*, Springer, New York, 2001.
22. V. Paxson, Fast, approximate synthesis of fractional Gaussian noise for generating self-similar network traffic, *Computer Communications Review*, 27(5) 1997, 5–18.
23. W. Willinger, and V. Paxson, Discussion of "heavy tail modeling and teletraffic data", *The Annals of Statistics*, 25(5) 1997, 1805–1869.
24. W. Willinger, and V. Paxson, Where mathematics meets the internet, *Notices of the American Mathematical Society*, 45(8) 1998, 961–970.
25. M. Li, An approach to reliably identifying signs of DDOS flood attacks based on LRD traffic pattern recognition, *Computers & Security*, 23(7) 2004, 549–558.
26. C. C. Wood, Why information security is now multi-disciplinary, multi-departmental, and multi-organizational in nature, *Computer Fraud & Security*, 2004(1) 2004, 16–17.
27. M. Li, An iteration method to adjusting random loading for a laboratory fatigue test, *International Journal of Fatigue*, 27(7) 2005, 783–789.
28. M. Li, W. Jia, and W. Zhao, Correlation form of timestamp increment sequences of self-similar traffic on ethernet, *Electronics Letters*, 36(19), 2000, 1168–1169.
29. M. Li, and C.-H. Chi, A correlation based computational model for synthesizing long-range dependent data, *Journal of the Franklin Institute*, 340(6–7), 2003, 503–514.
30. M. Li, and S. C. Lim, A rigorous derivation of power spectrum of fractional Gaussian noise, *Fluctuation and Noise Letters*, 6(4) 2006, C33–C36.
31. M. Livny, B. Melamed, and A. K. Tsiolis, The impact of autocorrelation on queuing systems, *Management Science*, 39, 1993, 322–339.
32. K. S. Fu, editor, *Digital Pattern Recognition*, 2nd ed., Springer, New York, 1980.
33. M. Basseville, Distance measure for signal processing and pattern recognition, *Signal Processing*, 18(4), 1989, 349–369.
34. J. S. Bendat, and A. G. Piersol, *Random Data: Analysis and Measurement Procedure*, 2nd ed., John Wiley & Sons, New York, 1991.
35. S. K. Mitra, and J. F. Kaiser, *Handbook for Digital Signal Processing*, John Wiley & Sons, New York, 1993.

CHAPTER 13

Change Trend of Hurst Parameter of Multi-Scale Traffic under DDOS Flood Attacks

Since arrival traffic pattern under distributed-denial-of-service (DDOS) flood attacks varies significantly from the pattern of normal traffic (i.e., attack-free traffic) at a protected site, anomaly detection plays a role in the detection of DDOS flood attacks. Hence, quantitatively studying statistics of traffic under DDOS flood attacks (abnormal traffic for short) is essential to anomaly detections of DDOS flood attacks.

Since DDOS attacking needs a certain period of time to make it effective, we can consider the Hurst parameter, H, of traffic as a feature. However, how H of traffic varies under DDOS flood attacks is rarely reported. This chapter shows that averaged H of abnormal traffic tends to be significantly smaller than that of normal one at the protected site. This abnormality of abnormal traffic is demonstrated with test data provided by MIT Lincoln Laboratory and explained from a view of Fourier analysis.

13.1 BACKGROUND

The Internet is the infrastructure that supports computer communications. It has actually become the "electricity" of the modern society because its use in modern society is so pervasive and many people rely on it so heavily. For instance, employees in the modern society would rather give up access to their telephone than give up their access to their email. Nevertheless, it is subject to electronic attacks, such as DDOS flood attacks [1–4]. The threats of DDOS attacks to the individuals are severe. For instance, any denial-of-service (DOS) of a bank server implies a loss of money, disgruntling or losing customers. DDOS flood attacks remain great threats to the Internet, though various approaches and

DOI: 10.1201/9781003354987-17

systems have been proposed to fight it [1–14]. Hence, intrusion detection system (IDS) and intrusion prevention system (IPS) are desired.

There are several categories of DOS attacks. One can divide DOS attacks into three categories: (1) flood (i.e., bandwidth) attacks, (2) protocol attacks, and (3) logical attacks. This chapter considers flood attacks.

A DDOS flood attack sends attack packets upon a site (victim) with a huge amount of traffic the sources of which are distributed over the world so as to effectively jam its entrance and block access by legitimate users or significantly degrade its performance. It never tries to break into the victim's system, making security defenses at the protected site irrelevant.

Usually, IDSs are classified into two categories. One is misuse detection and the other anomaly detection. Solutions given by misuse detection are primarily based on a library of known signatures to match against network traffic. Hence, unknown signatures from new variants of an attack mean 100% miss. Therefore, anomaly detectors play a role in detection of DDOS flood attacks. As far as anomaly detection is concerned, quantitatively characterizing abnormalities of statistics of abnormal traffic is fundamental.

A traffic stream is a packet flow. A packet consists of a number of fields, such as protocol type, source IP, destination IP, ports, flag setting in the case of TCP (Transmission Control Protocol) or UDP (User Datagram Protocol), message type in the case of ICPM (Internet Control Message Protocol), timestamp, and data length (packet size). Each may serve as a feature of a packet. The literature discussing traffic features is rich (see [15–30]). To the best of our knowledge, however, taking into account the Hurst parameter H in characterizing abnormality of traffic series in packet size under DDOS flood attacks is rarely seen. In Chapter 12, autocorrelation function (ACF) of traffic with long-range dependence (LRD) is taken as its statistical feature. As a supplementary to Chapter 12, this chapter specifically studies how H of traffic varies under DDOS flood attacks. In this regard, the following two questions are fundamental.

1. Whether H of traffic when a site is under DDOS flood attacks (abnormal traffic for short) is significantly different from that of normal one (i.e., attack free traffic)?

2. What is the change trend of H of traffic when a site suffers from DDOS flood attacks?

We will give the answers to the above questions from the point of views of processing data traffic and theoretic inference and analysis.

In the rest of chapter, Section 13.2 is about test data. We brief data traffic and use a series of normal traffic available from ref [31] to explain how its H normally varies in Section 13.3. The answer to the question (1) is given in Section 13.4. Then, in Section 13.5, we use a pair of series (one is normal traffic and the other abnormal traffic) that is provided by MIT Lincoln Laboratory [32] to demonstrate that averaged H of abnormal traffic tends to be significantly smaller than that of normal one and briefly discusses this abnormality of abnormal traffic from a view of Fourier analysis, giving the answer to the question (2). Section 13.6 gives the summary.

13.2 TEST DATA

Three series of test data are utilized in this chapter. The first one is an attack-free series measured at the Lawrence Berkeley Laboratory from 14:00 to 15:00 on Friday, January 29, 1994. It is named LBL-PKT-4, which has been widely used in the research of general (normal traffic) traffic pattern (see refs [15–30]). We use it to show a case how H of normal traffic varies. The second is Outside-MIT-week1-1-1999-attack-free (OM-W1-1-1999AF for short) [32]. It was recorded from 08:00:02, March 1 (Monday), to 06:00:02, March 2 (Tuesday), 1999. The third is Outside-MIT-week2-1-1999-attack-contained (OM-W2-1-1999AC for short) [32], which was collected from 08:00:01, March 8 (Monday), to 06:00:49, March 9 (Tuesday), 1999. Two MIT series are used to demonstrate a case how H of traffic varies under DDOS attacks. Though whether or not MIT test data are in the sense of standardization is worth further discussion as stated in [33], they are valuable and can yet be test data for the research of abnormality of abnormal traffic due to available data traffic under DDOS flood attacks being rare.

13.3 BRIEF OF DATA TRAFFIC

Let $x(t_i)$ be a traffic series, indicating the number of bytes in a packet at time t_i ($i = 0, 1, 2, ...$). From a view of discrete series, we write $x(t_i)$ as $x(i)$, implying the number of bytes of the ith packet. Let $r(k)$ be the ACF of $x(i)$. Then,

$$r_{xx}(k) \sim ck^{2H-2}, \tag{13.1}$$

where $c > 0$, $H \in (0.5, 1)$ is the Hurst parameter, and $\sim$ stands for the asymptotical equivalence under the limit $k \to \infty$.

The ACF in (13.1) is non-summable for $H \in (0.5, 1)$, implying LRD. Hence, H is a measure of LRD of traffic. According to the research in traffic engineering, fractional Gaussian noise (fGn) is an approximate model of traffic at large time scales [15]. The ACF of LRD fGn is given by

$$R(k;H) = \frac{\sigma^2}{2} \left[\left(|k|+1\right)^{2H} + \left\||k|-1\right\|^{2H} - 2|k|^{2H} \right], \quad 0.5 < H < 1, \tag{13.2}$$

where $\sigma^2 = \frac{\Gamma(2-H)\cos(\pi H)}{\pi H(2H-1)}$. Without the generality losing, we let $\sigma^2 = 1$.

By taking fGn as an approximate model of $x(i)$, we consider another series given by $x(i)^{(L)} = \frac{1}{L}\sum_{j=iL}^{(i+1)L-1} x(j)$. According to the analysis in self-similar processes, one has $\text{Var}[x^{(L)}] \approx L^{2H-2}\text{Var}(x)$, where Var implies the variance operator. Thus, traffic has the property of asymptotic self-similarity measured by H. Consequently, H characterizes the properties of both LRD and self-similarity of traffic of fGn type.

In practice, measured traffic is of finite length. Let x be a series of P length. Divide x into N non-overlapping sections. Each section is divided into M non-overlapping segments.

Divide each segment into K non-overlapping blocks. Each block is of L length. Let $x(i)_m^{(L)}(n)$ be the series with aggregated level L in the mth segment in the nth section ($m = 0, 1, ...,$ $M - 1$; $n = 0, 1, ..., N - 1$). Let $H_m(n)$ be the H value of $x(i)_m^{(L)}(n)$. Let $R(k; H_m(n))$ be the measured ACF of $x(i)_m^{(L)}(n)$ in the normalized case. Then,

$$R(k; H_m(n)) = \frac{1}{2}\left[\left(|k|+1\right)^{2H_m(n)} + \Big\||k|-1\Big\|^{2H_m(n)} - 2|k|^{2H_m(n)} \right]. \tag{13.3}$$

The above expression exhibits the multi-scale property of traffic.

Let $J[H_m(n)] = \sum_k \left[R(k; H_m(n)) - r(k) \right]^2$ be the cost function. Then, one has

$$H_m(n) = \arg \min J[H_m(n)]. \tag{13.4}$$

Averaging $H_m(n)$ in terms of index m yields

$$H(n) = \frac{1}{M} \sum_{m=0}^{M-1} H_m(n), \tag{13.5}$$

representing the H estimate of the series in the nth section. In practical terms, a normality assumption for $H(n)$ is quite accurate in most cases for $M > 10$ regardless of probability distribution function of H [34]. Thus,

$$H_x = \mathrm{E}\big[H(n)\big] \tag{13.6}$$

is taken as a mean estimate of H of x, where E is the mean operator.

Let σ_H be the standard deviation of $H(n)$. Then, $\mathrm{Prob}\left[z_{1-\alpha/2} < \frac{H(n)-H_x}{\sigma_H} \leq z_{\alpha/2} \right] = 1 - \alpha$, where $(1 - \alpha)$ is confidence coefficient. The confidence interval of $H(n)$ with $(1 - \alpha)$ confidence coefficient is given by $\left(H_x - \sigma_H z_{\alpha/2}, H_x + \sigma_H z_{\alpha/2} \right)$. The following demonstration exhibits $H(n)$ of traffic series LBL-PKT-4.

Demonstration 13.1. The first 1024 points of the series $x(i)$ of LBL-PKT-4 is indicated in Figure 13.1 (a). Consider the first 524,288 ($= P$) points of $x(i)$. The partition settings are as follows. $L = 32$, $K = 16$, $M = 32$, $N = 32$, and $J = 2048$. Computing H in each section yields $H(n)$ as shown in Figure 13.1 (b). Its histogram is indicated in Figure 13.1 (c).

According to (13.6), we have $H_x = 0.758$. The confidence interval with 95% confidence level is [0.750, 0.766]. Hence, we have 95% confidence to say that the H estimate in each section of that series takes $H_x = 0.758$ as its approximation with fluctuation not greater than 7.431×10^{-3}.

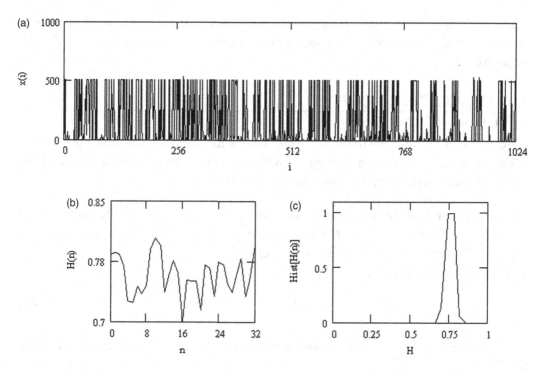

FIGURE 13.1 Demonstrating statistical invariable H. (a) A real traffic time series. (b) Estimate $H(n)$. (c) Histogram of $H(n)$.

13.4 USING H TO DESCRIBE ABNORMALITY OF TRAFFIC UNDER DDOS FLOOD ATTACKS

From the previous discussions, we see that H is a parameter to characterize the properties of both LRD and self-similarity of traffic. On the other hand, ACF is a statistical feature of a time series, which is used in queuing analysis of network systems [35, 36]. Hence, the following lemma.

Lemma 13.1 [6]. Let x and y be normal traffic and abnormal traffic, respectively. Let r_{xx} and r_{yy} be the ACFs of x and y, respectively. During the transition process of DDOS flood attacking, $\|r_{yy} - r_{xx}\|$ is noteworthy.

Proof. A network system is a queuing system. Arrival traffic x of a queuing system has its statistical pattern r_{xx} [35, 36]. Suppose the site suffers from DDOS flood attacks. Suppose that $\|r_{yy} - r_{xx}\|$ is negligible in this case. Then, the site would be overwhelmed at its normal state even if there were no DDOS flood packets. This is an obvious contradiction. □

For each values of $H \in (0.5, 1)$, there is exactly one ACF of fGn with LRD. Thus, a consequence of Lemma 13.1 is that $\|H_y - H_x\|$ is considerable, where H_y and H_x are average H

values of x and y, respectively. Hence, H is a parameter that can yet be used to describe abnormality of traffic under DDOS flood attacks. This gives the answer to the question (1) in Section 13.1.

13.5 CHANGE TREND OF H

13.5.1 Cases of Change Trend of H

Here, we show two demonstrations of $H(n)$. One is for normal traffic and the other for abnormal traffic. Two demonstrations show that average value of H of abnormal traffic tends to be significantly smaller than that of normal traffic.

Demonstration 13.2 (attack free traffic). The first 1024 points of the series $x(i)$ of attack free traffic OM-W1-1-1999AF is indicated in Figure 13.2 (a). Its $H(n)$ is plotted in Figure 13.2 (b) and histogram in Figure 13.2 (c).

By computation, we obtain

$$H_x = 0.895. \tag{13.7}$$

Its variance = 5.693×10^{-4} and the confidence interval with 95% confidence level [0.865, 0.895].

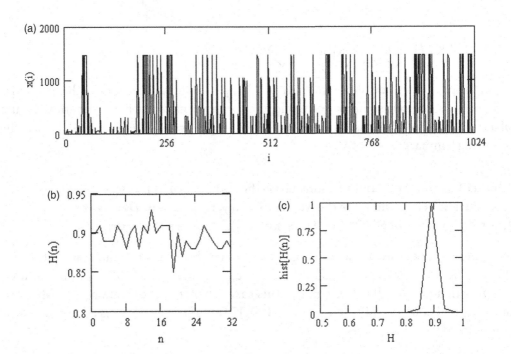

FIGURE 13.2 Demonstrating $H(n)$ of attack free traffic OM-W1-1-1999AF. (a) Time series of OM-W1-1-1999AF. (b) Estimate $H(n)$ of OM-W1-1-1999AF. (c) Histogram of $H(n)$ of OM-W1-1-1999AF.

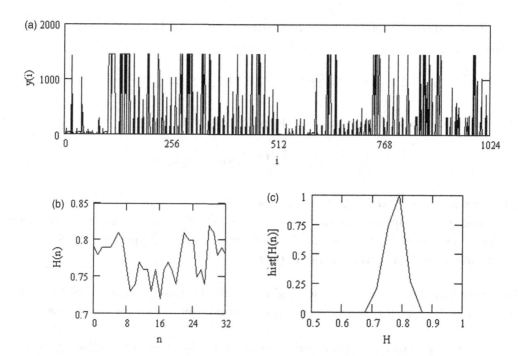

FIGURE 13.3 Demonstrating $H(n)$ of abnormal traffic OTM-W2-1999AC. (a) Time series of OM-W2-1-1999AC. (b) Estimate $H(n)$ of OM-W2-1-1999AC. (c) Histogram of $H(n)$ of OM-W2-1-1999AC.

Demonstration 13.3 (abnormal traffic). The first 1024 points of the series $x(i)$ of attack contained traffic OM-W2-1-1999AC is indicated in Figure 13.3 (a). Its $H(n)$ is plotted in Figure 13.3 (b) and histogram in Figure 13.3 (c).

By computation, we obtain

$$H_y = 0.774. \tag{13.8}$$

Its variance = 6.777×10^{-4} and the confidence interval with 95% confidence level [0.723, 0.825].

Comparing the means of H in the above two demonstrations, we see

$$H_y < H_x. \tag{13.9}$$

The above inequality exhibits a case of the change trend of H of traffic under DDOS flood attacks. It actually follows a general rule as can be seen from the following analysis.

13.5.2 Analysis of Change Trend of H of Traffic under DDOS Flood Attacks

In the case of multi-scale fGn, we let H represent the mean estimate of the Hurst parameter as that in (13.6) for the sake of simplicity. As $0.5\left[(\tau+1)^{2H} - 2\tau^{2H} + (\tau-1)^{2H}\right]$ is the finite second-order difference of $0.5\tau^{2H}$, approximating it with second-order differential of $0.5\tau^{2H}$ yields

$$0.5\left[(\tau+1)^{2H} - 2\tau^{2H} + (\tau-1)^{2H}\right] \approx H(2H-1)\tau^{2H-2}. \tag{13.10}$$

In the domain of generalized functions [44, p. 43], we obtain

$$F\left(|\tau|^{2H-2}\right) = 2\cos\frac{\pi(2H-1)}{2}(2H-2)!|\omega|^{-(2H-1)}, \tag{13.11}$$

where F is the operator of the Fourier transform.

As known, the frequency bandwidth of x is the width of its power spectrum $S(\omega)$, which is usually explained in the sense of the maximum effective frequency in engineering. Hence, the following is a consequence of (13.11).

Inference 13.1. Let B_1 and B_2 be the bandwidths of LRD fGn x_1 and x_2, respectively. Let mean estimates of H of x_1 and x_2 be H_1 and H_2, respectively. Then, $H_2 < H_1$ if $B_2 > B_1$. □

As known, the data rate of abnormal traffic is greater than that of attack free traffic [37]. Hence, the bandwidth of abnormal traffic is wider than that of attack-free traffic according to the relationship between data rate and bandwidth. Then, according to Inference 13.1, we see that average H of abnormal traffic is smaller than that of attack-free traffic, giving the answer to the question (2) in Section 13.1. Eq. (13.9) is a case about this rule. As the larger the H the stronger the LRD, we note that LRD of abnormal traffic become weaker than those of attack-free traffic.

13.6 SUMMARY

We have explained that average H of abnormal traffic significantly differs from that of normal one as a consequence of Lemma 13.1. We have given Inference 13.1 to show that average H of abnormal traffic is smaller than that of normal traffic. The results in theory are demonstrated and also validated with test data provided by MIT Lincoln Laboratory.

REFERENCES

1. R. A. Kemmerer, and G. Vigna, Intrusion detection: A brief history and overview, *Supplement to IEEE Computer, IEEE Security & Privacy*, 35(4) 2002, 27–30.
2. A. Householder, K. Houle, and C. Dougberty, Computer attack trends challenge internet security, *Supplement to IEEE Computer, IEEE Security & Privacy*, 35(4) 2002, 5–7.
3. E. Schultz, Intrusion prevention, *Computer & Security*, 23(4) 2004, 265–266.
4. S. E. Quincozes, C. Albuquerque, D. Passos, and D. Mossé, A survey on intrusion detection and prevention systems in digital substations, *Computer Networks*, 184, 2021, 107679.
5. R. Rasool, H. Wang, U. Ashraf, K. Ahmed, Z. Anwar, and W. Rafique, A survey of link flooding attacks in software defined network ecosystems, *Journal of Network and Computer Applications*, 172, 2020, 102803.

6. M. Li, An approach for reliably identifying signs of DDoS flood attacks based on LRD traffic pattern recognition, *Computer & Security*, 23(7) 2004, 549–558.

7. S. Q. A. Shah, F. Z. Khan, and M. Ahmad, Mitigating TCP SYN flooding based EDOS attack in cloud computing environment using binomial distribution in SDN, *Computer Communications*, 182, 2022, 198–211.

8. F. Zola, L. Segurola-Gil, J. L. Bruse, M. Galar, and R. Orduna-Urrutia, Network traffic analysis through node behaviour classification: A graph-based approach with temporal dissection and data-level preprocessing, *Computers & Security*, 115, 2022, 102632.

9. M. Li, and M. Li, An adaptive approach for defending against DDoS attacks, *Mathematical Problems in Engineering*, 2010, 2010, Article ID 570940 (15 pp).

10. S. H. Oh, and W. S. Lee, An anomaly intrusion detection method by clustering normal user behavior, *Computer & Security*, 22(7) 2003, 596–612.

11. M. Odusami, S. Misra, O. Abayomi-Alli, A. Abayomi-Alli, and L. Fernandez-Sanz, A survey and meta-analysis of application-layer distributed denial-of-service attack, *International Journal of Communication Systems*, 33(18) 2020, e4603.

12. S.-B. Cho, and H.-J. Park, Efficient anomaly detection by modeling privilege flows using hidden Markov model, *Computer & Security*, 22(1) 2003, 45–55.

13. S. Cho, and S. Cha, SAD: Web session anomaly detection based on parameter estimation, *Computer & Security*, 23(7) 2004, 312–319.

14. M. P. Singh, and A. Bhandar, New-flow based DDoS attacks in SDN: Taxonomy, rationales, and research challenges, *Computer Communications*, 154, 2020, 509–527.

15. V. Paxson, and S. Floyd, Wide-area traffic: The failure of Poisson modeling, *IEEE/ACM Transactions on Networking*, 3(3) 1995, 226–244.

16. J. Beran, *Statistics for Long-Memory Processes*, Chapman & Hall, New York, 1994.

17. W. Willinger, and V. Paxson, Where mathematics meets the internet, *Notices of the American Mathematical Society*, 45(8) 1998, 961–970.

18. W. Willinger, M. S. Taqqu, W. E. Leland, and D. V. Wilson, Self-similarity in high-speed packet traffic: Analysis and modeling of ethernet traffic measurements, *Statistical Science*, 10(10) 1995, 67–85.

19. H. Michiel, and K. Laevens, Teletraffic engineering in a broad-band era, *Proceedings of the IEEE*, 85(12) 1997, 2007–2033.

20. A. Adas, Traffic models in broadband networks, *IEEE Communications Magazine*, 35(7) 1997, 82–89.

21. E. Leland, M. Taqqu, W. Willinger, and D. V. Wilson, On the self-similar nature of ethernet traffic, (extended version), *IEEE/ACM Transactions on Networking*, 2(1) 1994, 1–15.

22. J. Beran, R. Shernan, M. S. Taqqu, and W. Willinger, Long-range dependence in variable bit-rate video traffic, *IEEE Transactions on Communications*, 43 (2–4) 1995, 1566–1579.

23. I. Csabai, 1/f noise in computer network traffic, *Journal of Physics A: Mathematical & General*, 27(12) 1994, L417–L421.

24. B. Tsybakov, and N. D. Georganas, Self-similar processes in communications networks, *IEEE Transactions on Information Theory*, 44(5) Sep. 1998, 1713–1725.

25. M. Li, *Fractal Teletraffic Modeling and Delay Bounds in Computer Communications*, 1st ed., CRC Press, Boca Raton, 2022.

26. M. Li, Modified multifractional Gaussian noise and its application, *Physica Scripta*, 96(12) 2021, 125002 (12 pp).

27. M. Li, Generalized fractional Gaussian noise and its application to traffic modeling, *Physica A*, 579, 2021, 1236137 (22 pp).

28. M. Li, Long-range dependence and self-similarity of teletraffic with different protocols at the large time scale of day in the duration of 12 years: Autocorrelation modeling, *Physica Scripta*, 95(4) 2020, 065222 (15 pp).

29. M. Li, Multi-fractional generalized Cauchy process and its application to teletraffic, *Physica A*, 550, 2020, 123982 (14 pp).
30. M. Li, Record length requirement of long-range dependent teletraffic, *Physica A*, 472, 2017, 164–187.
31. P. Danzig, J. Mogul, V. Paxson, and M. Schwartz, The internet traffic archive, 2000. ftp://ita. ee.lbl.gov/traces/. [dataset].
32. http://www.ll.mit.edu/IST/ideval.
33. J. McHugh, Testing intrusion detection systems: A critique of the 1988 and 1999 DARPA intrusion detection system evaluations as performed by Lincoln laboratory, *ACM Transactions on Information System Security*, 3(4) 2000, 262–294.
34. J. S. Bendat, and A. G. Piersol, *Random Data: Analysis and Measurement Procedure*, 2nd ed., John Wiley & Sons, New York, 1991.
35. M. Livny, B. Melamed, and A. K. Tsiolis, The impact of autocorrelation on queuing systems, *Management Science*, 39, 1993, 322–339.
36. S.-Q. Li, and C.-L. Hwang, Queue response to input correlation functions: Continuous spectral analysis, *IEEE/ACM Transactions on Networking*, 1(6) 1993, 678–692.
37. L. Garber, Denial-of-service attacks rip the internet, *IEEE Journal of Computer*, 33(4) 2000, 12–17.

Postscript

Having explained the theory of fractal time series in Chapters 1–5, the long-range dependence (LRD) and self-similarity (SS) of traffic in Chapter 6, data processing techniques of traffic in Chapters 7 and 8, multi-fractional/multi-fractal models of traffic in Chapters 9–11, applications to intrusion detections in computer networks in Chapters 12 and 13, I would like to talk about some related issues in this chapter.

14.1 LOCAL VERSUS GLOBAL OF FRACTAL TRAFFIC

We use the Hurst parameter H to measure global property of traffic time series $x(t)$, such as LRD, and the fractal dimension D to characterize local property of $x(t)$, such as SS. The two, namely, LRD and SS, are different concepts. However, there is no general rule about the relationship between LRD and SS. Whether LRD and SS are related or not relies on a concrete model of $x(t)$ one considers. For example (Mandelbrot [1], Li [2, 3]), the LRD of the fractional Gaussian noise (fGn) is completely and linearly related to its SS in the form

$$D_{fGn} = 2 - H_{fGn}. \qquad (14.1)$$

The LRD and SS of the generalized fractional Gaussian noise (gfGn) are coupled in the form (Li [2, 4])

$$D_{gfGn} = 2 - aH_{gfGn}, \, 0 < a \le 1, \qquad (14.2)$$

with the couple factor a. However, the LRD and SS of the generalized Cauchy (GC) process of traffic model are totally uncoupled (Li [2, 3], Gneiting and Schlather [5], Lim and Li [6], Li [7–9]).

When traffic $x(t)$ is multi-fractal, the measure of LRD or SS is time-varying. For the multi-fractional GC (mGC) process (Li [10]) or modified multi-fractional Gaussian noise model (mmfGn) (Li [11]), both LRD and SS are time-varying. The difference between two

DOI: 10.1201/9781003354987-18

is that the LRD of the mGC process is uncoupled with its SS while the SS of mmfGn is linearly related to its LRD in the form

$$D_{\text{mmfGn}}(t) = 2 - H_{\text{mmfGn}}(t). \tag{14.3}$$

14.2 STATIONARITY VERSUS MULTI-FRACTAL PROPERTY OF TRAFFIC

In Chapter 4, I have explained that traffic in the sense of mono-fractal is ergodic. However, since multi-fractal models of traffic, either mGC or mmfGn is with time-varying fractal dimension $D(t)$ or time-varying Hurst parameter $H(t)$, its statistical models, such as autocorrelation function (ACF) or power spectrum density (PSD), are time-varying. Thus, purely from the point of view of mathematics, multi-fractal models of traffic are non-stationary. In practice, however, a traffic series is of finite length. Consequently, whether a multi-fractal traffic with finite length is stationary or not depends on variation degree of $H(t)$ or $D(t)$ as well as observation time scales, see Chapters 7 and 10.

14.3 OPEN PROBLEMS

14.3.1 Open Problem of Traffic Bound

In computer science, a particular thing is about the bounding model of traffic introduced by Cruz [12] in the form

$$\int_0^t x(u)du = X(t) \leq \sigma + \rho t, \tag{14.4}$$

where σ and ρ are nonnegative constants (also see Li [2], Boudec [13], Jiang and Liu [14], Jiang [15, 16]). The quantity σ is a measure of burstiness, which is a local property of traffic, while ρ measures long-average rate, which is a global property of traffic. The pair (σ, ρ) is a kind of traffic model called Cruz bound [2]. Two parameters σ and ρ are independent of each other. For Li's bound [2] in the form

$$\int_0^t x(u)du = X(t) \leq r^{2D-5}\sigma + a^{-H}\rho t, \tag{14.5}$$

how to apply a model of multi-fractal traffic, for instance, the mGC one, to the above remains an open problem.

14.3.2 Open Problem with Respect to Traffic at Connection Level

Let $x_j(t)$ be the traffic at connection j ($j = 1, 2, ..., J$). Practically, it may be very difficult to demand that each traffic trace $x_j(t)$ has enough length for satisfactory estimation of its ACF as discussed in Chapter 8. Therefore, an open problem is how to obtain satisfactory estimations of D and H with short record length without the restrictions explained in Chapter 8.

14.3.3 Open Problem Regarding Refining the Stationarity Test of Traffic

We previously concluded that traffic is stationary at small time scales and non-stationary at large time scales in Chapters 7 and 10. However, what is the critical point or critical range between small time scales and large ones with respect to the stationarity issue of traffic is a problem unsolved.

14.3.4 Open Problem About Weight Function in Generalized Mean Square Error for Traffic Prediction

In Theorem 5.2, Chapter 5, we introduced a weight function $g(x) \in S$ in the mean square error of predicting an LRD series $x(n)$ so that $x(n)$ is predicable. However, what are possible function forms of $g(x)$ is a problem that remains unsolved.

REFERENCES

1. B. B. Mandelbrot, *Gaussian Self-Affinity and Fractals*, Springer, New York, 2001.
2. M. Li, *Fractal Teletraffic Modeling and Delay Bounds in Computer Communications*, CRC Press, Boca Raton, 2022.
3. M. Li, Record length requirement of long-range dependent teletraffic, *Physica A*, 472, 2017, 164–187.
4. M. Li, Generalized fractional Gaussian noise and its application to traffic modeling, *Physica A*, 579, 2021, 1236137 (22 pp).
5. T. Gneiting, and M. Schlather, Stochastic models that separate fractal dimension and Hurst effect, *SIAM Review*, 46(2) 2004, 269–282.
6. S. C. Lim, and M. Li, A generalized Cauchy process and its application to relaxation phenomena, *Journal of Physics A: Mathematical and General*, 39(12) 2006, 2935–2951.
7. M. Li, Long-range dependence and self-similarity of teletraffic with different protocols at the large time scale of day in the duration of 12 years: Autocorrelation modeling, *Physica Scripta*, 95(4) 2020, 065222 (15 pp).
8. M. Li, Self-similarity and long-range dependence in teletraffic, *Proc., the 9th WSEAS Int. Conf. on Multimedia Systems and Signal Processing*, Hangzhou, China, May 2009, 19–24.
9. M. Li, Fractal time series: A tutorial review, *Mathematical Problems in Engineering*, 2010, 2010, Article ID 157264 (26 pp).
10. M. Li, Multi-fractional generalized Cauchy process and its application to teletraffic, *Physica A*, 550, 2020, 123982 (14 pp).
11. M. Li, Modified multifractional Gaussian noise and its application, *Physica Scripta*, 91(12) 2021 (12 pp).
12. R. L. Cruz, A calculus for network delay, part I: Network elements in isolation, part II: Network analysis, *IEEE Transactions on Information Theory*, 37(1) 1991, 114–141.
13. J. Y. Le Boudec, Application of network calculus to guaranteed service networks, *IEEE Transactions on Information Theory*, 44(3) 1998, 1087–1096.
14. Y.-M. Jiang, and Y. Liu, *Stochastic Network Calculus*, Springer, New York, 2008.
15. Y.-M. Jiang, Per-domain packet scale rate guarantee for expedited forwarding, *IEEE/ACM Transactions on Networking*, 14(3) 2006, 630–643.
16. Y.-M. Jiang, A basic stochastic network calculus, *ACM SIGCOMM Computer Communication Review*, 36(4) 2006, 123–134.

Appendix

Convergence of Sample Autocorrelation of LRD Traffic

We depict my work on a fundamental issue in the theory of long-range-dependent traffic in the aspect of the convergence of sample autocorrelation function (ACF) of real traffic. The present results suggest that the sample ACF of traffic is convergent. In addition, we show that the sample size has considerable effects in estimating the sample ACF of traffic. More precisely, a sample ACF of traffic tends to be smoother when the sample size increases, which is also detailed in Chapter 8 from a view of measurement.

A.1 BACKGROUND

Let us start with the meaning of the convergence of a sample ACF of a stationary random function $x(t)$ for $t \in (0, \infty)$. Denote by $x_T(t)$ a sample record of $x(t)$ with length T, i.e., $x_T(t) = x(t)$ for $t \in (0, T)$. Let $r_T(\tau)$ be the ACF of $x_T(t)$, meaning $r_T(\tau) = E[x_T(t)x_T(t + \tau)]$ for $t, t + \tau \in (0, T)$. Then, we say that $r_T(\tau)$ is convergent if $\lim_{T \to} r_T(\tau) = r(\tau)$ exists, where $r(\tau)$ is the ACF of $x(t)$. Otherwise, $r_T(\tau)$ is divergent or does not exist. In the discrete case, $x(t)$ is replaced by $x(i)$ $(i = 0, \ldots, \infty)$, $x_T(t)$ is substituted by $x_I(i)$ for $i = 0, \ldots, I - 1$, and $r_T(\tau)$ is replaced with $r_I(k)$ $(k = 0, \ldots, I - 1)$. When $\lim_{I \to \infty} r_I(k) = r(k)$ exists, where $r(k)$ is the ACF of $x(i)$, we say that $r_I(k)$ is convergent. That is the meaning of the convergence or existence of a sample ACF of a random function.

The issue described above may be unnecessary to treat in the field of conventional second-order random functions because an ACF of a conventional second-order random function generally exists (Yaglom [1], Lindgren and McElrath [2], Papoulis and Pillai [3], Fuller [4], Bendat and Piersol [5], Beran [6]). However, the issue whether a sample ACF of teletraffic (traffic for short) time series with long-range dependence is convergent or not is worth discussing.

The research by Resnick et al. [7] stated that the sample ACF of stable random functions with infinite variance may be random when the sample size approaches infinity instead of being convergent. The example of such a type of random functions includes α-stable processes with infinite variance [7, p. 798]. Recall that traffic modeling using α-stable processes was described by reports, such as Karasaridis and Hatzinakos [8], Barbe and McCormick [9]. On the other side, the reports by Field et al. [10, 11] discussed the traffic modeling using the standard Cauchy distribution, which implied that traffic is of infinite variance because variance of a random function obeying the Cauchy distribution is infinite [2–4]. Therefore, a sample ACF of traffic may be divergent if it is of infinite variance. Nonetheless, correlations of traffic play a role in practice (see refs [12–23], just to cite a few). Consequently, the answer to the question whether a sample ACF of traffic is convergent or not is desired from the point of view of traffic theory. In this appendix, we aim at exhibiting that sample ACFs of real traffic data are convergent.

This appendix chapter is organized as follows: Problem statement is described in Section A.2. Results are explained in Section A.3. Discussions are given in Section A.4, which is followed by summary.

A.2 PROBLEM STATEMENTS AND RESEARCH THOUGHTS

In the field, the convergence of an ACF of a time series with LRD was challenged by Resnick et al. [7]. Hence, comes the problem whether an ACF of traffic with LRD convergences.

Note that we explained the egodicity of LRD series in Chapter 4. That implies that an ACF of an LRD series is convergent. However, we still describe more about the convergence of an ACF of LRD series using real traffic traces. In what follows, we assume that traffic $x(i)$ is causal. By causal we mean that $x(i)$ is defined for $i \geq 0$ and $x(i) = 0$ if $i < 0$. In the interval $[0, I]$, the sample ACF of $x_l(i)$ is given by

$$r_l(k) = \frac{1}{I} \sum_{i=0}^{I-1} x(i)x(i+k). \tag{A.1}$$

For $I_l \neq I_m$, assuming $I_l < I_m$, we have,

$$r_{I_l}(k) \neq r_{I_m}(k), k = 0, \ldots, I_l - 1. \tag{A.2}$$

The above may be expressed by

$$r_{I_l}(k) + \Delta_{I_l}(k) = r_{I_m}(k), k = 0, \ldots, I_l - 1, \tag{A.3}$$

where $\Delta_{I_l}(k)$ is a fluctuation noise. Further, For $I_l \neq I_m \neq I_n$, assuming $I_l < I_m < I_n$, we have,

$$r_{I_m}(k) + \Delta_{I_m}(k) = r_{I_n}(k), k = 0, \ldots, I_l - 1. \tag{A.4}$$

In general, if $r_l(k)$ is convergent, we have

$$\sum_{k=0}^{I_l-1} \left| \Delta_{I_m}(k) \right| \leq \sum_{k=0}^{I_l-1} \left| \Delta_{I_l}(k) \right|. \tag{A.5}$$

More precisely, in the case of $r_l(k)$ being convergent, we have

$$\lim_{I_l \to \infty} r_{I_l}(k) = \lim_{I_m \to \infty} r_{I_m}(k) = r(k). \tag{A.6}$$

Consequently, the above means that

$$\lim_{I_l \to \infty} \Delta_{I_l}(k) = 0. \tag{A.7}$$

Since a real traffic data series is always of finite length and because the true ACF of $x(i)$ is never achieved, the issue we treat in this research is to investigate whether $r_l(k)$ tends smoother when I becomes larger. If that is so, we infer that $r_l(k)$ is convergent. If not, it is divergent.

A.3 RESULTS

Before showing the results, we brief the real traffic data used in this section. Let $x(t(i))$ be a sample record of traffic, where $t(i)$ is the series of timestamps, indicating the timestamp of the ith packet arriving at a server. Thus, $x(t(i))$ represents the packet size of the ith packet recorded at time $t(i)$. Note that the pattern of $x(t(i))$ is consistent with that of $x(i)$ that represents the size of the ith packet on a packet-by-packet basis.

We use two real traffic traces. One is named BC-pOct89 that contains 1,000,000 packets. It was recorded on an Ethernet at the Bellcore Morristown Research and Engineering Facility [24], which is now known as Telcordia Technologies (http://en.wikipedia.org/wiki/Telcordia_Technologies), made by Will Leland and Dan Wilson. The other is DEC-PKT-1 with 3.3 million packets. It was made by Jeff Mogul of Digital's Western Research Lab [24]. The former was used in the pioneering research of the fractal statistics of traffic in [25, 26]. The latter was utilized by Paxson and Floyd in ref [27]. We cite the articles by Cao et al. [28], Resnick [29], D'Auria and Resnick [30] for the nice description of the basic statistics of traffic.

Figure A.1 is the plot of the first 2048 data points of traffic trace BC-pOct89. Figure. A.2 indicates the plots of sample ACFs of BC-pOct89 with different sample sizes as follows.

- Fig. A.2 (a): sample ACF of BC-pOct89 with the sample size $I = 2^{11} = 2048$.

- Fig. A.2 (b): sample ACF of BC-pOct89 with the sample size $I = 2^{12} = 4096$.

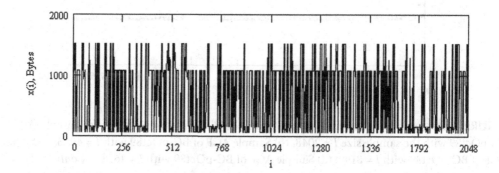

FIGURE A.1 Plot of packet-size series of traffic trace BC-pOct89 on a packet-by-packet basis.

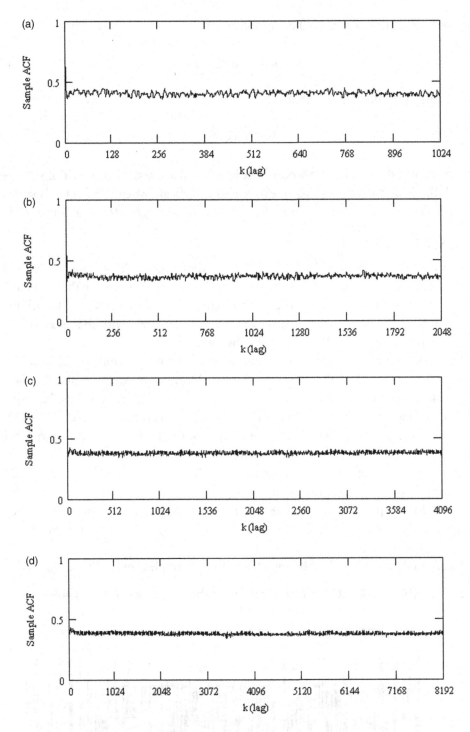

FIGURE A.2 Sample ACFs of BC-pOct89 with the different sample sizes. (a) Sample ACF of BC-pOct89 with the sample size $I = 2048$. (b) Sample ACF of BC-pOct89 with $I = 4096$. (c) Sample ACF of BC-pOct89 with $I = 8192$. (d) Sample ACF of BC-pOct89 with $I = 16384$. *Continued*

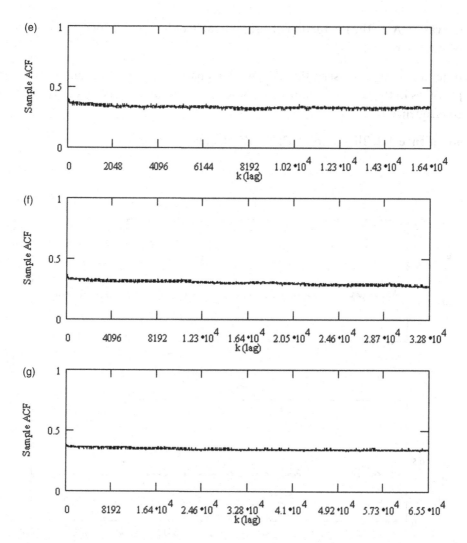

FIGURE A.2 *(Continued)* (e) Sample ACF of BC-pOct89 with $I = 32768$. (f) Sample ACF of BC-pOct89 with $I = 65536$. (g) Sample ACF of BC-pOct89 with $I = 131027$.

- Fig. A.2 (c): sample ACF of BC-pOct89 with the sample size $I = 2^{13} = 8192$.

- Fig. A.2 (d): sample ACF of BC-pOct89 with the sample size $I = 2^{14} = 16,384$.

- Fig. A.2 (e): sample ACF of BC-pOct89 with the sample size $I = 2^{15} = 32,768$.

- Fig. A.2 (f): sample ACF of BC-pOct89 with the sample size $I = 2^{16} = 65,536$.

- Fig. A.2 (g): sample ACF of BC-pOct89 with the sample size $I = 2^{17} = 131,027$.

From Figure A.2, we obtain the following observations:

Observation A.1. The sample ACF of BC-pOct89 with different sample sizes obeys a certain deterministic function.

Observation A.2. The fluctuation of a sample ACF of BC-pOct89 decreases as the sample size increases.

In order to refine the Observation A.2, we demonstrate the first 1024 points of the sample ACFs in the sense of zoom in as shown in Figure A.3. Thus, comes the following consequence.

Consequence A.1. The sample ACF of BC-pOct89 is convergent.

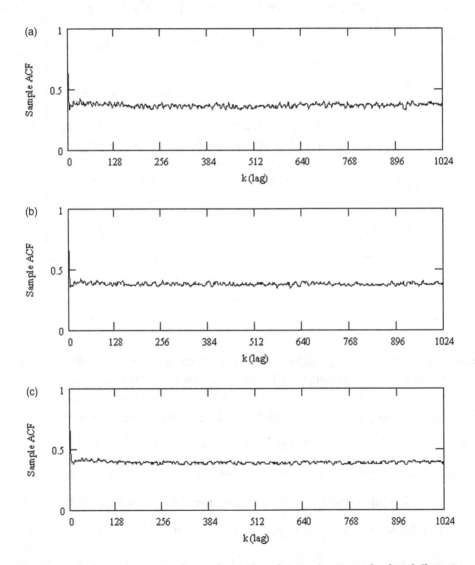

FIGURE A.3 First 1024 points of sample ACFs of BC-pOct89 with the different sample sizes. (a) First 1024 points of sample ACF of BC-pOct89 with the sample size $I = 4096$. (b) First 1024 points of sample ACF of BC-pOct89 with $I = 8192$. (c) Sample ACF of BC-pOct89 with $I = 16384$. *(Continued)*

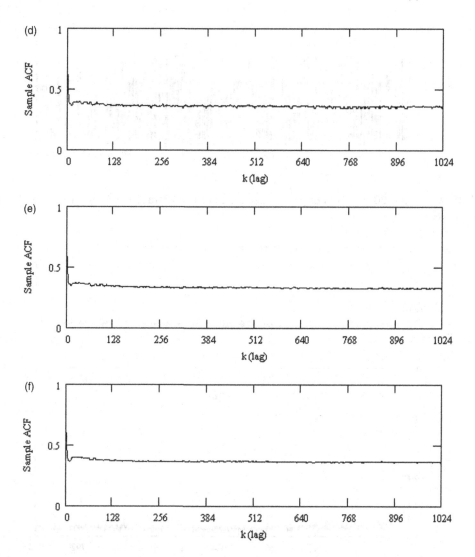

FIGURE A.3 *(Continued)* (d) First 1024 points of sample ACF of BC-pOct89 with $I = 32768$. (e) First 1024 sample ACF of BC-pOct89 with $I = 65536$. (f) First 1024 sample ACF of BC-pOct89 with $I = 131027$.

The demonstrations of the sample ACF of DEC-PKT-1 are given in Figures A.4–A.6. The first 2048 data points of DEC-PKT-1 are plotted in Figure A.4. Figure A.5 shows its plots of sample ACFs with different sample sizes from $I = 2^{11} = 2048$ to $I = 2^{17} = 131027$.

Figure A.5 implies that the sample ACF of DEC-PKT-1 with different sample sizes is a deterministic function. Besides, the fluctuation of its sample ACF decreases as the sample size increases. We refine the observation that the fluctuation of its sample ACF decreases as the sample size increases in Figure A.6 by indicating the first 1024 points of each sample ACF, making the result that the sample ACF of DEC-PKT-1 is convergent more clearly.

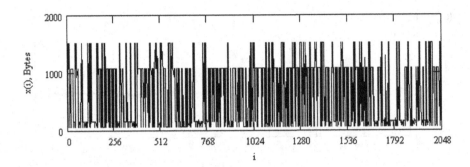

FIGURE A.4 First 2048 points of packet-size series of traffic trace DEC-PKT-1 on a packet-by-packet basis.

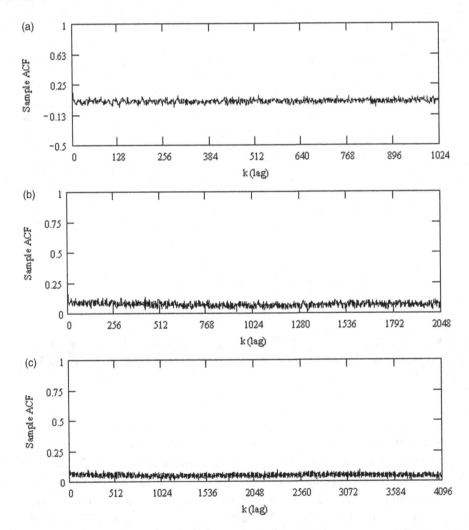

FIGURE A.5 Plots of sample ACFs of DEC-PKT-1 with the different sample sizes. (a) Sample ACF of DEC-PKT-1 with the sample size $I = 2048$. (b) Sample ACF of DEC-PKT-1 with $I = 4096$. (c) Sample ACF of DEC-PKT-1 with $I = 8192$. (d) Sample ACF of DEC-PKT-1 with $I = 16384$. (e) Sample ACF of DEC-PKT-1 with $I = 32768$. (f) Sample ACF of DEC-PKT-1 with $I = 65536$. (g) Sample ACF of DEC-PKT-1 with $I = 131027$. *(Continued)*

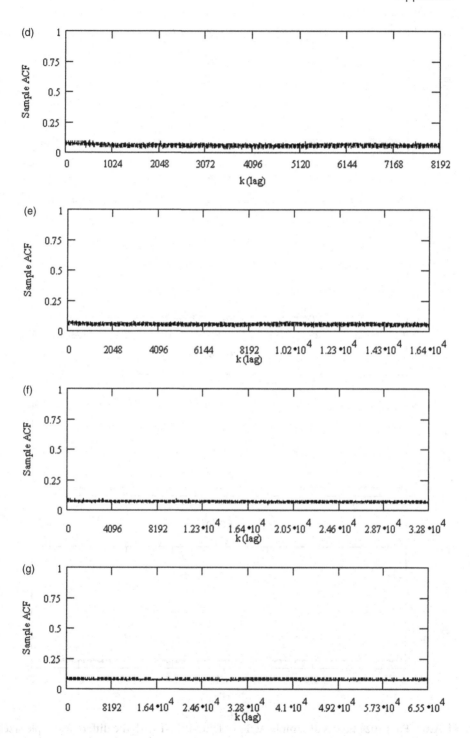

FIGURE A.5 *(Continued)*

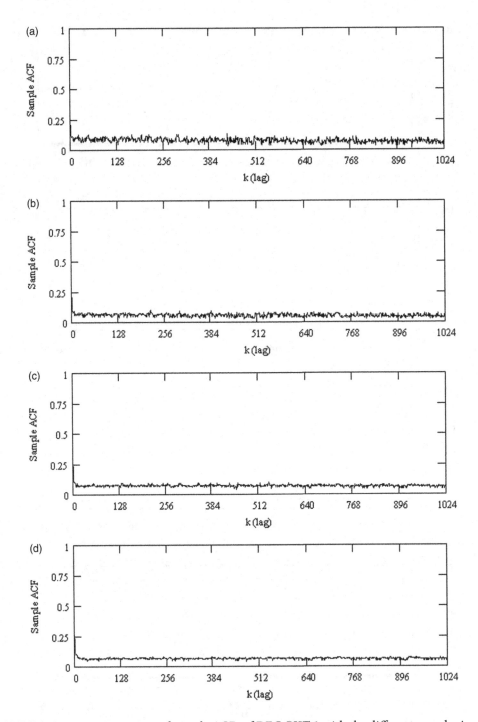

FIGURE A.6 First 1024 points of sample ACFs of DEC-PKT-1 with the different sample sizes. (a) First 1024 points of sample ACF of DEC-PKT-1 with the sample size $I = 4096$. (b) First 1024 points of sample ACF of DEC-PKT-1 with $I = 8192$. (c) Sample ACF of DEC-PKT-1 with $I = 16384$. (d) First 1024 points of sample ACF of DEC-PKT-1 with $I = 32768$. (e) First 1024 sample ACF of DEC-PKT-1 with $I = 65536$. (f) First 1024 sample ACF of DEC-PKT-1 with $I = 131027$. *(Continued)*

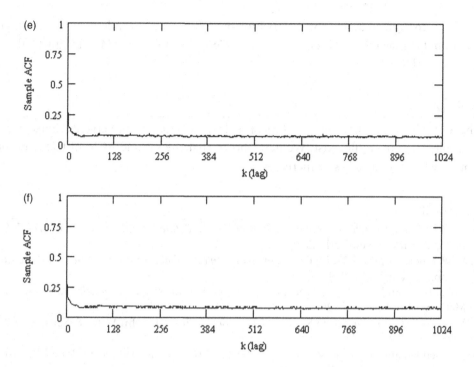

FIGURE A.6 *(Continued)*

A.4 DISCUSSIONS

The modeling of the sample ACF of BC-pOct89 may be fractional Gaussian noise (fGn), which is appropriate when the time scaling is large (Feldmann and et al. [31], Li et al. [32]), or the generalized fGn (Li [33]). Both BC-pOct89 and DEC-PKT-1 may be well described by the generalized Cauchy (GC) correlation structure (Li and Lim [34]). The theme of this research is the convergence of sample ACF of traffic instead of correlation modeling of traffic. From the present results, nevertheless, one may infer that Eqs. (A.5), (A.6), and (A.7) hold.

The data used on this research were measured in the last century, more precisely, BC-pOct89 in 1989 and DEC-PKT-1 in 1995. However, the research by Borgnat et al. [35], Abry et al. [36], and Li [37] exhibited that statistics of traffic remain identical from the early age of the Internet to present time. Therefore, we infer that sample ACF of traffic today is convergent.

Though we utilized traffic data in this research, the methodology described in this research might be possible for investigating the convergence of sample ACFs of other data series, such as those in marine science [38–41]. Finally, we note that this research does not imply something that might deviate from the point of view in traffic modeling using Levy stable structure, such as stable Lévy motion (Mikosch et al. [42]) or Levy flights (Terdik and Gyires [43]) or α-stable models [8] or the standard Cauchy distribution [10, 11] when those correspond to random functions with infinite variance such that they may yield divergent sample ACF. Rather, we suppose that it seems still faraway to thoroughly understand the statistics of traffic, without contradictions, with its commonly used models that may not be enough as Mandelbrot stated for modeling fractal random functions in general [44], but things may need developing (see Cohen and Lindner [45], Denby et al. [46], Lazarou et al. [47]).

A.5 SUMMARY

We have exhibited that the sample ACFs of real traffic data are convergent. Consequently, the sample ACFs of traffic exist. We have shown that the sample ACFs of traffic become smoother when the sample size increases.

REFERENCES

1. A. M. Yaglom, *Correlation Theory of Stationary and Related Random Functions, Vol. I: Basic Results*, Springer, New York, 1987.
2. B. W. Lindgren, and G. W. McElrath, *Introduction to Probability and Statistics*, The Macmillan Company, New York, 1959.
3. A. Papoulis, and S. U. Pillai, *Probability, Random Variables, and Stochastic Processes*, 3rd ed., McGraw-Hill, New York, 2002.
4. W. A. Fuller, *Introduction to Statistical Time Series*, 2nd ed., John Wiley & Sons, New York, 1996.
5. J. S. Bendat, and A. G. Piersol, *Random Data: Analysis and Measurement Procedure*, 3rd ed., John Wiley & Sons, New York, 2000.
6. J. Beran, *Statistics for Long-Memory Processes*, Chapman & Hall, New York, 1994.
7. S. Resnick, G. Samorodnitsky, and F. Xue, How misleading can sample ACFs of stable MAs be? (Very!), *The Annals of Applied Probability*, 9(3) 1999, 797–817.
8. A. Karasaridis, and D. Hatzinakos, Network heavy traffic modeling using α-stable self-similar processes, *IEEE Transactions on Communications*, 49(7) 2001, 1203–1214.
9. P. Barbe, and W. P. McCormick, Heavy-traffic approximations for fractionally integrated random walks in the domain of attraction of a non-Gaussian stable distribution, *Stochastic Processes and Their Applications*, 122(4) 2012, 1276–1303.
10. A. J. Field, U. Harder, and P. G. Harrison, Measurement and modelling of self-similar traffic in computer networks, *IEE Proceedings-Communications*, 151(4) 2004, 355–363.
11. T. Field, U. Harder, and P. Harrison, Network traffic behaviour in switched ethernet systems, *Performance Evaluation*, 58(2–3) 2004, 243–260.
12. Y. Afek, A. Bremler-Barr, and Y. Koral, Space efficient deep packet inspection of compressed web traffic, *Computer Communications*, 35(7) 2012, 810–819.
13. V. N. G. J. Soares, F. Farahmand, and J. J. P. C. Rodrigues, Traffic differentiation support in vehicular delay-tolerant networks, *Telecommunication Systems*, 48(1–2) 2011, 151–162.
14. A. H. Taherinia, and M. Jamzad, A two-step watermarking attack using long-range correlation image restoration, *Security and Communication Networks*, 5(6) 2012, 625–635.
15. Y. Cai, P. P. C. Lee, W. Gong, and D. Towsley, Analysis of traffic correlation attacks on router queues, *Computer Networks*, 55(3) 2011, 734–747.
16. Y. Zhu, X. Fu, B. Gramham, R. Bettati, and W. Zhao, Correlation-based traffic analysis attacks on anonymity networks, *IEEE Transactions on Parallel and Distributed Systems*, 21(7) 2010, 954–967.
17. S. Lee, B. Chung, H. Kim, Y. Lee, C. Park, and H. Yoon, Real-time analysis of intrusion detection alerts via correlation, *Computers & Security*, 25(3) 2006, 169–183.
18. J. Zhou, M. Heckman, B. Reynolds, A. Carlson, and M. Bishopm, Modeling network intrusion detection alerts for correlation, *ACM Transactions on Information and System Security*, 10(1) 2007, 1–31.

19. G. Min, and M. Ould-Khaoua, Prediction of communication delay in torus networks under multiple time-scale correlated traffic, *Performance Evaluation*, 60(1–4) 2005, 255–273.

20. B. Balcıoglu, D. L. Jagerman, and T. Altıok, Merging and splitting autocorrelated arrival processes and impact on queueing performance, *Performance Evaluation*, 65(9) 2008, 653–6669.

21. L. B. Lim, L. Guan, A. Grigg, I. W. Phillips, X. G. Wang, and I. U. Awan, Controlling mean queuing delay under multi-class bursty and correlated traffic, *Journal of Computer and System Sciences*, 77(5) 2011, 898–916.

22. M. Ashour, and T. Le-Ngoc, Priority queuing of long-range dependent traffic, *Computer Communications*, 31(17) 2008, 3954–3963.

23. M. Livny, B. Melamed, and A. K. Tsiolis, The impact of autocorrelation on queuing systems, *Management Science*, 39, 1993, 322–339.

24. P. Danzig, J. Mogul, V. Paxson, and M. Schwartz, The internet traffic archive, 2000. ftp://ita.ee.lbl.gov/traces/. [dataset].

25. W. E. Leland, M. S. Taqqu, W. Willinger, and D. V. Wilson, On the self-similar nature of ethernet traffic (extended version), *IEEE/ACM Transactions Networking*, 2(1) 1994, 1–15.

26. H. J. Fowler, and W. E. Leland, Local area network traffic characteristics, with implications for broadband network congestion management, *IEEE JSAC*, 9(7) 1991, 1139–1149.

27. V. Paxson, and S. Floyd, Wide area traffic: The failure of Poisson modeling, *IEEE/ACM Transactions on Networking*, 3(3) 1995, 226–244.

28. J. Cao, W. S. Cleveland, and D. X. Sun, Bandwidth estimation for best-effort internet traffic, *Statistical Science*, 19(3) 2004, 518–543

29. S. Resnick, On the foundations of multivariate heavy-tail analysis, *Journal of Applied Probability*, 41, 2004, 191–212.

30. B. D'Auria, and S. I. Resnick, The influence of dependence on data network models, *Advances in Applied Probability*, 40(1) 2008, 60–94.

31. A. Feldmann, A. C. Gilbert, W. Willinger, and T. G. Kurtz, The changing nature of network traffic: Scaling phenomena, *ACM SIGCOMM Computer Communication Review*, 28(2) 1998, 5–29.

32. M. Li, W. Zhao, W. Jia, D. Y. Long, and C.-H. Chi., Modeling autocorrelation functions of self-similar teletraffic in communication networks based on optimal approximation in Hilbert space, *Applied Mathematical Modelling*, 27(3) 2003, 155–168.

33. M. Li, Modeling autocorrelation functions of long-range dependent teletraffic series based on optimal approximation in Hilbert space: A further study, *Applied Mathematical Modelling*, 31(3) 2007, 625–631.

34. M. Li, and S. C. Lim, Modeling network traffic using generalized Cauchy process, *Physica A*, 387(11) 2008, 2584–2594

35. P. Borgnat, G. Dewaele, K. Fukuda, P. Abry, and K. Cho, Seven years and one day: Sketching the evolution of Internet traffic, *Proc. the 28th IEEE INFOCOM 2009*, Rio de Janeiro, Brazil, May 2009, pp. 711–719.

36. R. Fontugne, P. Abry, K. Fukuda, D. Veitch, K. Cho, P. Borgnat, and H. Wendt, Scaling in internet traffic: A 14 year and 3 day longitudinal study, with multiscale, analyses and random projections *IEEE/ACM Transactions on Networking*, 25(4) 2017, 2152–65.

37. M. Li, Long-range dependence and self-similarity of teletraffic with different protocols at the large time scale of day in the duration of 12 years: Autocorrelation modeling, *Physica Scripta*, 95(4) 2020, 065222 (15 pp).

38. J. Y. He, G. Christakos, J. P. Wu, M. Li, and J. X. Leng, Spatiotemporal BME characterization and mapping of sea surface chlorophyll in Chesapeake Bay (USA) using auxiliary sea surface temperature data, *Science of the Total Environment*, 794, 2021, 148670.

39. X. Xiao, C. C. Xu, Y. Yu, J. Y. He, M. Li, and C. Cattani, Fat tail in the phytoplankton movement patterns and swimming behavior: New insights into the prey-predator interactions, *Fractal and Fractional*, 5(2) 2021, 49.

40. M. Li, and J.-Y. Li, Generalized Cauchy model of sea level fluctuations with long-range dependence, *Physica A*, 484, 2017, 309–335.
41. M. Li, Y. C. Li, and J. X. Leng, Power-type functions of prediction error of sea level time series, *Entropy*, 17(7) 2015, 4809–4837.
42. T. Mikosch, S. Resnick, H. Rootzén, and A. Stegeman, Is network traffic approximated by stable Lévy motion or fractional Brownian motion? *The Annals of Applied Probability*, 12(1) 2002, 23–68.
43. G. Terdik, and T. Gyires, Levy flights and fractal modeling of internet traffic, *IEEE/ACM Transactions on Networking*, 17(1) 2009, 120–129.
44. B. B. Mandelbrot, *Multifractals and 1/f Noise*, Springer, New York, 1998.
45. S. Cohen, and A. Lindner, A central limit theorem for the sample autocorrelations of a Lévy driven continuous time moving average process, *Journal of Statistical Planning and Inference*, 143(8) 2013, 1295–1306.
46. L. Denby, J. M. Landwehr, C. L. Mallows, J. Meloche, J. Tuck, B. Xi, G. Michailidis, and V. N. Nair, Statistical aspects of the analysis of data networks, *Technometrics*, 49(3) 2007, 318–334.
47. G. Y. Lazarou, J. Baca, V. S. Frost, and J. B. Evans, Describing network traffic using the index of variability, *IEEE/ACM Transactions on Networking*, 17(5) 2009, 1672–1683.

Index